Quantum Processes and Measurement

This accessible and self-contained text presents the essential theoretical techniques developed to describe quantum processes, alongside a detailed review of the devices and experimental methods required in quantum measurement. Ideal for advanced undergraduate and graduate students seeking to extend their knowledge of the physics underlying quantum technologies, the book develops a thorough understanding of quantum measurement theory, quantum processes, and the evolution of quantum states. A wide range of basic quantum systems are discussed, including atoms, ions, photons, and more complex macroscopic quantum devices such as opto-mechanical systems and superconducting circuits. Quantum phenomena are also covered in detail, from entanglement and quantum jumps, to quantum fluctuations in optical systems. Numerous problems at the end of each chapter enable the reader to consolidate key theoretical concepts and to develop their understanding of the most widely used experimental techniques.

Claude Fabre Claude Fabre is Emeritus Professor in Physics at Sorbonne University at the Kastler–Brossel Laboratory in Paris, and has made pioneering contributions within the field of quantum optics, especially concerning the manipulation of quantum fluctuations and correlations in light and quantum parameter estimation. He is Fellow of the Optical Society of America and an honorary member of the Institut Universitaire de France. He was awarded the Grand Prix Léon Brillouin. He coauthored the successful textbook *Introduction to Quantum Optics* (2010) with Gilbert Grynberg and Alain Aspect.

Rodrigo G. Cortiñas is a postdoctoral researcher at Yale University and a Fellow of the Yale Quantum Institute. His research is currently focused on experimental quantum error correction in superconducting circuits. He obtained his PhD from the *École Normale Supérieure de Paris* (ENS) for experimental work on Rydberg atomic physics for quantum simulation. As a teaching assistant, he has led several exercise classes at the graduate level at the ENS and at the *Universidad de Buenos Aires*, from where he graduated.

Quantum Processes and Measurement

Theory and Experiment

CLAUDE FABRE
Sorbonne University

WITH CONTRIBUTIONS FROM
RODRIGO G. CORTIÑAS
Yale University

Shaftesbury Road, Cambridge CB2 8EA, United Kingdom

One Liberty Plaza, 20th Floor, New York, NY 10006, USA

477 Williamstown Road, Port Melbourne, VIC 3207, Australia

314–321, 3rd Floor, Plot 3, Splendor Forum, Jasola District Centre,
New Delhi – 110025, India

103 Penang Road, #05-06/07, Visioncrest Commercial, Singapore 238467

Cambridge University Press is part of Cambridge University Press & Assessment,
a department of the University of Cambridge.

We share the University's mission to contribute to society through the pursuit of education,
learning and research at the highest international levels of excellence.

www.cambridge.org
Information on this title: www.cambridge.org/9781108477772

DOI: 10.1017/9781108774918

First published 2023

A catalogue record for this publication is available from the British Library.

Library of Congress Cataloging-in-Publication Data
Names: Fabre, Claude, 1951- author. | Cortiñas, Rodrigo G., contributor.
Title: Quantum processes and measurement : theory and experiment / Claude Fabre,
Sorbonne University ; with contributions from Rodrigo G. Cortiñas.
Description: Cambridge, United Kingdom ; New York, NY : Cambridge University Press,
2023. | Includes bibliographical references and index.
Identifiers: LCCN 2022057768 (print) | LCCN 2022057769 (ebook) | ISBN 9781108477772
(hardback) | ISBN 9781108774918 (epub)
Subjects: LCSH: Quantum theory–Textbooks. | Quantum measure theory–Textbooks.
Classification: LCC QC174.12 .F325 2023 (print) | LCC QC174.12 (ebook) | DDC
530.12–dc23/eng20230506
LC record available at https://lccn.loc.gov/2022057768
LC ebook record available at https://lccn.loc.gov/2022057769

ISBN 978-1-108-47777-2 Hardback

Contents

Introduction

Un coup de dés jamais n'abolira le hasard
A throw of the dice never will abolish chance

Stéphane Mallarmé

Historical Context

The study of quantum physics started at the very beginning of the 20th century with the explanation by Planck [146] of the frequency spectrum of the light emitted by heated objects, and with its interpretation in terms of light quanta by Einstein [64]. Later that century, Bohr [22] succeeded in calculating the spectrum of light emitted by hydrogen. Elaborating on an intuition by Louis de Broglie that matter could behave like a wave, quantum mechanics was developed in the 1920s by Heisenberg, Schrödinger, Dirac, and many others as a global theory. They were able to explain the properties of more and more complicated objects (multi-electron atoms, ions, molecules, solid state...), and with greater and greater accuracy. We recall these rules in Appendix A. In particular, they allowed physicists to calculate quantities measured on a great number of quantum objects, like macroscopic samples of atoms, molecules, and solid state systems [57, 172].

The newborn quantum theory introduced new paradigms, like the superposition principle and the possibility of quantum interferences, wave-particle duality, the existence of conjugate quantities and the Heisenberg inequality, the special properties of the measurement process, and the quantum state collapse. Furthermore, it showed the necessity of introducing a dose of irreducible randomness into the theory, a true revolution in physics. All these quantum features looked strange to physicists acquainted with classical physics, especially when it came to describing the behavior of individual quantum objects and their uncontrollable quantum jumps between different states. These questions were the object of lively debates in the physics community, reaching a climax at the occasion of N. Bohr and A. Einstein's discussions.

In the 1930s, Einstein [62] and Schrödinger [158] pinpointed another "strangeness," linked with the nonlocal and paradoxical nature of quantum correlations between remote parties, that Schrödinger named "Verschränkung" (entanglement in English, intrication in French). Later,

Bell [15] made a big step forward by showing that the interpretation of entanglement in terms of local "hidden" and unpredictable supplementary variables could contradict in some situations the predictions of quantum mechanics. Experimentalists [8], using more and more sophisticated experimental set-ups, showed that Nature indeed followed the laws of quantum mechanics and not the ones of local hidden variables approach, pinpointing the radically new nature of quantum correlations, which are in no way reducible to the classical ones, and the validity of quantum mechanics to describe them. Entanglement, and the other specific quantum features that we have just quoted, gradually became a resource, enabling physicists to go further in the exploration of quantum strangeness and to imagine applications: This is the domain of quantum technologies. For example, quantum entanglement is exploited in cryptography to make intrinsically secure quantum key distribution devices, and the quantum superposition principle is harnessed to make computation tasks in a highly parallel way in the future quantum computer, the Holy Grail of quantum physics. To account for these new situations, theoreticians developed a full frame of novel concepts and new theoretical tools.

Meanwhile, the experimental techniques underwent a tremendous development in terms of sensitivity, quality of state preparation, and detection of the physical systems and data analysis capacity. New experimental devices and techniques were developed: single particle manipulation and detection, use of highly controlled lasers, nanotechnologies, powerful computers, etc. In many experiments, physicists are now able to precisely master and observe *single quantum objects*, to make them evolve in a well-defined way and to accumulate in computer memories data about millions of successive individual quantum objects. Not only are mean values of measurements now subject to measurement, but also all the details of the statistical distribution of *quantum fluctuations*. In particular, sudden and unpredictable changes, called *quantum jumps*, which particularly puzzled Schrödinger, are now easily observed in individual quantum objects. One is now able to describe in a general way the sequence preparation -> evolution -> detection of single quantum objects, which is often called a *quantum process*.

Aim of the Book

The aim of this textbook is to provide a self-contained and accessible introduction to the physics underlying quantum technologies by presenting typical experimental achievements and by introducing theoretical tools that have been developed to precisely characterize quantum processes in more complex situations than the ones covered by basic quantum mechanics: Let us quote, among many others, Kraus operators, POVMs, symplectic groups,

Gaussian and non-Gaussian states, and entanglement witnesses. We leave aside the subject of *quantum information*, which has already been covered by several remarkable textbooks [13, 136, 147].

We also detail the characteristics and nature of the measurement process of quantum objects by introducing the concept of positive operator measures, and post-measurement operators. We will describe the generation and evolution of quantum states that have been post-selected, i.e. conditioned by the result of a measurement.

Finally, we consider the domain of quantum metrology and quantum estimation by determining the optimized quantum limits of high precision and/or high sensitivity estimations of physical parameters within the constrains of the Heisenberg inequality and of the back action of the measurement on the physical system. For this purpose we will introduce and study the "quantum Cramér–Rao bound" in different physical situations.

This short textbook is based on lectures given at the Ecole Normale Supérieure in the frame of the Masters degree of the specialty "International Center for Fundamental Physics" (ICFP). It constitutes a concise approach of this subject that is accessible to graduate and PhD students. In order to not extend too much the size of this textbook, most demonstrations and mathematical developments are skipped. The reader will find more detailed accounts of the different subjects treated in this book in other textbooks and review papers [11, 21, 44, 45, 57, 75, 82, 88, 107, 112, 136, 139, 142, 147, 161, 172, 173, 179].

Instructions for Use

This textbook addresses the subjects we have just listed from different points of view that complement each other:

1. Five chapters (1, 3, 6, 8, and 10) are experimental. They briefly describe typical and illustrative experiments performed on different experimental platforms: atoms, ions, photons, and macroscopic devices such as optomechanical systems and superconducting circuits. They have been chosen by the authors in a way that is personal and non-exhaustive. They set the stage for the theoretical chapters.
2. Six chapters (2, 4, 5, 7, 9, and 11) constitute the core of this textbook. They present various theoretical tools and use them to describe different aspects of quantum physics, especially its most recent developments.
3. At the end of the book, the appendices (A to K) present brief descriptions of physical quantum systems that are currently encountered in quantum physics: light–matter interaction, light-induced motion, single particle trapping, and superconducting circuits. They display in a concise way important results that are useful to understand the quantum

machinery presented in the experimental and theoretical chapters. They serve as reminders of useful results for some students, and of introductions to the basic features for others.

4. At the end of chapters and appendices, exercises of various levels are included, from calculations of orders of magnitude to advanced theoretical developments, that enable the reader to assimilate more deeply and personally the new concepts and tools described in the book. The detailed solutions of these exercises can be found on the web at https://cambridge.org/9781108477772.

1 Experiment: Detecting Single Quantum Objects

In 1952, E. Schrödinger, one of the fathers of quantum mechanics and a Nobel Prize winner, wrote [160]:

We never experiment with just one electron or atom or (small) molecule. In thought experiments, we sometimes assume that we do; this invariably entails ridiculous consequences.

In 2012, S. Haroche and D. Wineland were awarded the Nobel prize

... *for ground-breaking experimental methods that enable measuring and manipulation of individual quantum systems*, according to the Nobel committee citation.

The juxtaposition of these two statements illustrates the tremendous progress of the experimental techniques in 60 years. They have allowed physicists to progress from gedanken experiments to feasible experiments. It also highlights the conceptual change that this progress induced. Physicists are now able to investigate the consequences of the single particle experiments envisioned by Schrödinger, but also to harness them to develop applications that are now known as quantum technologies.

With the exception of high energy physics experiments, for which the energy of the particles is so high that a single particle provides measurable macroscopic effects, like bubble formation, at Schrödinger time, most measurements were made on samples containing macroscopic quantities of quantum objects, of the order of the Avogadro number. For these experiments, the relevant quantities in the quantum mechanical description of the studied systems are *ensemble averaged quantities*, i.e. mean values, and sometimes variances and correlations. This approach has lead to great successes, in particular the ab initio determination, from a quantum description of the microscopic world, of the value of many macroscopic parameters such as thermal and electrical conductivity, light absorption and scattering parameters.

The situation has evolved with the development over the years of increasingly sophisticated and sensitive experimental techniques: it is now possible to measure various properties of single quantum systems having an energy at the electron Volt level or even at the meV level: single atoms, ions, molecules, or photons are now the object of experimental studies. This allowed physicists to investigate new properties of the quantum world that may look strange or paradoxical (hence the word ridiculous used by Schrödinger), because they are far away from our intuition based on the macroscopic world. Nowadays, there are many experimental techniques permitting to detect and manipulate single low energy quantum systems. In the remainder of this chapter, we give some examples of these techniques.

1.1 Destructive Detection of a Single Photon

A first category of photodetectors is based on the photoelectric effect: a photon that hits their surface has some probability to expel from it an electron that is then free. In photomultipliers the ejected electron is accelerated and hits another metallic surface from which it expels several electrons. The avalanche process is repeated several times and gives rise at the end of the series of metal surfaces to a macroscopic bunch of electrons, with a gain of up to 10^8. This avalanche process produces a measurable pulsed electric signal that we call it a "click". It witnesses the arrival of a single photon. Channel electron multipliers, or channeltrons, are similar devices that are able to deliver a macroscopic electric pulse from a single incident electron.

Photomultipliers need vacuum to operate. Most convenient devices are the solid-states avalanche photodiodes, made of a semiconductor junction. The photon incident on it creates an electron-hole pair. An avalanche process taking place in the high field region of the junction produces an electrical current in the external circuit. If the detector operates above the breakdown voltage (Geiger mode), the gain can be very high, of the order of 10^9, so that a single photon gives rise to a measurable pulse with a high probability (up to 70%).

Bolometers constitute a second category of photodetectors. They record the temperature rise of an opaque body due to the absorption of light energy. Some of them are able to measure very high luminous powers. Others are extremely sensitive to very weak powers. The best ones consist of superconducting thin (5 nm) and narrow (100 nm) nanowires of very small heat capacity and cooled to very low temperatures: they are able to detect the transition from superconductivity to normal conduction induced by the heating induced by the energy carried by a single photon in the visible or infrared range! They give the same pulse height when more than one photon is present, so that they do not count the number of photons impinging on the detector, but detect only the presence of at least one photon.

Other kinds of superconducting devices, called transition edge sensors (TESs), are able to give different signals when 1, 2, 3 ... photons are incident on the detector (see Figure 1.1): they are called photon number resolving detectors [126]. They have a negligible dark count and a very high quantum efficiency, up to 90% that can be optimized in a wide spectral range, from mm waves to X-rays. They also have defects: they are slow and require a dilution refrigerator to operate.

Using more than one photon detector allows physicists to extract more information: with two detectors, one has access to photon correlations by *photon coincidence measurements*, which are of paramount importance in quantum physics (Section F.5); with the help of linear arrays of

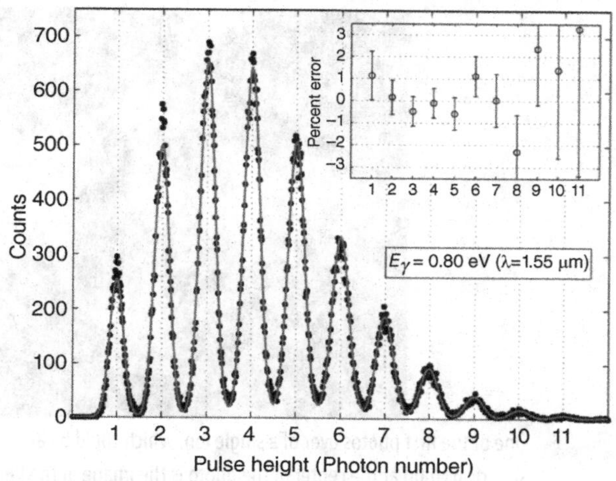

Fig. 1.1 Measured pulse-height distribution of a very weak laser pulse in a coherent state containing an average photon number of four photons using a superconductor calorimetric photon counter. The signals corresponding to 1–9 photons are well resolved. From A. Miller et al. Demonstration of a low-noise near-infrared photon counter with multiphoton discrimination, *Applied Physics Letters*, 2003. Reprinted from [126] with the permission of AIP.

photodiodes, or of charge coupled devices (CCD cameras) that are sensitive to single photons, experimentalists have access to the 1D or 2D position of the recorded single photons.

Electron multipliers and semiconductor devices are also sensitive to other kinds of incident particles, provided these particles bring an energy larger than the work function of the metallic electrode, or the energy gap of the semiconductor. This is, for example, the case of helium atoms. They can be easily detected individually when they are excited to the metastable state because of its high energy (20 eV). These atoms can also be precisely localized using a microchannel plate, which is an array of channeltrons.

1.2 Nondestructive Detection of the Presence of a Single Trapped Particle

Appropriate configurations of electric and magnetic fields, described in Appendix J, are able to create a harmonic potential around a chosen point $\mathbf{r}_0$ of the empty space so that a charged particle, like an electron or an ion, feels in its neighborhood a restoring force towards $\mathbf{r}_0$ likely to trap single electrons or ions.

Trapped charged particles, as any moving charged objects, induce currents in the electrodes. This current can be detected and witnesses the presence of the particle without destroying it. These induced currents are damped by the circuit's electrical resistance. This constitutes a relaxation process for the particle that efficiently reduces its energy in the trap. Because

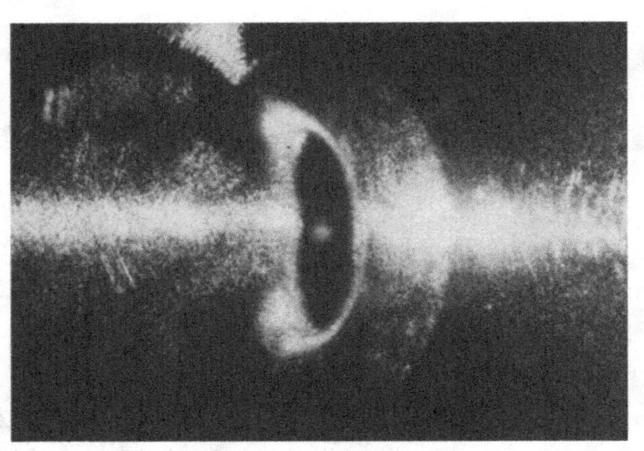

Fig. 1.2 One of the first photos ever of a single ion, which could be also seen by the naked eye. The small spherical cloud at the center of the photo is the image of the light scattered by the single Ba ion, whereas the surrounding light is the light scattered by the close-by trapping electrodes. From W. Neuhauser et al. Visual observation and optical cooling of electrodynamically contained ions, *Applied Physics*, 1978 [135].

of collisions with other particles present in the evacuated chamber in which the experiment is made, the charged particles can be expelled one by one from the trap. Their number in the trap decreases with time, and can be equal to exactly one during a time that ranges from seconds to days or even months.

To measure the presence of the single trapped ion, one shines on the point $\mathbf{r}_0$ a laser beam that is resonant on the transition linking the ion ground state $|G\rangle$ to an excited state $|B\rangle$ of short lifetime T: the laser light excites the ion to level $|B\rangle$, which then decays back to $|G\rangle$ by emitting a spontaneous emission photon in a random direction: this is the well-known phenomenon of fluorescence. A proportion of these photons are collected by a large aperture lens and monitored on a photodetector. Due to the continuous laser irradiation the particle is quickly excited back to $|G\rangle$, and the cycle can be repeated. The flux of photons is thus of the order of 1 photon per lifetime, i.e. typically 10^8 photons per second for a 10 ns lifetime. This is an easily detectable bunch of photons, even by the naked eye: it is possible to actually see and to take a picture of a single ion. Figure 1.2 shows the fluorescence of a single Ba^+ ion. Note that the recorded fluorescence spot has a diffraction limited size of the order of the wavelength (1 μm), which is much bigger than the ion size itself (roughly 50 pm).

The dipole force, described in Section J.1, induced by a highly focused off-resonant intense laser, provides a trap for polarizable objects like atoms. As with ions, a single atom can be detected by its fluorescence induced by another laser beam resonant on a closed transition, which is able to give rise to a large number of fluorescent photons. The fact that the trapping is due to a laser beam and not to fixed electrodes adds a lot of flexibility: using laser standing waves one can create an optical lattice of periodically disposed traps (Section 3.4 and Figure 1.3). Arrays of single atoms, candidates to be quantum registers for qubits, can also be produced in this way.

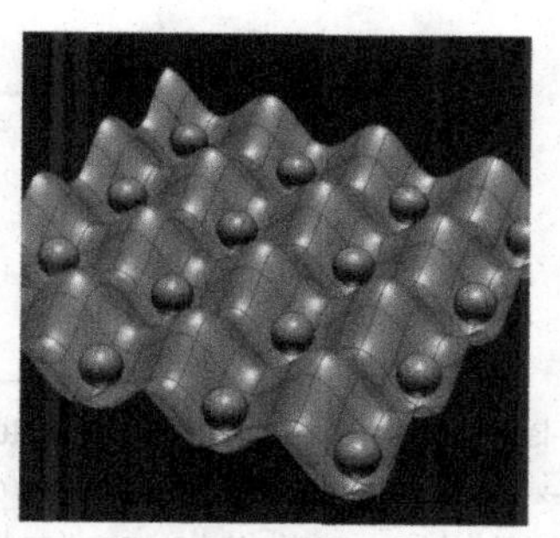

Fig. 1.3 Spatial shape of the dipole force potential experienced by atoms and created by interfering laser beams. Atoms are trapped at the periodic minima of this potential. From M. Takamoto et al. An optical lattice clock, *Nature*, 2005 [166].

1.3 Nondestructive Detection of the Energy Level of a Single Trapped Particle

Let us now describe a technique used to monitor in a nondestructive way, not only the presence of a single particle, but also its energy level. It is called the shelving technique, which can be used with ions and atoms as well.

We consider a three-level particle, which has a second excited state $|D\rangle$, that has a very long lifetime, in the configuration sketched in Figure 1.4. If the particle is in the ground state $|G\rangle$, a laser, resonant on the transition $|G\rangle \rightarrow |B\rangle$, will induce a strong fluorescence, as explained in Section 1.2, which can be easily recorded. If some process excites the atom to the level $|D\rangle$ (for example a weak lamp illumination), the cycle stops immediately and the fluorescence signal drops to zero. So the level of fluorescence is a direct measurement of the energy of the particle.

Figure 1.5 shows one example of a temporal sequence of the fluorescence signal from a single ion: its switch-off indicates a passage from $|G\rangle$ to $|D\rangle$ due to some excitation process whereas its switch-on indicates a passage from $|D\rangle$ to $|G\rangle$ due to spontaneous emission. The fluorescence signal of Figure 1.5 is characterized by the presence of on- and off- switches that occur randomly. They are the famous and fascinating *quantum jumps* that are distinctive features of the behavior of single quantum objects. *The exact time of such jumps cannot be controlled or predicted* by the experimentalist, to his great despair, as the main aim of physicists is to find ways to predict all physical phenomena. We will consider these quantum jumps in more detail in Section 4.5.

It is possible to record over very long times fluorescence signals that display many on- and off-quantum jumps. From this recorded data one can build the histogram of the dwell times t_D in the excited state $|D\rangle$ (Figure 1.6). This histogram can be fit by an exponential function, $e^{-t_D/T}$, which yields the lifetime T of the excited state $|D\rangle$. We see in this simple experimental example the very root of quantum physics: even though a

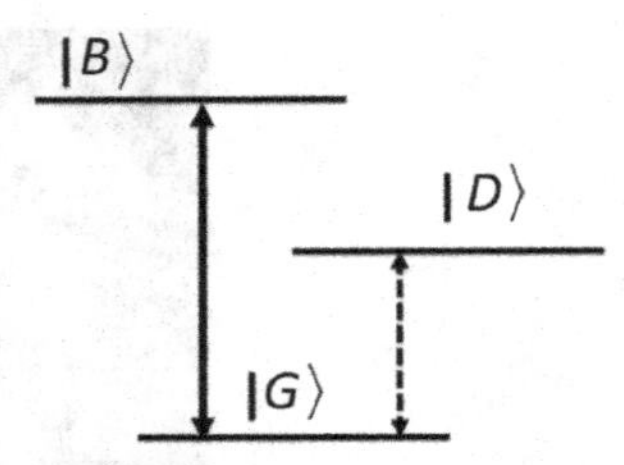

Fig. 1.4 Three-level particle (ion or atom) submitted to a laser light resonant with the transition $|G\rangle \longrightarrow |B\rangle$ ("bright transition") producing a detectable fluorescent signal, and to a weak incoherent light resonant with the transition $|G\rangle \longrightarrow |D\rangle$ ("weak transition").

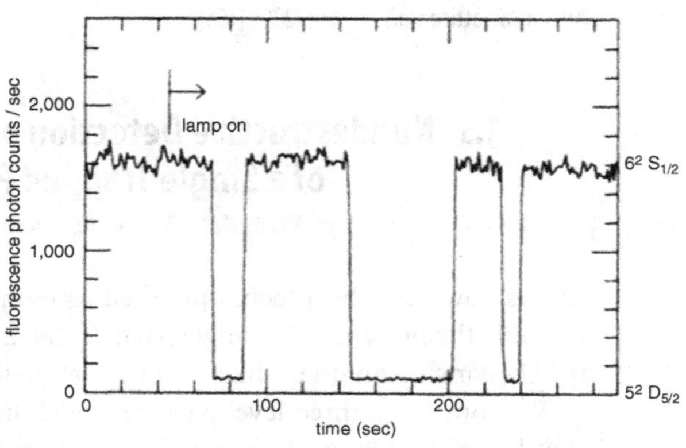

Fig. 1.5 Fluorescent signal emitted by a single ion submitted to a strong resonant laser field and a weak incoherent pumping light. The sudden changes in the signal reveal the *quantum jumps* experienced by the ion. From W. Nagourney et al. Shelved optical electron amplifier: Observation of quantum jumps, *Phys. Rev. Letters*, 1986 [132].

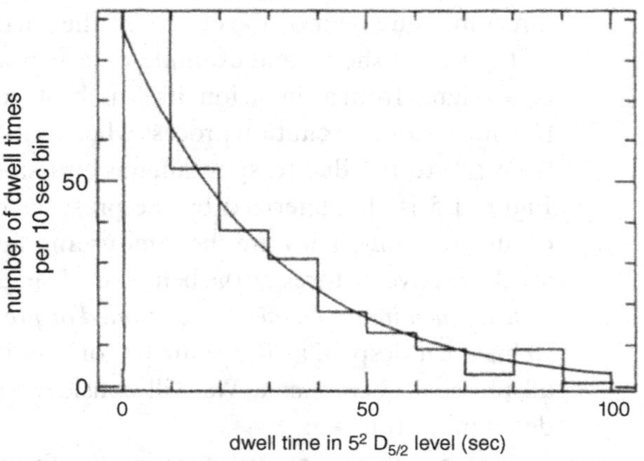

Fig. 1.6 Histogram of times spent in the excited state $|D\rangle$ (dwell times) corresponding to the periods of switched-off fluorescence in Figure 1.5. From W. Nagourney et al. Shelved optical electron amplifier: Observation of quantum jumps, *Phys. Rev. Letters*, 1986 [132].

single event on a single quantum object occurs randomly and cannot be predicted, the analysis of many successive single particle events allows us to accurately retrieve its properties such as its lifetime. This parameter could also have been directly measured on the fluorescence signal from a macroscopic sample of particles.

1.4 Destructive Detection of the Energy Level of a Single Atom

Highly excited states of atoms called Rydberg states (with high principal quantum numbers n, of the order of 50) have the interesting property that they can be counted one by one because their valence electron is weakly bound to the atomic core: an electrostatic field of 100 V/cm is enough to overcome the binding Coulomb field of the atomic core and therefore to ionize it: this is the field ionization phenomenon, which cannot be used for low-lying levels because it requires huge electrostatic fields for these levels. The ejected electron can then be individually counted by an electron multiplier, or channeltron.

To detect the presence of less excited atoms than Rydberg states, field ionization can be replaced by a suitable laser irradiation that is able to photo-ionize the atom.

The field ionization technique can even be used to distinguish atoms in Rydberg states with different quantum numbers n using the fact that

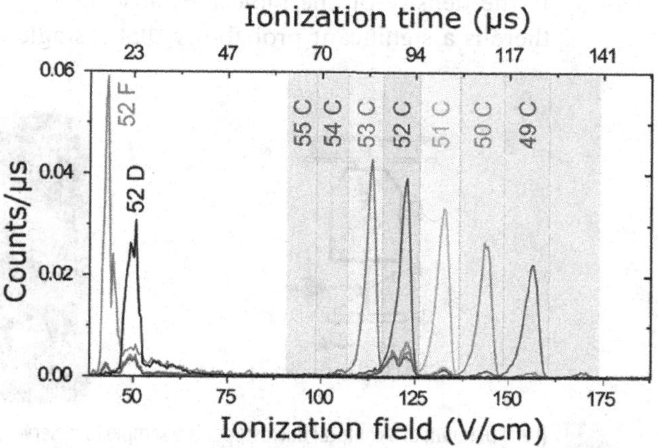

Fig. 1.7 Ionization spectrum of single highly excited rubidium atoms, which are prepared in different circular states of principal quantum number ranging from 49 to 54. The ionizing field increases with time and the different levels, ionized for different values of the field, appear well separated: for well chosen values of the ionizing field, one circular level can be ionized and the next one not. From T. Cantat-Moltrecht et al. Long-lived circular Rydberg states of laser-cooled rubidium atoms in a cryostat, *Physical Review Research*, 2020 [32], CC BY 4.0.

the minimum electrostatic field needed to ionize such atoms varies with n (it is actually proportional to n^{-4}). The set-up is sketched in Figure 3.2 of Chapter 3. The atom excited to a Rydberg state $|e_n\rangle$ of principal number n travels in the vacuum of an evacuated chamber between the two plates of plane capacitors, inside which exists an electrostatic field that is just enough to ionize level $|e_n\rangle$ but not the more bound state $|e_{n-1}\rangle$. The level $|e_{n-1}\rangle$ is ionized when the atom travels in a second capacitor placed downstream, inside which exists a stronger electrostatic field that can ionize the atom when it is in state $|e_{n-1}\rangle$.

In Figure 1.7, the state-selective field-ionization signal for Rydberg atoms of principal quantum numbers around 50 is shown. The measurement is done by ramping up the voltage of the ionization electrode and recording in real time the ionization signal from single Rydberg atoms using a channeltron. One notes that each Rydberg level has a characteristic ionization feature that is well separated from the next one.

1.5 Nondestructive Detection of Single Particles on a Surface

It is possible to detect a single molecule in a dilute sample of molecules located on a surface. This is done by using confocal microscopy (Figure 1.8). It involves a highly focused excitation beam that induces the fluorescence of the molecule, and a confocal pinhole that rejects the out-of-focus light and selects a small detection region of the image plane. If the density of the molecules adsorbed on the surface is low enough, there is a significant probability that a single molecule is in the detection

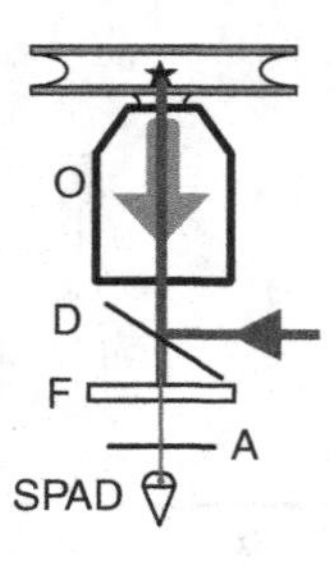

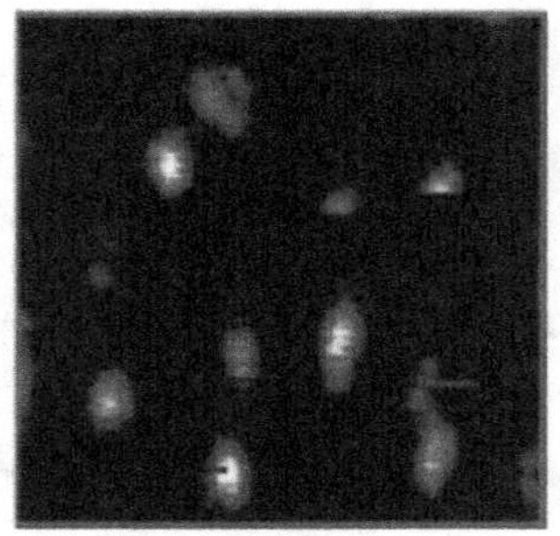

Fig. 1.8 (Left) Diagram of a confocal microscope. The sample to observe is at the top. O: high aperture objective; D: dichroic mirror transmitting the fluorescent light; F: filter; A: confocal aperture that restricts the axial extent of the observation volume; SPAD: single photon avalanche detector. (Right) Confocal image of isolated single molecules trapped on the surface (field: 10 μm $\times$ 10 μm). From W. E. Moerner and D. P. Fromm. Methods of single-molecule fluorescence spectroscopy and microscopy, *Review of Scientific Instruments*, 2003 [129].

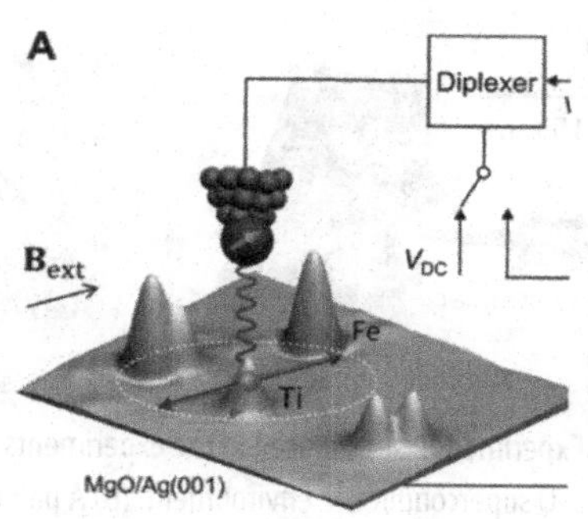

Fig. 1.9 Diagram of a STM imaging set-up: Ti atoms are individually positioned on a metallic surface with the help of the metallic tip of a scanning tunneling microscope. The spin states of these atoms are coherently manipulated at the single atom level by the tip playing the role of a microwave antenna. From Kai Yang et al. Coherent spin manipulation of individual atoms on a surface, *Science*, 2019 [181]. CC BY 4.0.

area. The laser beam is scanned over the surface, and searches for the location of the molecule on the surface. Usually, there are still many other background species in the detection area. One eliminates their fluorescence by the selective resonant character of the excitation of the desired molecule. It is often necessary to cool down the sample in order to get narrow enough transition linewidths and reduce the migration of the molecule.

In conventional spectroscopy, the inhomogeneity of the environment of the molecules located at different places induces a broadening of the transition called *inhomogeneous broadening*. This is not the case for single molecule spectroscopy, as there is no longer an ensemble averaging. The inhomogeneity now translates into a position-dependent shift of the resonance frequency, from one detected single molecule to another.

Single particles atoms lying on the surface of a solid can be also detected by scanning tunneling microscopy (STM), which allows us to scan a metallic surface with a nanometric resolution and detect single atoms that are adsorbed on it. By changing the voltage applied to the cantilever electrode that explores the surface, one can create a force that is sufficient to attach a chosen atom to the cantilever, to remove it from the surface and transport it to another chosen location, always with a single atom resolution. Figure 1.9 shows an STM image of Ti and Fe isolated atoms on MgO. The cantilever tip senses the different Ti spin orientations and their quantum superpositions. They constitute magnetic qubits that can be controlled by the microwave field emitted by the cantilever tip.

1.6 Quantum Jumps in Superconducting Circuits

Superconducting circuits are objects of several millimeters in size that can easily be seen with bare eyes (Figure 1.10) and look very classical.

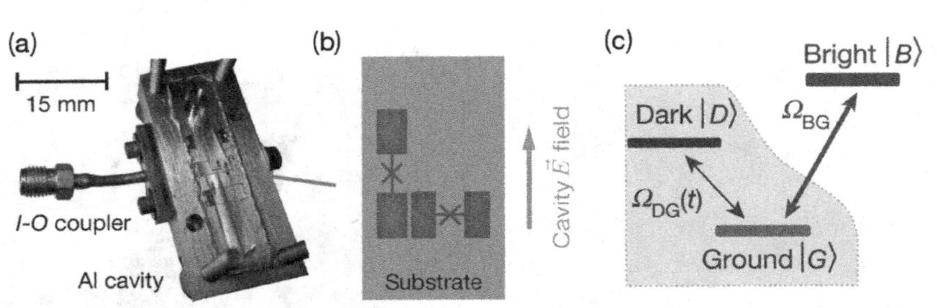

Fig. 1.10 (a) Experimental set-up used in the experiments from [185]. An aluminium cavity provides the high-Q superconductive environment. (b) A pair of artificial atoms (transmons on a sapphire chip) creates an artificial three-level molecule (c) used in the same way as the three-level system in ion jumps (Section 1.3). With permission from Z. Minev. *Catching and Reversing a Quantum Jump Mid-Flight*, PhD thesis, Yale University, 2019 [185].

They are indeed macroscopic objects acting as microwave antennas. These antennas produce a microwave field that can be measured with ordinary electronics and usual data processing techniques. Nonetheless, as described in Appendix K, they can be brought to a fully quantum regime where they admit a description in terms of a single pair of conjugate quantum variables ruled by the Schrödinger equation [24, 133]. Just as with atoms, one can neglect their many-body "subatomic" structure in their description and treat them like indivisible quantum objects [55], hence their nickname "artificial atoms." Their control and use is possible thanks to the Josephson tunnel junction, a nonlinear circuit element that is nondissipative and exhibits strong nonlinearity at the single photon level [56]. The nonlinearity allows the isolation of the two lower levels of an otherwise harmonic LC oscillator and their use as two-level systems [19]. Superconducting circuits can achieve a high quantum efficiency since the microwave environment can be engineered for the photons to leak though a single mode superconducting waveguide channeling them to an amplifier and subsequent detector. Thanks to the availability of cryogenic and microwave commercial technologies coherent Rabi and Ramsey oscillations (see Section G.3) are observed in these qubit systems with relative ease. With specially engineered quantum limited amplifiers the single quantum trajectories of these systems can be tracked, allowing quantum jumps to be observed in these macroscopic systems [171].

In Figure 1.10, we show a set-up [127] that was capable of observing quantum jumps in a macroscopic circuit imitating the "three level atom" already discussed in this chapter (see Section 1.3). Figure 1.10(a) shows the aluminium cavity with its microwave port (an SMA connector). A sapphire substrate with two aluminum artifical atoms, or "transmons" printed in orthogonal directions is in its interior and sketched in Figure 1.10(b) (see Appendix K). One of the transmons is aligned with the electromagnetic field of the fundamental mode of the cavity. This makes its antenna "bright" (B) since it can easily emit radiation into this mode. The other transmon is normal to the same electric field making it a "dark" (D) antenna which is weakly coupled to the fundamental mode. In Figure 1.10(c) we reproduce

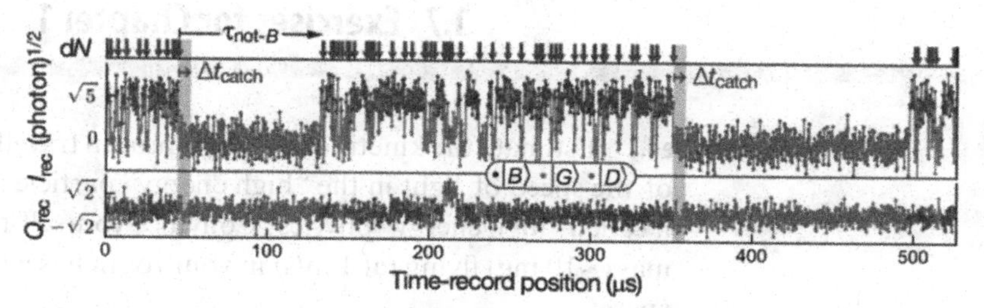

Fig. 1.11 Quantum jumps in a superconducting circuit experiment. Quadratures of the current I (middle recording) and charge Q (lower recording) of a probe microwave pulse are measured by homodyne detection to infer the state of the artificial atom. The vertical arrows of the upper recording correspond to "clicks" in the detection. The variable Δt_{catch} is the time period after the last "click" before a series of "no-clicks". From Z. Minev et al. To catch and reverse a quantum jump mid-flight, *Nature*, 2019 [127].

the three main levels of importance for this experiment. The ground state (G) is coupled to the bright state and to the dark state by two coherent drive fields. The field connecting to the bright state is strong and produces a fast Rabi oscillation, while the Rabi frequency towards the dark state is orders of magnitude weaker. The detection scheme relies on the fact that when the atom is in B state the frequency of the fundamental cavity mode (line width ~3 MHz) is dispersively shifted by ~5 MHz, and thus a probe field can provide information over the atomic state. This is reminiscent of a standard quantum optics measurement technique that is explained in detail in Appendix K: by measuring the quadratures of the reflected probe one can measure the resonance frequency of the resonator, which in turns reveals the atomic state in a quantum non-demolition way. Importantly in the experiment discussed, the D state is engineered to have a dispersive shift of ~0.3 MHz, making it indistinguishable from the signal corresponding to G. The binary measurement performed is then B, not-B. In Figure 1.11 we reproduce an experimental jump trace. The arrows in the top part of the panel represent "clicks" at the times at which the bright state is observed. Whenever that bright state is not detected the atom is most likely in the ground state and a sudden bright click is expected soon after. Rarely, though, the clicks are interrupted for a longer period corresponding to the atom visiting the dark state. After being "shelved" for a while in D a jump back to the ground state restarts the bright state "clicking."

Even more remarkably, the high controllability offered by superconducting circuits allows us to highlight the difference in nature between the "up-jumps" (from G to D, during Δt_{catch}, see Figure 1.11) from the "down-jumps" (from G to D), events that were previously regarded jointly as discontinuous "jumps" [33, 48, 88]. We discuss this subject in more detail in Section 5.8.

1.7 Exercises for Chapter 1

Exercise 1.1 Estimate the kinetic energy of a proton travelling at 99.999999% of the speed of light in the "high energy" particle accelerator. Why is it called "high energy" physics? Convince yourself that a mosquito (of mass ~10 mg) flying (at 1 m/s) in your room has a comparable kinetic energy.

Exercise 1.2 "Low energy" experiments on single atoms or single ions experiments require laser cooling. Compute the mean kinetic energy and the velocity of Rubidium atoms at 1 μK. Compute the recoil velocity of a Rubidium atom after the absorption of a single λ =780 nm photon.

Exercise 1.3 Using that in the Bohr atomic model the radius of the nth orbit is $r_n = a_0 n^2$, show that the dependence of the ionization field for circular Rydberg atoms with the principal quantum number is n^{-4}.

Exercise 1.4 What is the average number of photons in a ~51 GHz cavity if the environment is at 0.8 K? Compare to the expected value at room temperature. How is the argument modified for a resonator at 5 GHz?

Exercise 1.5 The photon of Figure 3.3 was observed to live in the cavity during $T \sim 0.476$ s. The lifetime of the cavity was measured independently to be $T_{\text{cav}} \sim 0.13$ s. Compute the probability of observing a photon living during $\geq 0.476\text{s} \simeq 3.7 \times T_{\text{cav}}$. What is the probability for a human to live 3.7 times (or more) its life expectancy? Discuss.

2 Description of Quantum Systems in Terms of the Density Matrix

In Appendix A, the reader will find a short reminder of the basic rules of quantum mechanics in terms of state vectors. These rules have a limited range of validity: the system must be "closed," i.e. isolated or submitted to externally fixed forces, prepared in a totally controlled way. The case where the quantum system has these ideal properties is called a *pure case*.

As we will see in the Section 2.1, these basic rules can be reformulated in terms of another mathematical object, the density matrix ρ, which was introduced in the early days of quantum mechanics [108, 173]. The advantage of the new formulation is that it has a much greater domain of applicability, including in open systems, and is indeed able to set the frame allowing us to describe almost any physical situation.

2.1 The Density Matrix as an Alternative Description of Basic Quantum Mechanics

The density matrix is the operator acting on the Hilbert space $\mathcal{H}$ of state vectors given by

$$\rho = |\Psi\rangle\langle\Psi|, \tag{2.1}$$

where $|\Psi\rangle$ is the state vector. In this textbook, we will omit the hat on ρ to alleviate further expressions.

2.1.1 Basic Properties

The density matrix is Hermitian and is a positive operator:

$$\rho = \rho^\dagger; \quad \forall\,|\phi\rangle,\ \langle\phi|\rho|\phi\rangle = \langle\phi|\Psi\rangle\langle\Psi|\phi\rangle \geq 0. \tag{2.2}$$

Two state vectors differing by a phase factor $e^{i\alpha}$ describe the same physical situation. This freedom of choice for the global phase does not exist for the density matrix.

Let us recall two very useful relations concerning traces:

$$Tr|\phi_1\rangle\langle\phi_2| = \langle\phi_2|\phi_1\rangle; \quad Tr\hat{A}\hat{B} = Tr\hat{B}\hat{A}. \tag{2.3}$$

As a consequence, $\mathrm{Tr}\rho = 1$ and $\rho^2 = \rho$: the density matrix is a projector and its eigenvalues are 0s and a single 1. The eigenvector associated with the eigenvalue 1 is precisely the state vector of the system.

In the pure state case, state vectors and density matrices are two equivalent mathematical objects that one can use to describe a given physical system.

2.1.2 Basic Rules of Quantum Mechanics in the Pure State Case in Terms of the Density Matrix

We recall in Appendix A the rules of elementary quantum mechanics in terms of the state vector. We now show that it is possible to rewrite them in terms of the density matrix:

(S) *Description of physical system*: an isolated physical system, or a system submitted to externally fixed forces, prepared in a totally controlled way (pure state case), is described by a single mathematical object, the *density matrix*, or *density operator* ρ, which is a linear operator acting in a well-defined Hilbert space $\mathcal{H}$ of dimension N (finite or infinite). The density matrix is a projector and is normalized to 1:

$$\rho^2 = \rho; \quad Tr\rho = 1. \tag{2.4}$$

(E) *Evolution*: the density matrix $\rho(t)$ at any time t is given by

$$\rho(t) = \hat{U}(t)\rho(0)\hat{U}^{\dagger}(t), \tag{2.5}$$

where the unitary evolution operator $\hat{U}(t)$ is ruled by the operatorial Schrödinger equation:

$$i\hbar\frac{d}{dt}\hat{U}(t) = \hat{H}(t)\hat{U}(t), \tag{2.6}$$

where $\hat{H}$ is the Hamiltonian of the system, which is the observable associated with its energy.

(M) *Measurement*:

(M1) *Quantization*: the measurement of the physical quantity A can only give as a result one of the eigenvalues a_n of the associated Hermitian operator $\hat{A}$, called an Observable.

(M2) *Born's rule*: in general the result of the measurement cannot be predicted with certainty. The conditional probability $Proba(a_n|\rho)$ of obtaining the result a_n given that the system is in the quantum state described by the density matrix ρ is given by:

$$Proba(a_n|\rho) = Tr\hat{P}_n\rho, \tag{2.7}$$

where $\hat{P}_n$ is the projector on the eigensubspace associated with the eigenvalue a_n.

(M3) *State collapse*: the conditional state of the system just after the measurement and *in the subset of cases where the measurement has given the result* a_n is described by the density matrix

$$\rho^{after|a_n} = \frac{\hat{P}_n\rho\hat{P}_n}{Proba(a_n|\rho)}. \tag{2.8}$$

2.1.3 Examples

Let us consider two simple examples. More detailed information can be found in Appendix B.

Qubit Case

A two-level system, or qubit, is described by the 2-dimensional state vector $|\psi(t)\rangle = \alpha e^{-iE_- t/\hbar}|-\rangle + \beta e^{-iE_+ t/\hbar}|+\rangle$. It can be also described by the 2×2 density matrix

$$\rho = \begin{pmatrix} |\alpha|^2 & \alpha^* \beta e^{-i\omega_0 t} \\ \alpha \beta^* e^{i\omega_0 t} & |\beta|^2 \end{pmatrix}, \tag{2.9}$$

with $\hbar\omega_0 = E_+ - E_-$, the first column being associated with the basis state $|-\rangle$ of the qubit, and the second with the state $|+\rangle$. The diagonal terms are real, positive, and time independent. They are called *populations*. They give the probability of being in the state $|-\rangle$ or $|+\rangle$, respectively. The off-diagonal terms oscillate in time at the Bohr frequency ω_0 and are complex conjugate to each other. They are called *coherences* and are more subtle to understand. As for all oscillating quantities, not only their amplitude but also their phase plays an important role. Coherences contain most of the wave aspects of the quantum system.

1D Particle Case

A 1D particle described by the wavefunction $\psi(x)$ is also described by the two variable density matrix $\rho(x_1, x_2)$:

$$\rho(x_1, x_2) = \psi^*(x_2)\psi(x_1). \tag{2.10}$$

The diagonal part $x_1 = x_2 = x$ gives the probability density for the particle to be at point x, whereas the off-diagonal part gives information about the correlations between position measurements made at two different points. Note that when the dimension of the Hilbert space $\mathcal{H}$ is infinite, the trace is not always a well-defined object.

2.2 First Extension: Open Systems

Let us consider the very frequent case of a system S_A that is not isolated or submitted to externally fixed forces, that we call "open system." It is the case, for example, of the single trapped ions described in Section 1.2 and Figure 1.2. A trapped ion cannot be considered as alone in the universe. It has a long history before it is trapped and submitted to the measurements: the ion comes from a chemical reaction and has been heated, evaporated, and captured. In the trap it is submitted to collisions with the residual gas in the vacuum chamber, and its motion induces currents in the electrodes which create the trapping potential.

So, in addition to the Hilbert space $\mathcal{H}_A$ of the studied system S_A, one has to consider the Hilbert space $\mathcal{H}_B$ of the system S_B with which S_A has interacted beforehand, or with which it is presently interacting. In this situation one can consider that the total Hilbert space $\mathcal{H}_A \otimes \mathcal{H}_B$ is isolated. We will denote by $|u_i\rangle$ $(i = 1, \ldots, d_A)$ a basis of $\mathcal{H}_A$, by $|v_j\rangle$ $(j = 1, \ldots, d_B)$ a basis of $\mathcal{H}_B$. Therefore, $|u_i\rangle \otimes |v_j\rangle = |u_i, v_j\rangle$ is a basis of $\mathcal{H}_A \otimes \mathcal{H}_B$ (for a precise introduction of tensor products of spaces and states, see Appendix A and chapter 2F of reference [45]).

2.2.1 Reduced Density Matrix

The whole system is isolated, and therefore is in a pure state described by the state vector $|\Psi_{tot}\rangle$ or by the total density matrix $\rho_{tot} = |\Psi_{tot}\rangle\langle\Psi_{tot}|$ acting on $\mathcal{H}_A \otimes \mathcal{H}_B$.

We now want to measure a physical quantity A (like the position of the ion in the trap) that is defined only inside system S_A. It is associated with an observable $\hat{A} = \sum_n a_n \hat{P}_n$ operating only in $\mathcal{H}_A$, where $\hat{P}_n$ is the projector on the eigenspace associated to eigenvalue a_n. The probability of measuring the eigenvalue a_n is, according to Equation (2.7),

$$Proba(a_n|\rho_{tot}) = Tr\rho_{tot}\hat{P}_n = \sum_{i,j} \langle u_i, v_j|\rho_{tot}\hat{P}_n|u_i, v_j\rangle. \tag{2.11}$$

Inserting the closure relation $\sum_{i',j'} |u_{i'}, v_{j'}\rangle\langle u_{i'}, v_{j'}| = \hat{1}$, where $\hat{1}$ is the identity matrix, and taking into account the fact that the projector $\hat{P}_n$ acts only in the Hilbert space $\mathcal{H}_A$, one obtains,

$$Proba(a_n|\rho_{tot}) = \sum_{i,j,i',j'} \langle u_i, v_j|\rho_{tot}|u_{i'}, v_{j'}\rangle\langle u_{i'}|\hat{P}_n|u_i\rangle\delta_{j,j'} \tag{2.12}$$

$$= \sum_{i,j,i'} \langle u_i, v_j|\rho_{tot}|u_{i'}, v_j\rangle\langle u_{i'}|\hat{P}_n|u_i\rangle. \tag{2.13}$$

Let us now introduce the operator $\rho_{reduced}$ acting only in space $\mathcal{H}_A$, called *the reduced density matrix*, which is defined by its matrix elements,

$$\langle u_i|\rho_{reduced}|u_{i'}\rangle = \sum_j \langle u_i, v_j|\rho_{tot}|u_{i'}, v_j\rangle. \tag{2.14}$$

Using the closure relation inside the Hilbert space $\mathcal{H}_A$, *Sigma* $\sum_{i'} |u_{i'}\rangle\langle u_{i'}| = \hat{1}_{\mathcal{H}_A}$ we can write

$$Proba(a_n, \rho_{tot}) = \sum_{i,i'} \langle u_i|\rho_{reduced}|u_{i'}\rangle\langle u_{i'}|\hat{P}_n|u_i\rangle \tag{2.15}$$

$$= \sum_i \langle u_i|\rho_{reduced}\hat{P}_n|u_i\rangle = Tr\rho_{\text{reduced}}\hat{P}_n.$$

One finally retrieves the usual Born's rule relative to S_A only, provided that one uses the reduced density matrix defined in Equation (2.14), which can be written in a more symbolic way as

$$\rho_{reduced} = \sum_j \langle v_j | \rho_{tot} | v_j \rangle = Tr_{\mathcal{H}_B} \rho_{tot}. \tag{2.16}$$

This operation is called a *partial trace*, as the trace is made by summing the diagonal elements only in the Hilbert space $\mathcal{H}_B$ (the partial trace over S_B does not yield a number, but an operator acting in the space S_A).

Born's rule is therefore valid for a non-isolated system S_A provided that one uses in the formula the density matrix (2.14) or (2.16), obtained by making a partial trace of the total density matrix over the part of the system that one wants to ignore.

What about the state collapse postulate M3? The conditional state of the system just after the measurement on S_A when the measurement has given the result a_n is described by the total density matrix

$$\rho_{tot}^{after|a_n} = \frac{\hat{P}_n \rho_{tot} \hat{P}_n}{Tr(\rho_{tot} \hat{P}_n)}. \tag{2.17}$$

The reduced density matrix $\rho_{reduced}^{after|a_n}$ is obtained by the partial trace operation over $\mathcal{H}_B$, which commutes with the projection operator $\hat{P}_n$ of $\mathcal{H}$. Normalization to 1 of the reduced density matrix implies that

$$\rho_{reduced}^{after|a_n} = \frac{\hat{P}_n \rho_{reduced} \hat{P}_n}{Tr(\rho_{reduced} \hat{P}_n)}. \tag{2.18}$$

The postulates M_2 and M_3 given at the beginning of this chapter in terms of the density matrix for ideal measurements are also valid for open systems described in terms of reduced density matrices. Note that the previous demonstration does not use the fact that ρ_{tot} is a pure state. It also applies to the case of an initial non-isolated system (and/or imperfectly prepared; see Section (2.3) on which one wants to make measurements only in a smaller subspace).

2.2.2 Non-locality of Quantum Physics?

When one wants to describe a non-isolated quantum system, does one need to consider it within the whole Hilbert space describing all the quantum systems with which it has interacted, even though these systems are now far away from S, so that no interaction is possible between them? In Chapter 8, we will see in great detail the case where the whole system is, or is not, in a tensor product of quantum states defined in S_A and S_B. When the whole system is in a non-factorizable state (called an *entangled state*), the state collapse provoked by a measurement in S_A affects S_B even when these two systems have ceased to physically interact. This situation is an example of non-locality in quantum physics.

We saw in the previous section that *a complete local quantum mechanical description of a system S_A is possible* in terms of reduced density matrices acting in the Hilbert space $\mathcal{H}_A$ only, even if S_A is a part of a larger ensemble with which it interacts. This is of course true *provided that one restricts* *oneself to extracting information from measurements on system S_A only*. The puzzling nonlocal properties of quantum systems appear only when one is making correlation measurements on systems S_A and S_B.

2.2.3 Basic Properties of the Reduced Density Matrix

Note the following useful relation:

$$Tr_{\mathcal{H}_B}|u_{i'}, v_{j'}\rangle\langle u_i, v_j| = \langle v_j|v_{j'}\rangle|u_{i'}\rangle\langle u_i|. \tag{2.19}$$

Using Equation (2.14), one easily shows that $\rho_{reduced}$ has the regular properties of a density matrix, i.e. the unit trace,

$$Tr\rho_{reduced} = \sum_i \langle u_i|\rho_{reduced}|u_i\rangle = \sum_{i,j} \langle u_i, v_j|\rho_{tot}|u_i, v_j\rangle = Tr\rho_{tot} = 1, \tag{2.20}$$

and positivity,

$$\langle\phi|\rho_{reduced}|\phi\rangle = \langle\phi|\left(\sum_j \langle v_j|\rho_{tot}|v_j\rangle\right)|\phi\rangle = \sum_j \langle\phi, v_j|\rho_{tot}|v_j, \phi\rangle \geq 0, \tag{2.21}$$

as a sum of positive numbers whatever $|\phi\rangle \in \mathcal{H}$.

2.2.4 Examples

Qubit Coupled to Another

Let us consider the simple case where S_A is a qubit that is coupled to another qubit S_B, and assume that the total system of both qubits is described by the state vector

$$|\Psi_-\rangle = \frac{1}{\sqrt{2}}(|A:+, B:-\rangle - |A:-, B:+\rangle), \tag{2.22}$$

where A and B is a short notation for the two systems S_A and S_B. This state is called a Bell state. It contains strong correlations between the two qubits, as we will see in Chapter 7. The corresponding total density matrix, written on the basis $|A:-, B:-\rangle$, $|A:-, B:+\rangle$, $|A:+, B:-\rangle$, $|A:+, B:+\rangle$, is

$$\rho_{tot} = \frac{1}{2}\begin{pmatrix} 0 & 0 & 0 & 0 \\ 0 & 1 & -1 & 0 \\ 0 & -1 & 1 & 0 \\ 0 & 0 & 0 & 0 \end{pmatrix}. \tag{2.23}$$

The reduced density matrix can be calculated using Equation (2.14):

$$\rho_{reduced} = \begin{pmatrix} 1/2 & 0 \\ 0 & 1/2 \end{pmatrix}. \tag{2.24}$$

Being proportional to the identity matrix, it has the same expression in any qubit basis. It therefore corresponds to a totally randomized qubit. We will come back to this example in Chapter 7 (which is about entanglement).

Spontaneous Emission

Let us consider a two-level atom coupled to the electromagnetic field (see Appendix G). Let us suppose that at time $t = 0$ it is in the excited state $|+\rangle$ without any photon present $|\Psi(0)\rangle = |+\rangle \otimes |0\rangle$, where $|0\rangle$ is the vacuum state. From this state, the system spontaneously decays to the ground state by emitting a single photon in an electromagnetic field mode ℓ, written $|1 : \ell\rangle$, these modes differing by the direction of emission, polarization, and frequency. The complete theory, found for example in [82] and recalled in Appendix G.2, shows that the total system, which is isolated, is described by the wavevector $|\Psi(t)\rangle$ given at any time by expression (G.9),

$$|\Psi(t)\rangle = \gamma_0(t)|+\rangle \otimes |0\rangle + \sum_\ell \gamma_\ell(t)|-\rangle \otimes |1 : \ell\rangle, \tag{2.25}$$

with $\gamma_0(t)$ and $\gamma_\ell(t)$ given by Equations (G.18) and (G.19):

$$\gamma_0(t) = e^{-\frac{\Gamma}{2}t}e^{-i(\omega_0+\delta\omega)t}, \tag{2.26}$$

$$\gamma_\ell(t) = \frac{V_\ell^*}{i\hbar}\frac{1 - e^{-\frac{\Gamma}{2}t}e^{-i(\omega_\ell-\omega_0+\delta\omega)t}}{\frac{\Gamma}{2} - i(\omega_\ell - \omega_0 - \delta\omega)}, \tag{2.27}$$

where Γ is the spontaneous decay constant.

If one is only interested in the fate of the atom, one will use the atomic reduced density matrix obtained by tracing over the field degrees of freedom:

$$\rho_{at} = \langle 0|\Psi(t)\rangle\langle\Psi(t)|0\rangle + \sum_\ell \langle 1 : \ell|\Psi(t)\rangle\langle\Psi(t)|1 : \ell\rangle \tag{2.28}$$

$$= |\gamma_0(t)|^2|+\rangle\langle+| + \left(1 - |\gamma_0(t)|^2\right)|-\rangle\langle-|. \tag{2.29}$$

If one is only interested in the fluorescent light emitted by the decaying atom, one will use the photon reduced density matrix obtained by tracing over the atomic degree of freedom:

$$\rho_{photon} = \langle +|\Psi(t)\rangle\langle\Psi(t)|+\rangle + \langle -|\Psi(t)\rangle\langle\Psi(t)|-\rangle \tag{2.30}$$

$$= |\gamma_0(t)|^2|0\rangle\langle 0| + \sum_{\ell,\ell'} \gamma_\ell(t)\gamma_{\ell'}(t)|1 : \ell\rangle\langle 1 : \ell'|. \tag{2.31}$$

These two reduced density matrices ignore, of course, the strong correlations existing between the atom and the field: if the atom is in the excited state, one is sure that no photon is present. If the atom is in the ground state, one is sure that a photon has been emitted.

At long times $\left(t \gg \frac{1}{\Gamma}\right)$, $\gamma_0(t) = 0$ and $\gamma_1(t) \neq 0$, the state vector is factorizable in an atomic part (the atomic ground state) and a field part:

$$|\psi_R\rangle = \sum_\ell \frac{V_\ell^*}{i\hbar}\frac{e^{-i\omega_\ell t}}{\frac{\Gamma}{2} - i(\omega_\ell - \omega_0 - \delta\omega)}|1_\ell\rangle. \tag{2.32}$$

One has $\widehat{N}|\psi_R\rangle = |\psi_R\rangle$: the field state is an eigenstate of the photon number operator with eigenvalue 1. It is therefore a single photon state, sometimes called a single photon wavepacket.

The probability of finding a single photon in mode ℓ at long times is

$$Proba(1_\ell) = \frac{|V_\ell|^2}{\hbar^2} \frac{1}{\frac{\Gamma^2}{4} + (\omega_\ell - \omega_0 - \delta\omega)^2}. \tag{2.33}$$

It is significant only when ω_ℓ is close to $\omega_0 + \delta\omega$ in a frequency band of width Γ: the excited atomic level, which has acquired a finite lifetime, has therefore been broadened and shifted by its interaction with the quantized electromagnetic field in the vacuum mode.

Spontaneous emission of a single atom (or single qubit) is a very efficient way to create a single photon state. It has been experimentally implemented in many two-level systems. The drawback of this technique is that this state exists only when $t \gg \Gamma^{-1}$. One must then wait for some time after the atom excitation in order to be sure to get such a single photon: such a single photon source is therefore not very intense.

Thermal Quantum State

In the thermodynamical system named a canonical ensemble, one has a small system that is in contact with the reservoir with which it is at thermal equilibrium at temperature T. By tracing over the reservoir variables, one obtains a reduced state for the small system, described by a density matrix ρ_T, which can be shown to be

$$\rho_T = \frac{e^{-\hat{H}/k_B T}}{Z}, \tag{2.34}$$

where $\hat{H}$ is the Hamiltonian of the small system, k_B is the Boltzmann constant, and $Z = Tre^{-\hat{H}/k_B T}$ the partition function, well known in thermodynamics.

2.3 Second Extension: Imperfect State Preparation

2.3.1 Example

Let us consider a spin 1/2 particle. Its two basis states, $|+\rangle$ and $|-\rangle$, correspond to two opposite orientations of its magnetic moment. A spin pointing in the direction of polar angles (θ, ϕ) is described by the state vector

$$|\psi(\theta, \phi)\rangle = \cos(\theta/2)|-\rangle + e^{i\phi}\sin(\theta/2)|+\rangle. \tag{2.35}$$

In the Stern–Gerlach experiment, the particle (more precisely the silver atom) is produced in a heated oven. As a result, the atom exiting the oven has a spin pointing in a direction (θ, ϕ) that is not controlled, and therefore its exact state vector is not known.

2.3.2 Statistical Mixture

More generally, let us consider the situation where the quantum system has a probability p_1 to be prepared in state $|\psi_1\rangle, \ldots, p_i$ to be prepared in state $|\psi_i\rangle$, with $i = 1, \ldots, N$ and $\sum_i p_i = 1$ (these states are not necessarily orthogonal). One can define an ensemble averaged density matrix, or *statistical mixture*, ρ_M, by

$$\rho_M = \sum_{i=1}^{N} p_i |\psi_i\rangle\langle\psi_i|. \tag{2.36}$$

Here, ρ_M, also called *mixed state*, is of trace 1, and, because $p_i \geq 0$, remains a positive operator.

2.3.3 Measurement on a Statistical Mixture

The probability of obtaining the result a_n in the measurement of A is, in the present situation, the average of the probabilities that one obtains when the system is in one of the prepared states $|\psi_i\rangle$, i.e.

$$Proba(a_n|\rho_M) = \sum_{i=1}^{N} p_i Tr\hat{P}_n|\psi_i\rangle\langle\psi_i| = Tr\hat{P}_n\rho_M. \tag{2.37}$$

One sees that the usual Born's rule can be extended to the case of a statistical mixture in the density matrix formalism. This is not possible in the wave vector formalism. The reason for this is that Born's expression for the probability is quadratic in $|\psi\rangle$ but linear in ρ.

What about the state collapse postulate M3? The conditional state of the system when the measurement has given the result a_n is described by the statistical mixture of collapsed states $\hat{P}_n|\psi_i\rangle\langle\psi_i|\hat{P}_n/\langle\psi_i|\hat{P}_n|\psi_i\rangle$, weighted by the probability of getting such a state, $p_i\langle\psi_i|\hat{P}_n|\psi_i\rangle/\sum_i p_i\langle\psi_i|\hat{P}_n|\psi_i\rangle$, giving the normalized product of the probability p_i of state preparation and the probability $\langle\psi_i|\hat{P}_n|\psi_i\rangle$ of getting the result a_n:

$$\rho^{after|a_n} = \sum_i \frac{p_i}{\sum_i p_i\langle\psi_i|\hat{P}_n|\psi_i\rangle}\hat{P}_n|\psi_i\rangle\langle\psi_i|\hat{P}_n = \frac{\hat{P}_n\rho_M\hat{P}_n}{Tr\hat{P}_n\rho_M}. \tag{2.38}$$

The state collapse postulate can therefore be extended to statistical mixtures.

2.3.4 Back to the Example

In the spin 1/2 case, the density matrix describing a spin pointing in the direction of polar angles (θ, ϕ) is, from relation (D.6) of Appendix D,

$$\rho(\theta,\phi) = \begin{pmatrix} \cos^2(\theta/2) & \cos(\theta/2)\sin(\theta/2)e^{i\phi} \\ \cos(\theta/2)\sin(\theta/2)e^{-i\phi} & \sin^2(\theta/2) \end{pmatrix}. \tag{2.39}$$

When such a matrix is averaged over all the possible values of the angles $d\Omega = \sin\theta d\theta d\phi$, supposed to have the same probability in all directions, one gets the statistical mixture

$$\rho_M = \frac{1}{2}\begin{pmatrix} 1 & 0 \\ 0 & 1 \end{pmatrix} = \hat{1}/2. \tag{2.40}$$

We see that the process of averaging has completely washed out the coherences and equalized the population of the two levels. Being proportional to the identity operator, such a density matrix *carries minimal information about the quantum system* as it has exactly the same form in any basis of the two-level system. We also note that the same density matrix could have been obtained in another set-up where one has probability 1/2 of preparing the state $|+\rangle$ and probability 1/2 of preparing the state $|-\rangle$. Different physical situations may lead to the same statistical mixture density matrix, and therefore to identical measurement results.

This example allows us to highlight the role of the coherences, the off-diagonal elements of the density matrix. Let us consider a spin pointing in the $\theta = \pi/2, \phi = 0$ direction. It is described by the state vector $(|+\rangle + |-\rangle)/\sqrt{2}$ or by the density matrix

$$\rho(\theta = \pi/2, \phi = 0) = \frac{1}{2}\begin{pmatrix} 1 & 1 \\ 1 & 1 \end{pmatrix}. \tag{2.41}$$

Matrices (2.40) and (2.41) differ only by the coherences; $\rho(\theta = \pi/2, \phi = 0)$ describes a pure state, coherent superposition of the two basis states with equal coefficients, representing a system which is at the same time in states $|-\rangle$ AND $|+\rangle$. In contrast, ρ_M describes a statistical mixture representing a system which is (in a random way) in state $|-\rangle$ OR $|+\rangle$.

2.3.5 Two Kinds of Randomness

Let us go back to expression (2.37) for the probability of a measurement on the system in a mixed state:

$$Proba(a_n|\rho_M) = \sum_{i=1}^{N} p_i Tr\hat{P}_n|\psi_i\rangle\langle\psi_i|. \tag{2.42}$$

We have in this formula a probability that is the result of two uncertainties of radically different natures: the uncertainty described by the probabilities p_i is due to the experimentalist's ignorance or laziness. It can be reduced and even suppressed by using more sophisticated experimental set-ups able to generate a unique quantum state. In contrast, the uncertainty described by Born's rule $Tr\hat{P}_n|\psi_i\rangle\langle\psi_i|$ is of a fundamental nature. It exists even when everything is perfectly controlled in the experiment and it cannot be completely reduced. The existence of such a fundamental, uncontrollable, quantum uncertainty makes some physicists feel uncomfortable with this situation. Einstein is the most famous of them. Let us quote him:

(Quantum) theory brings us many things, but it hardly allows us to get closer to the secret of the Old Man. Anyway, I am convinced that at least He does not play dice (A. Einstein, letter to M. Born 1926)

Physicists have wondered whether the quantum uncertainty is due to our partial ignorance of the quantum world. One can suppose that there exist supplementary degrees of freedom of the system under study that we do not know yet, and therefore do not master in the experiments: the uncertainty in the control of these hidden variables would induce a corresponding uncertainty in the measurement results that we phenomenologically describe presently by the Born's rule. This was, in particular, the hope of Einstein, hence the title of his famous paper [62]: *Can quantum-mechanical description of physical reality be considered complete?* We will examine this issue further in Section 7.4. Let us just say here that that the existence of *local* supplementary unknown degrees of freedom leads to conclusions (the famous Bell inequality) which are in contradiction with experimental results.

2.4 General Properties of the Set $\mathcal{D}_{\mathcal{H}}$ of Density Matrices

We have shown that the density matrix formalism is able to describe extended categories of situations encountered in quantum mechanics. We will now consider the properties of these matrices from a more general perspective.

2.4.1 Postulate S for General Density Matrices

Let us begin by reformulating the postulate (S) of Section 2.1 that deals with the description of a physical system, now in a way which is valid in all cases, open and/or imperfectly prepared systems:

(S) *Description of physical system*: a physical system is described by a single mathematical object, the *density matrix* ρ, **which is a linear operator, acting in a well-defined Hilbert space $\mathcal{H}$, that is positive and normalized to 1**:

$$\forall\, |\psi\rangle \in \mathcal{H} \quad \langle\psi|\rho|\psi\rangle > 0 \quad ; \quad Tr\rho = 1. \tag{2.43}$$

We will deal with the rules (E) and (M1, M2, M3) about evolution and measurement in non-ideal situations in Chapters 4 and 5.

2.4.2 Mathematical Properties of ρ

Let us call $\mathcal{D}_{\mathcal{H}}$ the set of density matrices over $\mathcal{H}$, i.e. the set of linear operators acting on $\mathcal{H}$ that are positive and of unit trace.

The definition of ρ implies that $Tr\rho$ is well defined (trace-class operator), even in Hilbert spaces of infinite dimension. It can be shown that density matrices are Hermitian: ρ can then be diagonalized, with real eigenvalues p_i and orthonormal eigenstates $|\phi_i\rangle$. Because $Tr\rho = 1$ these eigenvalues are comprised between 0 and 1 of sum 1, like probabilities. One can always write any density matrix ρ in diagonal form, i.e. as the statistical mixture

$$\rho = \sum_i p_i |\phi_i\rangle\langle\phi_i|. \tag{2.44}$$

A positive operator such as ρ can always be written as

$$\rho = \hat{A}^\dagger \hat{A}, \tag{2.45}$$

where $\hat{A}$ is any linear operator, Hermitian or non-Hermitian, which is not unique. For example, for a pure state $\rho = |\psi\rangle\langle\psi|$, $\hat{A} = |\phi\rangle\langle\psi|$ whatever the value of $|\phi\rangle$.

This positivity implies for the matrix elements that $|\rho_{ij}|^2 \leq \rho_{ii}\rho_{jj}$. Therefore, if the coherence ρ_{ij} is nonzero, then the product of the corresponding populations is not zero: coherences exist only when the two states connected by the coherence are populated.

2.4.3 Physical Properties of $\mathcal{D}_\mathcal{H}$

The set $\mathcal{D}_\mathcal{H}$ of density matrices over $\mathcal{H}$ is not a vector space, because a linear combination of positive operators is not, in general, a positive operator. This implies that there is no superposition principle for the density matrix: in a Young slit experiment, for example, if ρ_1 describes the state of the particle going through slit 1, and ρ_2 the state of the particle going through slit 2, the whole system is not described by $\rho = (\rho_1 + \rho_2)/2$, which would rather describe the statistical mixture when the particle has a 50% chance of going through slit 1 and 50% chance of going through slit 2. The interference pattern arises from the linear superposition of the corresponding wave vectors $|\psi_1\rangle$ and $|\psi_2\rangle$, so that the density matrix describing the whole system is actually $\rho_{tot} = (\rho_1 + \rho_2 + |\psi_1\rangle\langle\psi_2| + |\psi_2\rangle\langle\psi_1|)/2$: the information about the interference is therefore more in the coherence operator $|\psi_1\rangle\langle\psi_2|$ than in the states ρ_1 and ρ_2 describing the two possible paths.

The set $\mathcal{D}_\mathcal{H}$ is actually a *convex space*, because of the following property:

$$\forall\, \rho_1, \rho_2 \in \mathcal{D}_\mathcal{H}, \forall\, p \in [0, 1], \qquad p\rho_1 + (1-p)\rho_2 \in \mathcal{D}_\mathcal{H}, \tag{2.46}$$

meaning that one can superpose density matrices only in a probabilistic way.

As explained in Appendix D, any mixed density matrix ρ of a qubit can be written as:

$$\rho = p\rho_1 + (1-p)\rho_2, \tag{2.47}$$

where ρ_1 and ρ_2 are pure states, eigenvectors of the density matrix, and $0 < p < 1$ is a probability.

Pure orthogonal states like ρ_1 and ρ_2 in relation (2.47) belong to the surface of the Bloch sphere (see Appendix D) at the extremities of a diameter, whereas the mixed state ρ in relation (2.47) lies inside the Bloch sphere on this diameter. The center of the sphere is associated with the minimum information state $\hat{1}/2$.

2.5 Purity

2.5.1 Definitions, Examples

Let us recall that one calls a *pure state* a quantum state that is described by a density matrix ρ that can be written as $|\psi\rangle\langle\psi|$, and *mixed state* a quantum state for which this reduction is not possible.

Let us now calculate ρ^2, using expression (2.44):

$$\rho^2 = \sum_{i,j} p_i p_j |\phi_i\rangle\langle\phi_i|\phi_j\rangle\langle\phi_j| = \sum_i p_i^2 |\phi_i\rangle\langle\phi_i|. \tag{2.48}$$

This quantity is different from ρ unless the sum reduces to its first term, i.e. when the density matrix is equal to $|\phi_1\rangle\langle\phi_1|$ and corresponds to a pure state. Consequently, *the density matrix of a mixed state is not a projector.*

One can then define the *purity* P_u of a state described by a density matrix ρ by:

$$P_u = Tr\rho^2. \tag{2.49}$$

One has $P_u = \sum_i p_i^2$ (with $\sum_i p_i = 1$). This last expression implies that $P_u \leq 1$ and that the state is pure if and only if $P_u = 1$. Its smallest value is $1/d$, where d is the dimension of the Hilbert space. This value is obtained when the density matrix ρ is proportional to the identity $\hat{1}$, i.e. when $\rho = \hat{1}/d$, where d is the dimension of the Hilbert space. In this case, Born's rule implies that all measurements have equal probabilities $1/d$ whatever the physical quantity A and the measurement result a_n: It is a situation of minimal control over the state preparation. The only piece of information about the prepared state is that it belongs to a given Hilbert space, which is already important: it tells us that other possible parameters of the system must not be taken into account. Let us take the example an hydrogen atom in the mixed state,

$$\rho = \frac{1}{3}(|2,1,-1\rangle\langle 2,1,-1| + |2,1,0\rangle\langle 2,1,0| + |2,1,1\rangle\langle 2,1,1|), \tag{2.50}$$

where $|2,1,-1\rangle$, for example, means a hydrogen state of electronic quantum numbers $n = 2, l = 1, m = -1$. Here, ρ is proportional to the identity in the subspace of states of $n = 2, l = 1$. It describes an atom where the orientation of the orbital angular momentum of its electron is not mastered by the preparation. But using such a state implies that the other degrees of freedom of the hydrogen atom, like the electron spin and nuclear spin, are not interacting with the electron charge (no spin-orbit coupling or hyperfine interaction), and therefore that the electronic and nuclear spin part of the hydrogen quantum state can be factored out and are not affected by the electronic motion.

Let us consider in more detail the two examples described in Section 2.2.

In the case of a qubit coupled to another qubit, the total system is in a pure state, whereas the reduced state is in the maximally mixed state $\hat{1}/2$ of minimum purity $P_u = 1/2$ in which all information is lost, meaning that all the information was contained in the correlation between the two qubits. We will come back to this example in Chapter 7.

In the case of spontaneous emission, one notes that the density matrix of the atom alone (Equation (2.29)) is pure for $t = 0$ and $t \to \infty$, and that the purity reaches its minimum value 1/2 of a fully randomized state for $\Gamma t = ln\,2$. The purity properties of the photonic density matrix (2.30) turn out to be the same as for the atomic density matrix.

The purity of a thermal state of a qubit (Equation (2.34)) can be shown to be equal to $P_u = \tanh(\hbar\omega_0/2k_BT)$: it is close to a pure state at low temperatures or high frequencies, i.e. $k_BT \ll \hbar\omega_0$.

2.5.2 Purification Theorem

We now show that any density matrix acting in a Hilbert space $\mathcal{H}$ can always be considered as the reduced density matrix of a pure state in a larger Hilbert space $\mathcal{H}_{large}$

Let us consider a density matrix $\rho \in \mathcal{D}_{\mathcal{H}}$. As stated before, it can always be written as

$$\rho = \sum_i p_i |\phi_i\rangle\langle\phi_i|. \tag{2.51}$$

Now let us introduce another Hilbert space, called the ancilla space, $\mathcal{H}_a$, of dimension equal to or larger than the number of nonzero eigenvalues of ρ, and its basis $\{|v_j\rangle\}$. Let us consider the state vector defined in $\mathcal{H}_{large} = \mathcal{H} \otimes \mathcal{H}_a$:

$$|\psi_{tot}\rangle = \sum_i \sqrt{p_i}|\phi_i, v_i\rangle. \tag{2.52}$$

The corresponding density matrix is

$$\rho_{tot} = \sum_{i,j} \sqrt{p_i p_j}|\phi_i, v_i\rangle\langle\phi_j, v_j|. \tag{2.53}$$

It is, by construction, the density matrix of a *pure state*. It is easy to check that the partial trace over $\mathcal{H}_a$ of the pure case ρ_{tot} gives precisely the mixed state ρ.

The conclusion is that one can always purify a mixed state by going in a Hilbert state of larger dimension. The purification can be actually performed in many different ways. We will use this property in the following chapters.

2.6 Exercises for Chapter 2

Exercise 2.1 Assume one will perform experiments over an ensemble of quantum particles and that these particles are prepared with probability π_i in the *pure* state $|\psi_i\rangle$. Such an ensemble is a statistical mixture of pure states and we refer to it as being in a *mixed* quantum state.

(a) Derive the density matrix ρ and the generalized Born's rule $\langle\hat{A}\rangle = Tr[\rho\hat{A}]$ from Born's rule for kets, $\langle\hat{A}\rangle = \langle\psi_i|\hat{A}|\psi_i\rangle$, and probability arguments.

(b) Can one obtain a given density operator from different sets of preparation choices of π_i and $|\psi_i\rangle$? Give an example.

(c) Can one distinguish, by performing experiments on the system, two ensembles that have the same density operator but that were prepared from different choices of p_i and $|\psi_i\rangle$?

(d) Consider these two state preparation protocols for a spin-1/2 system:

- In a stream of particles the 1st, 3rd, 5th, ... (odd) will be prepared in the "up" state $|\uparrow\rangle$ and the 2nd, 4th, ... (even) particles will be prepared in the "down" state $|\downarrow\rangle$.

- In a stream of particles the 1st, 3rd, 5th, ... (odd) will be prepared in the "right" state state $|\rightarrow\rangle = (|\uparrow\rangle + |\downarrow\rangle)/\sqrt{2}$ and the 2nd, 4th, ... (even) particles will be prepared in the "left" state $|\leftarrow\rangle = (|\uparrow\rangle - |\downarrow\rangle)/\sqrt{2}$.

Can one distinguish these two preparation protocols by only performing measurements on the stream of particles? Is this result in conflict with the correct answer to the previous question?

Exercise 2.2 The expression for Pauli matrices is recalled in Appendix D.

(a) Find the eigenvalues of the Pauli matrices $\hat{\sigma}_j$.

(b) Find an expression for the commutator $[\hat{\sigma}_i, \hat{\sigma}_j]$ and the anticommutator $\{\hat{\sigma}_i, \hat{\sigma}_j\}$.

(c) Show that

$$\hat{\sigma}_i^2 = \hat{1},\ det(\hat{\sigma}_i) = -1,\ Tr(\hat{\sigma}_i) = 0,\ \hat{\sigma}_i\hat{\sigma}_j = \delta_{ij}\hat{1} + i\epsilon_{ijk}\hat{\sigma}_k, \quad (2.54)$$

$$(\vec{a}\cdot\vec{\hat{\sigma}})(\vec{b}\cdot\vec{\sigma}) = \vec{a}\cdot\vec{b},\hat{1} + i(\vec{a}\times\vec{b})\cdot\vec{\sigma}. \quad (2.55)$$

Here, ϵ_{ijk} and δ_{ij} are the Levi-Civita and Kronecker tensors, respectively.

Exercise 2.3 Consider the density matrix $\hat{\rho}$ of a qubit. Show that it can be written in terms of the Pauli operators as

$$\rho = \frac{1}{2}\left(\hat{1} + U_x\hat{\sigma}_x + U_y\hat{\sigma}_y + U_z\hat{\sigma}_z\right),$$

where the real numbers U_i are the Cartesian components of a vector $\vec{U}$, called the Bloch vector, defined in an abstract three-dimensional space. Defining $\vec{\hat{\sigma}} = (\hat{\sigma}_x, \hat{\sigma}_y, \hat{\sigma}_z)$ one has $U_x\hat{\sigma}_x + U_y\hat{\sigma}_y + U_z\hat{\sigma}_z = \vec{U}\cdot\vec{\sigma}$.

Exercise 2.4 Show that the Bloch vector is also defined by $U_i = \langle \hat{\sigma}_i \rangle$.

Exercise 2.5 Consider the operator $\hat{P}_n = \frac{1}{2}(\hat{1} + \vec{n} \cdot \vec{\sigma})$, where $\vec{n}$ is a real unit vector.

(a) Show that it is a projector.
(b) Show that $\vec{n} \cdot \vec{\sigma} \hat{P}_n = \hat{P}_n$.
(c) Show that $\hat{P}_n|\psi\rangle$ is eigenvector of $\vec{n} \cdot \vec{\sigma}$ with eigenvalue $+1$ (assume $\hat{P}_n|\psi\rangle \neq 0$). We call the normalized version of this eigenvector $|+\vec{n}\rangle = \hat{P}_n|\psi\rangle / \|\hat{P}_n|\psi\rangle\|$ (independent of ψ).
(d) Convince yourself that one can write $\hat{P}_n = |+\vec{n}\rangle\langle +\vec{n}|$ and that $\hat{P}_{-n} = (\hat{1} - |+\vec{n}\rangle\langle +\vec{n}|)$ is the projector over the eigenspace of eigenvalue -1 of $\vec{n} \cdot \vec{\sigma}$. Show that by defining $|-\vec{n}\rangle$ by the relation $\vec{n} \cdot \vec{\sigma}|-\vec{n}\rangle = -|-\vec{n}\rangle$, the second relation can be written as $\hat{P}_{-n} = |-\vec{n}\rangle\langle -\vec{n}|$.
(e) What is the physical meaning of $\vec{n}$, $\vec{n} \cdot \vec{\sigma}$, and $\hat{P}_n$?

Exercise 2.6 Determine the direction of the Bloch vector $\vec{U}$ for a general pure state,

$$|\psi\rangle = \cos\frac{\theta}{2}|0\rangle + e^{i\phi}\sin\frac{\theta}{2}|1\rangle, \quad \theta \in [0, \pi], \ \phi \in [0, 2\pi).$$

Compute the purity in this case.

Exercise 2.7 Overlap of states.

(a) We consider two density matrices $\hat{\rho}_1$ and $\hat{\rho}_2$. We define the overlap between them as $O_{12} = \mathrm{Tr}(\hat{\rho}_1\hat{\rho}_2)$. Show that this definition is consistent with the usual definition of the overlap $|\langle\psi_2|\psi_1\rangle|^2$ between two pure states $|\psi_1\rangle$ and $|\psi_2\rangle$.
(b) Show that, for qubits, the overlap O_{12} is given by $O_{12} = \frac{1}{2}\left(1 + \vec{U}_1 \cdot \vec{U}_2\right)$. Relate this result with that of Exercise 2.3.

Exercise 2.8 Purity

Show that purity does not change with time if the evolution is driven by the Schrödinger equation (ie: the evolution is unitary). What kind of evolution does it change then?

Exercise 2.9 Modification of purity under the operation of partial trace

We want to compare the purity of a state before and after the partial trace operation. We will consider successively the following density matrices:

– $(|u_1, v_1\rangle\langle u_1, v_1| + |u_1, v_2\rangle\langle u_1, v_2|)/2$;
– $(|u_1, v_1\rangle\langle u_1, v_1| + |u_2, v_2\rangle\langle u_2, v_2|)/2$;

Determine the reduced states, traced over the $|v_i\rangle$ basis, then calculate the purity of the full states and of the reduced states.

Exercise 2.10 Show that for an ensemble of states $\frac{1}{\sqrt{2}}(|+\rangle + e^{i\theta}|-\rangle)$, where θ is a Gaussian random variable, the purity reads $P_u = \frac{1}{2}(1 + |\langle e^{i\theta}\rangle_j|) = \frac{1}{2}(1 + e^{-\frac{1}{2}\sigma_\theta^2})$, where $\sigma_\theta^2 = Var(\theta)$. Here, j enumerates the realizations of the noise.

3 Experiment: Quantum Processes

In Chapter 1 we described in a nonexhaustive and simplified way some of the techniques allowing experimentalists to detect single quantum objects in well-defined states and to record their quantum jumps. Here, in this experimental chapter, we give a few examples of situations often encountered in quantum physics, including the experiments performed by D. Wineland [178] and S. Haroche [87] for which they were awarded the 2012 Nobel Prize in Physics.

In many experiments, one finds the following sequence: a single quantum system is prepared in a state that is experimentally controlled as exhaustively as possible. The system then interacts with the outer world and undergoes a well-defined evolution. It is finally detected. The whole sequence is repeated and its successive results stored. This sequence (*preparation → evolution → detection*) is often called a *quantum process*. We will go into more detail about the three steps of quantum processes in Chapters 4 and 5.

3.1 Quantum Control and Detection of a Trapped Ion State

The experiment (see Figure 3.1), described in D. Wineland's Nobel lecture [178], is performed on a single mercury ion trapped in a Penning trap (Section J.2). The three-level electronic structure given in Figure 3.1 is the same as in Figure 1.3. One uses the shelving technique described in Chapter 1: A laser resonant on the transition $|G\rangle \rightarrow |B\rangle$ ($S_{1/2}$ to $P_{1/2}$ in the diagram) induces a strong fluorescent signal displaying quantum jumps, which depend on the populated state of the ion, $|G\rangle$ or $|D\rangle$ ($S_{1/2}$ or $D_{5/2}$ in the diagram). The detection laser can be stopped at any time, allowing for a high fidelity detection of the internal states $|G\rangle$ or $|D\rangle$ of the ion, which is then left unperturbed. Let ω_0 be the frequency of the qubit transition $|G\rangle \rightarrow |D\rangle$. The ionic harmonic motion, at a frequency ω_v of a few MHz, is quantized, with external motional states $|n_v\rangle$ that are thermally excited. When one applies a classical field on the ion, the coupling between internal and external degrees of freedom, described by the Hamiltonian given in Section J.2, induces the presence of two well-separated sidebands at frequencies $\omega_0 \pm \omega_v$ in the Lamb–Dicke regime, corresponding to transitions $|G, n_v\rangle$ to $|D, n_v \pm 1\rangle$ coupling motional and internal states.

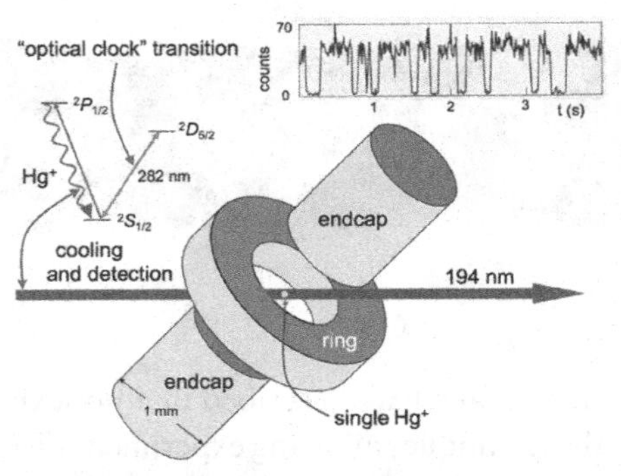

Fig. 3.1 Diagram of the electrode configuration that creates a potential trapping the ion in the center of the ring, together with the relevant level diagram and a recording of quantum jumps measured on the fluorescence. From D. Wineland, Superposition, entanglement and raising Schrödinger's cat, Nobel lecture, *Reviews of Modern Physics*, 2013 [178].

In the first step of the experiment one generates a pure ground state $|G, n_v = 0\rangle$ using the technique of sideband cooling described in Appendix J: A laser at frequency $\omega_0 - \omega_v$ cools down, after many cycles of excitation and spontaneous emission, the ion to the ground trapping state $|G, n_v = 0\rangle$ from which it cannot escape. The motional ground state of the ion $|n_v = 0\rangle$ is affected by zero point motion and has a spatial extension of tens of nanometers.

From this, the application of sequences of classical field pulses tuned at the resonant frequencies ω_0, $\omega_0 - \omega_v$, and $\omega_0 + \omega_v$ is a convenient way to manipulate in a well-controlled way the internal-motional quantum states of the ion, and its motional state when it is possible to factorize out the internal part. They allow us to generate any desired motional state $\alpha|G\rangle + \beta|D\rangle$ with coherence times around 1 s. For example, motional Fock states $|n_v\rangle$ have been generated [124], as well as motional coherent states $|\alpha\rangle$, which are oscillating wavepackets with the same spatial extension as the ground state $|n_v = 0\rangle$. Entangled motional-internal states can also be generated. For example, the generation of the Schrödinger cat state [130]:

$$|\Psi\rangle = \frac{1}{\sqrt{2}}(|G\rangle \otimes |\alpha\rangle + |D\rangle \otimes |-\alpha\rangle), \tag{3.1}$$

where $|\alpha\rangle$ and $|-\alpha\rangle$ are motional coherent states oscillating with opposite phases.

The detection of the exact motional state that has been generated by the experiment is done by analyzing many sequences of fluorescence measurements. They give the populations of the levels, but not the coherences, which are derived from a recording of Rabi oscillations as a function of the pulse durations.

3.2 Continuous Nondestructive Monitoring of a Single Microwave Photon

We turn now to an experiment, described in S. Haroche's Nobel lecture [88], that is complementary to the previous one: Instead of working with a single trapped ion submitted to an appropriate electromagnetic wave to nondestructively create nonclassical states of the ion motion, one considers a single trapped photon interacting with a series of single atoms to create in the trapping cavity, and in a nondestructive way, nonclassical states of the electromagnetic field in the cavity mode.

The experimental set-up is sketched in Figure 3.2: The horizontal arrow gives the path of a very weak beam of circular rubidium highly excited states (Rydberg states). They first cross the box R_1, in which the circular Rydberg states $|+\rangle = |n = 50, l = 49, m = 49\rangle$ or $|-\rangle = |n = 49, l = 48, m = 48\rangle$, of lifetime in the 30 ms range, are generated. A $\pi/2$ pulse of microwave light resonant with the transition $|-\rangle \rightarrow |+\rangle$ is applied to the atoms in cavity R_1. It induces a rotation by $\pi/2$ of the Bloch vector (see Section G.3), and creates a superposition of the two atomic states. The atoms enter then into an ultra-high finesse microwave cavity C (called photon box), made of superconducting mirrors, that keeps microwave photons trapped over very long times, of the order of 0.1 s. Single rubidium atoms interact one by one with the cavity mode. At the exit of the cavity C, they are subject to a second $\pi/2$ pulse in cavity R_2 that transforms back coherences into populations. The quantum state of the atom is detected downstream in the condenser D in either state $|+\rangle$ or $|-\rangle$, as described in Section 1.4.

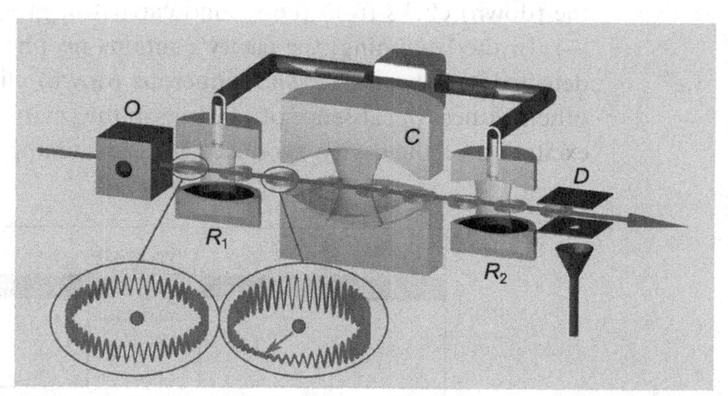

Fig. 3.2 Diagram of the experimental set-up. Single circular Rydberg atoms, created in O, cross the cavity C one by one and interact with the quantized microwave field trapped in C. Cavities R_1 and R_2 modify the quantum state of the atom in a controlled way (represented by the oscillating wave functions in the inserts). From S. Haroche, Controlling photons in a box and exploring the quantum to classical boundary, Nobel lecture, *Reviews of Modern Physics*, 2013 [87].

Using this set-up, reminiscent of the Ramsey configuration (Appendix G.2.1), it is possible to detect the presence of a single microwave photon in the cavity without destroying it by using a nonresonant interaction between the atom and the cavity. We know from Appendix G.3.3 that such a nonresonant interaction does not induce transitions between the atomic levels but affects the phase of the qubit wavefunction components by a factor $e^{\pm i\phi}/2$, corresponding to a rotation by ϕ in the equatorial plane of the Bloch vector. In addition, the quantum state phase shift ϕ depends in a detectable way on the presence of a single microwave photon in the cavity (Equation (G.41)). In the experiment, the atom-cavity coupling is so large that a single photon induces a complete 2π rotation of the atomic Bloch vector. If the cavity is empty, the atom stays in its initial state. The transformation provoked by the high Q cavity on the atom in state $|-\rangle$ is therefore

$$|-\rangle \otimes |1\rangle \rightarrow -|-\rangle \otimes |1\rangle; \quad |-\rangle \otimes |0\rangle \rightarrow |-\rangle \otimes |0\rangle, \tag{3.2}$$

taking into account the fact that a 2π rotation induces a sign change on the state vector (see Section G.2). Note that the intracavity photon state is left invariant by the interaction, making it a nondemolition interaction. The $\pi/2$ pulse in the second auxiliary cavity R_2 transforms the coherences back into populations, so that the phase change is converted into a change of the final state of the atom that is measured by the detector D. The present set-up is therefore able to detect the presence of a single photon in the cavity without destroying it.

Figure 3.3 shows a temporal sequence of single Rydberg atom detections. The upper vertical bars are the (up) clicks from the first capacitor, produced by ionization of atoms in state $|+\rangle$. The lower bars of the upper panel are the (down) clicks from the second capacitor, produced by atoms in state $|-\rangle$. In the beginning, the cavity contains no photons, so that atoms are detected mostly in $|-\rangle$. The numerous (down) clicks, very close to each other, witness the absence of photons in the cavity. A very weak microwave excitation of the cavity, produced by a low temperature thermal source S,

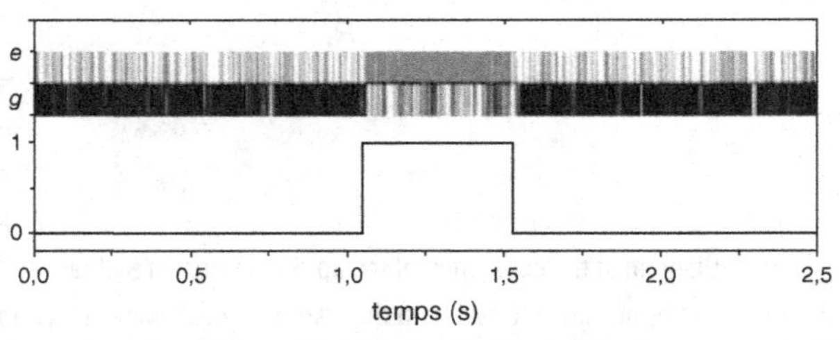

Fig. 3.3 (Top) Temporal sequence of recordings of field-ionized Rydberg atoms in state $e = |+\rangle$ (upper series of vertical segments) and $g = |-\rangle$ (lower series) by the set-up of Figure 3.2. (Bottom) Time evolution of the number of photons in the cavity, jumping from the values 0 to 1, and back, by quantum jumps occurring at random times. From S. Haroche, Controlling photons in a box and exploring the quantum to classical boundary, Nobel lecture, *Reviews of Modern Physics*, 2013 [87].

provokes at some random time a quantum jump to a single photon state in the cavity, the presence of which is witnessed by the detection of atoms in level $|+\rangle$ ((up) clicks). This photon is monitored many times until it disappears suddenly, after a fraction of a second, by a new quantum jump occurring at an unpredictable time, so that atoms are then mostly detected afterwards in level $|+\rangle$ and again give rise to a great number of (down) clicks. The statistical analysis of many traces like this one allows us to determine the cavity lifetime. Note that in the upper trace the presence of less dense stray clicks that correspond to false counts due to imperfections in the state-selective field ionization technique (see Section 5.4).

This technique can be extended to count larger numbers of quanta by using a phase shift per photon of $\pi/4$, so that photon numbers from 1 to 8 can be experimentally distinguished. One first excites the cavity to the coherent state $|\alpha = 2\rangle$ by a very small classical field sent between the two mirrors of the cavity, then, using a Bayesian approach of repeated measurements (see Section 5.7) that update step by step the knowledge of the photon number distribution, one conditionally prepares the Fock state $|n = 4\rangle$, which then slowly decays to the lower state by losing photons one by one at random times, as can be seen in Figure 3.4, which

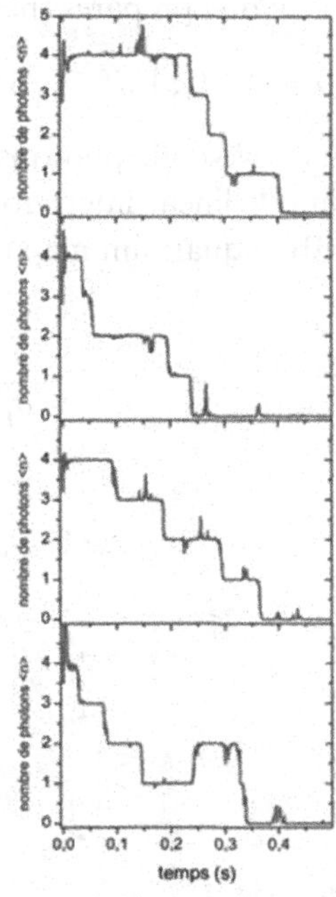

Fig. 3.4 Different recorded real time trajectories of the temporal evolution of the intracavity photon number for an initially conditionally prepared Fock state $|n = 4\rangle$. The photon number decays to lower values by random quantum jumps, but sometimes increases because of thermal re-excitation induced by blackbody light. With permission from C. Guerlin. *Mesure quantique non-destructive répétée de la lumière: états de Fock et trajectoires quantiques*. Université Pierre et Marie Curie Paris, PhD thesis, VI, 2007 [83].

displays in real time different individual trajectories showing randomly distributed quantum jumps. Note on the last trace a re-excitation of the cavity mode induced by a blackbody photon emitted by the surroundings of the cavity mode.

One can also conditionally generate Schrödinger cats $|+\rangle \otimes |\alpha\rangle + |-\rangle \otimes |-\alpha\rangle$, where $|\alpha\rangle$ is a photonic coherent state. They are the cousins of the Schrödinger cat states of a trapped ion motion mentioned in the Section 3.1; see Equation (3.1).

3.3 Photons in a Multimode Linear Interferometer: Boson Sampling

One of the important alleys of research in the direction of universal quantum computing is to find physical systems that unambiguously display a *quantum supremacy*, i.e that are able to perform a given computational task faster than classical computers. "Boson sampling" is one of them [1, 74].

The boson- (in our case photon-) sampling experimental set-up is shown in Figure 3.5: one prepares the input state,

$$|\psi_{in}\rangle = |1 : \ell_1, 1 : \ell_2, ..., 1 : \ell_N, 0 : \ell_{N+1}, 0 : \ell_{N+2}, ..., 0 : \ell_M\rangle, \tag{3.3}$$

consisting of N single photons in M different modes. One sends this state in a multimode linear interferometer having M inputs and M outputs. The corresponding quantum map (see Chapter 4), written in terms of creation operators is

$$\hat{a}_i^{\dagger out} = \sum_{i=1}^{M} S_i^j \hat{a}_j^{\dagger in}, \tag{3.4}$$

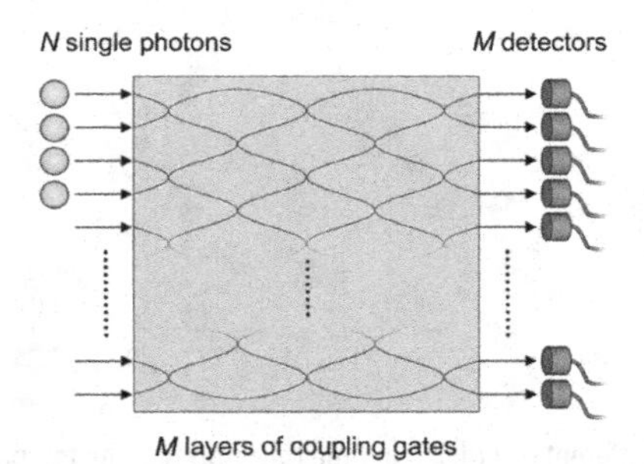

Fig. 3.5 Photon sampling set-up. The gray box is a linear interferometer consisting of a great number of beamsplitters that mix the different input modes in which one injects single photons. One records by an array of single photon detectors the photons exiting the device. From R. Garcia-Patron et al. Simulating boson sampling in lossy architectures, *Quantum*, 2019 [73], CC BY 4.0.

where $S = [S_i^j]$ is the symplectic matrix characterizing the interferometer, the extension of the two-mode matrix (9.71) to M modes. The output state is a superposition of the different configurations describing how the N photons are distributed over the M output modes. An array of M single photon detectors placed at the output repeatedly sample these distributions. One can show that the probability of measuring at the output a given configuration, which is the result of the interference between all the paths linking the input configuration to the output configuration, is proportional to the square of the *permanent* of the $N \times N$ submatrix deduced from the $M \times M$ matrix S.[1]

Various experiments of photon sampling have been implemented in the last decade [28]. The N single photons are generally obtained by a series of two-photon generation by spontaneous parametric down conversion where the idler photon is detected, and the signal photon is sent in the interferometer. The N photons are heralded by the simultaneous detection of the N idler photons, which is a rare event when N is larger than a few units. The use of semiconductor quantum dots that emit photons in a well-controlled way allows for a more efficient generation of the input state. The interferometer is implemented using an architecture of multiport waveguides, which allow for a simple control of the corresponding transformation S using tunable guided-wave directional couplers and phase shifters. The single photon detectors are avalanche photodiodes or superconducting nanowire single photon detectors (see Section 1.1).

So far, the number of photons and modes is too small to be competitive with the determination of the probability distribution of output photons configurations calculated by classical computers. Technological advances are needed to increase these numbers, as well as a clear and nonambiguous criterion characterizing the onset of the searched quantum supremacy.

3.4 Manipulation of Single Atoms Trapped in an Optical Lattice

As described in the first section of Appendix J, atoms can be trapped using the dipole force induced by intense blue detuned light, which creates an effective potential for the atoms, proportional to the light intensity, that traps atoms at their minima. Many optical techniques allow experimentalists to create light patterns likely to trap atoms in the regions of minimum intensity: two or more intersecting light beams create by interference *optical*

[1] The exact expression can be found in [74]. The determinant of a square matrix is calculated by summing cofactors with alternating + and − signs. The permanent of the same matrix is obtained by summing the same cofactors, but all with a + sign. The permanent turns out to be hard to compute when N and M are very large.

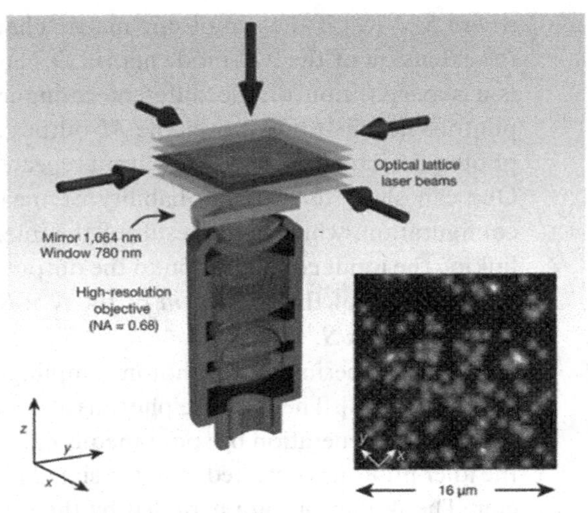

Fig. 3.6 (Left) Experimental set-up, with co-propagating beams creating a two-dimensional lattice. The atoms are individually detected using fluorescence imaging with the help of a high-resolution microscope objective. (Right) Site resolved fluorescence image of a thermal cloud of single atoms. From Sherson et al. Single-atom resolved fluorescence imaging of an atomic Mott insulator, *Nature*, 2010 [163].

lattices with periodic maxima and minima, while spatial light modulators and properly designed phase plates can generate arbitrary geometries of light patterns and wavefronts compatible with the laws of diffraction.

Figure 3.6 shows a typical experimental set-up [163]: The light pattern created by the counter-propagating beams generate a two-dimensional optical lattice that traps the atoms in its minima. The fluorescent light emitted by the atoms is collected by a high-resolution microscope and imaged on a CCD camera. Atoms from a very dilute sample fall in the periodic minima of the structure. When more than one atom is trapped in the same site, light assisted collisions expel both. This set-up allows us to trap and detect many single atoms simultaneously with a single atom single site resolution. One reaches an interesting regime, inaccessible in condensed matter experiments, of a very dilute, strongly interacting gas that can be characterized and studied atom by atom.

By changing the light intensities and detunings, one can tailor the interaction between the atoms and study the phase transitions between different regimes and their different collective excitations at the microscopic level. This interaction may also generate entangled states of many atoms, such as the *cluster states* that have been proposed to be the initial block of measurement-based quantum computing [149]. On such a platform, one can also "mimic" quantum Hamiltonians and quantum states of more complicated and difficult-to-handle quantum systems: This is the rising domain of quantum simulation, with possible applications to quantum information processing.

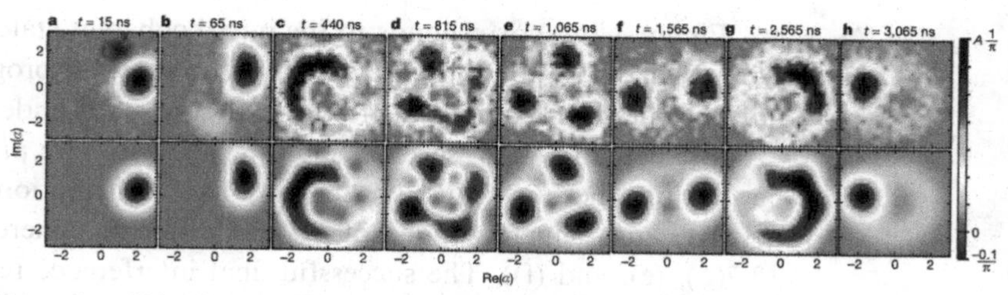

Fig. 3.7 Generation of Schrödinger cats by Kerr effect. The plots correspond to the (upper plots) experimental and (lower plots) theoretical Husimi Q-function as a function of time under Kerr evolution. The initial state is a coherent state represented by a phase space Gaussian. "Two-, three-, and four-legged cats" are observed. The quantum state Q-function finally goes back to its initial shape. Figure from G. Kirchmair et al. Observation of quantum state collapse and revival due to the single-photon Kerr effect, *Nature*, 2013 [103].

3.5 Evolution in a Kerr Medium: A Quantum Optics Experiment in Superconducting Circuits

Coherent states $|\pm\beta\rangle$ of the electromagnetic field (Section F.4) are robust against decoherence and are thus a good choice to encode quantum information [148]: The so-called cat-qubits are encoded in superpositions of coherent states $|\psi\rangle = a|\beta\rangle \pm b|-\beta\rangle$. As originally proposed by Yurke and Stoler [182] in the context of nonlinear optics, one can bring a coherent state $|\psi(t = 0)\rangle = |\beta\rangle$ to a coherent Schrödinger cat superposition using the free evolution in a Kerr medium during a well-defined time [80]. Superconducting circuits provide a highly nonlinear medium with low enough dissipation. The quantum process is extremely simple conceptually: The preparation consists of displacing the nonlinear circuit oscillator from its vacuum state to a coherent state $|\beta\rangle$. The subsequent evolution takes place under the Kerr Hamiltonian (see also Equation (K.9)),

$$\hat{H} = K\hat{a}^{\dagger 2}\hat{a}^2, \tag{3.5}$$

At time $t = \pi/K$ the system is found in state $|\psi(\pi/K)\rangle \propto |\beta\rangle + i|-\beta\rangle$. This is a $\pi/2$ gate. After twice this time ($t' = 2\pi/K$) the state refocuses to $|\psi(2\pi/K)\rangle = |-\beta\rangle$, implementing a full "$\pi$" gate over the cat qubit.

In Figure 3.7 we reproduce the experimental observations (upper row) and the theoretical predictions (lower row) for the temporal evolution of the $Q(\alpha)$ Husimi phase-space distribution $Q(\alpha, t) = \frac{1}{\pi}|\langle\alpha|\psi(t)\rangle|^2$ (see Section 9.1) induced by the Kerr Hamiltonian [103]. The measurement is made by applying a final displacement operator $\mathcal{D}$ to the field of amplitude $-\alpha$ and measuring the probability of finding the field in vacuum that directly yields the Q-function.

Figure 3.7(a) corresponds to the initial coherent state represented by a Gaussian distribution. Figure 3.7(b) and (c) show the progressive distortion of the quasi-probability distribution of the state under Kerr evolution. As predicted in [182], several "multi-legged cats" are produced along the evolution: the non-Gaussian "banana-like" distribution (c) progressively evolves into superpositions of several different coherent states (Figure 3.7(d), (e), and (f)). The successful final interference refocusing the field into a single Gaussian proves that the quantum state coherence is preserved at all intermediate times, evidence for a pure Hamiltonian evolution. This spectacular manipulation of the microwave field quantum state is used to work out logical gates acting on the superconducting qubits [80].

3.6 Exercises for Chapter 3

Exercise 3.1 In many experiments one can only measure the populations of the excited and ground states. This is the case of the Rydberg atoms experiments discussed in this chapter: Ionization spectroscopy only can resolve $|g\rangle = |50c\rangle$ from $|e\rangle = |51c\rangle$. How does one get to measure the coherence among them, then?

Exercise 3.2 What is the expression of the probability per unit time that a thermal excitation appears the cavity as in Figure 3.3 (see also Figure 3.4)?

Exercise 3.3 How to generalize the single photon nondemolition measurement to the measurement of $0, 1, 2 \ldots, n$ photons? How does the population in state $n + 1$ manifest in such a measurement?

Exercise 3.4 Discuss the experimental traces in Figure 3.4. Is the nondemolition photon measurement causing a projection? What sets the timescale of the plateaus, and what sets the timescale of the transition between them?

Exercise 3.5 Consider two laser beams of wavevectors $|\vec{k}_i| = 2\pi/\lambda$ intersecting at an angle θ. What is the spacing of the intensity minima that one can use to make an optical lattice?

Exercise 3.6 Show that the map $\hat{K}\left(\alpha_0|0\rangle + \alpha_1|1\rangle + \alpha_2|2\rangle\right) = \alpha_0|0\rangle + \alpha_1|1\rangle - \alpha_2|2\rangle$ is equivalent to the unitary transformation generated by $\hat{H}t = \frac{\pi\hbar}{2}\left(5\hat{n} - \hat{n}^2\right)$. The map can be conditionally achieved by measurement in linear systems and it produces an effective nonlinearity much stronger than the one produced by an optically passive Kerr media [109].

Exercise 3.7 Compute the permanent $\text{Per}(M)$ of the 4×4 matrix M defined by $M_{ij} = i + j$. Compare with the calculation of the determinant of the same matrix.

Exercise 3.8 Consider a boson sampling scenario where the mode couplings are given by the symplectic map $\hat{U}\hat{a}_i^\dagger\hat{U}^\dagger = \sum_{j=1}^m U_{i,j}\hat{a}_j^\dagger$. Convince yourself that the input state $|\psi_{in}\rangle = |1_1, 1_2, \ldots, 1_n, 0_{n+1}, 0_{n+2}, \ldots, 0_m\rangle$ transforms into $|\psi_{out}\rangle = \sum_l \gamma_l \left|n_1^{(l)}, \ldots, n_m^{(l)}\right\rangle$, where l enumerates all the possible outcomes with

$$\gamma_l = \frac{\text{Per}(U_l)}{\sqrt{n_1^{(l)}! \ldots n_m^{(l)}!}}.$$

Here, $n_k^{(l)}$ represents the amount of photons in mode k in the l possible outcomes.

Exercise 3.9 We define a phase space half-rotation as the transformation $x \to -x$ and $p \to -p$.

(a) Show that, at the level of the state vector, this corresponds to a sign flip over odd Fock state components. Show that a way to implement this is to let a harmonic motion ($\hat{H}/\hbar = -\delta \hat{a}^\dagger \hat{a}$) take place during half a period.
(b) Show that the equal weight real superposition of the initial and final states of this rotation have only even Fock components.
(c) Show that a generalized Kerr evolution ($\hat{H}/\hbar = -K(\hat{a}^\dagger \hat{a})^N$) will refocus any initial state into itself every time $2\pi/K$. Show that for Kerr evolution ($N = 2$) at the intermediate times $t = \pi/K$, the state is a superposition of the initial state and its phase space half-rotation. Write down explicitly the case where the initial state is a coherent state and show that one gets a cat state at the intermediate time.

Exercise 3.10 A "q-legged cat" is a pure state, quantum superposition of q different coherent states. We consider more precisely the following "q-legged cat":

$$\left|c_q^k\right\rangle = \frac{1}{\sqrt{q}} \sum_{p=0}^{q-1} e^{-i2pk\frac{\pi}{q}} \left|\alpha e^{i2p\frac{\pi}{q}}\right\rangle. \tag{3.6}$$

Show that it is a superposition of Fock states $|n\rangle$ with all the n equal modulo q. Use that $\frac{1}{q}\sum_{p=0}^{q-1} e^{-2ip(k-k')\frac{\pi}{q}} = \delta_{k,k'}$. Compute the overlap $\langle\alpha'|\alpha\rangle$ between two coherent states. What is the condition to neglect their overlap? What is the condition that α and q must fulfill in order to be able to distinguish the different coherent states? Why are no more than four-legged cats are shown in the experiment in Figure 3.7?

4 Evolution

We are now concerned with the evolution of a quantum system. We begin by recalling the main properties of Hamiltonian evolution. We then extend the description of quantum evolutions to more general situations in terms of a Kraus sum.

4.1 Hamiltonian Evolution of the Density Matrix

When a physical system is submitted to external forces that do not depend on the system state, but may depend on time, we know from Chapter 2 that its evolution is ruled by a time-dependent Hamiltonian $\hat{H}(t)$. The time dependence of the density operator in the pure state case, $\rho_p(t) = |\psi(t)\rangle\langle\psi(t)|$, is given by

$$\rho_p(t) = \hat{U}(t)\rho_p(0)\hat{U}^\dagger(t), \tag{4.1}$$

where the unitary evolution operator $\hat{U}(t)$ obeys the operatorial differential equation

$$i\hbar\frac{d}{dt}\hat{U}(t) = \hat{H}(t)\hat{U}(t). \tag{4.2}$$

We will call *Hamiltonian evolution* the one ruled by Equation (4.1).

When the Hamiltonian is time-independent, which corresponds to a system submitted to fixed forces, or not submitted from any influence coming from the outer world, Equation (4.2) has a simple solution:

$$\hat{U}(t) = e^{-i\hat{H}t/\hbar}, \quad \rho(t) = e^{-i\hat{H}t/\hbar}\rho(t=0)e^{i\hat{H}t/\hbar}. \tag{4.3}$$

When the preparation of the quantum system is not perfectly controlled, as already considered in Section 2.3, we can describe it by the statistical mixture:

$$\rho_M(t) = \sum_{i=1}^{N} p_i|\psi_i(t)\rangle\langle\psi_i(t)|. \tag{4.4}$$

Because of the linearity of Equation (4.1), one has

$$\rho_M(t) = \hat{U}(t)\rho_M(0)\hat{U}^\dagger(t). \tag{4.5}$$

So the relation (4.1) is valid for both pure states and statistical mixtures.

The Hamiltonian evolution of the density matrix can also be written as a differential equation, called von Neumann equation, which is written in the general case as

$$i\hbar\frac{d}{dt}\rho(t) = [\hat{H}(t), \rho(t)]. \tag{4.6}$$

What about the Hamiltonian evolution of the purity of the system? One has

$$\rho^2(t) = \hat{U}(t)\rho^2(0)\hat{U}^\dagger(t) \quad \text{and} \quad Tr\rho^2(t) = Tr\rho^2(0). \tag{4.7}$$

In a Hamiltonian evolution, the purity is therefore a constant of motion: a pure (mixed) state will stay pure (mixed). In addition, the Hamiltonian evolution of an isolated system is *reversible*, as shown in Section A.3.

4.2 Evolution of the Density Matrix in the General Case

4.2.1 Physical Evolution of an Open System

We consider a system S_A that is not isolated, but in contact with a system S_B. As we saw in Section 2.2, its state is described by a reduced density matrix $\rho_A(t)$, which is the partial trace over the system S_B of the total density matrix ρ_{tot}. Assuming that the total system $S_A + S_B$ is isolated, its evolution is Hamiltonian and ruled by Equation (4.1):

$$\rho(t) = \hat{U}_{tot}(t)\rho_{tot}(0)\hat{U}^\dagger_{tot}(t). \tag{4.8}$$

Assuming that at time $t = 0$, $\rho_{tot}(0)$ is the factorized state $\rho_A(0) \otimes \rho_B(0)$, and writing the density matrix ρ_B as the sum over pure states $|\psi_k\rangle$ of the reservoir $\sum_k p_k|\psi_k\rangle\langle\psi_k|$, the state of the system S_A at later time t, $\rho_A(t)$, is then given by

$$\rho_A(t) = Tr_B\left[\hat{U}_{tot}(t)\rho_A(0) \otimes \rho_B(0)\hat{U}^\dagger_{tot}(t)\right] \tag{4.9}$$

$$= \sum_{j,k} p_k\langle v_j|\hat{U}_{tot}(t)\rho_A(0) \otimes |\psi_k\rangle\langle\psi_k|\hat{U}^\dagger_{tot}(t)|v_j\rangle, \tag{4.10}$$

where $\{|v_j\rangle\}$ is a basis of S_B. Because of the presence of interactions, the evolution operator $\hat{U}_{tot}(t)$ is not factorizable into a part acting on $\mathcal{H}_A$ and a part acting on $\mathcal{H}_B$, and therefore the partial trace operation does not simply remove the reservoir part: One cannot write the reduced density matrix $\rho_A(t)$ as $\hat{U}'(t)\rho_A(0)\hat{U}'^\dagger(t)$, where $\hat{U}'(t)$ is some unitary operator acting on $\mathcal{H}_A$. Introducing the operators $\hat{\mathcal{M}}_{j,k}$ that act on the Hilbert space $\mathcal{H}_A$ of system S_A, given by

$$\hat{\mathcal{M}}_{j,k} = \sqrt{p_k}\langle v_j|\hat{U}_{tot}(t)|\psi_k\rangle, \tag{4.11}$$

the density matrix $\rho_A(t)$ can be simply written as the following "operator sum" representation:

$$\rho_A(t) = \sum_{j,k} \hat{\mathcal{M}}_{j,k} \rho_A(0) \hat{\mathcal{M}}^\dagger_{j,k}, \tag{4.12}$$

which is a generalization of the Hamiltonian evolution (4.1).

4.2.2 Quantum Map

In order to get the expression (4.12) in the previous section, we have used very reasonable assumptions: Hamiltonian evolution for the total system and factorization of the initial state. We will now envision the situation from a more formal point of view and determine in the most general case possible, the final quantum state ρ_f of a system knowing its initial state ρ_{in}. The superposition principle imposes that such an evolution is necessarily *linear in* ρ_{in}. We can therefore formally write

$$\rho_f = \hat{\hat{\mathcal{L}}}[\rho_{in}], \tag{4.13}$$

where $\hat{\hat{\mathcal{L}}}$ is a linear "superoperator," or "Liouvillian," that acts on the set $\mathcal{D}_\mathcal{H}$ and linearly transforms a density matrix into another one. This evolution, considered as directly connecting the initial state to the final state *without being interested in the intermediate values* is called a *quantum map*. This concept of a quantum map can be applied to the temporal evolution, and also to the characterization of the *spatial propagation* of the quantum state, for example, for states of light beams.

Because of the linearity of $\hat{\hat{\mathcal{L}}}$ and of the convex nature of density matrices, one must have, for any $0 \leq p \leq 1$,

$$\hat{\hat{\mathcal{L}}}[\rho_1 + ((1-p)\rho_2)] = p\hat{\hat{\mathcal{L}}}[\rho_1] + (1-p)\hat{\hat{\mathcal{L}}}[\rho_2]. \tag{4.14}$$

Equation (4.14), looking like the one for a classical average, also implies that it transforms a density matrix into another density matrix, i.e. during evolution, it preserves the positivity of the operator and the value 1 of the trace. This imposes strong constraints on the form of the quantum map that we consider in the next section.

4.2.3 Kraus Theorem

We will not give a demonstration of this theorem here; that can be found, for example, in [88] or [13].

Actually, it turns out that the requirements of trace preservation and positivity that characterize density matrices are not enough for the demonstration. One must use a stronger, but quite natural, requirement, named *complete positivity*, linked to the fact that the evolution of the system S_A is the result of interactions inside the Hilbert space $\mathcal{H}_A$ of the system S_A, but also of interactions with system S_B, characterized by a Hilbert space $\mathcal{H}_B$.

The hypothesis of complete positivity is the requirement that the operator $\hat{\hat{\mathcal{L}}} \otimes \mathcal{I}_B$ (where $\mathcal{I}_B$ is the identity in space $\mathcal{H}_B$) maps a positive operator of space $\mathcal{H}_A \otimes \mathcal{H}_B$ into another positive operator.

Note that all positive maps are not complete positive maps: For example, the transpose operation, exchanging the off-diagonal elements of a density matrix, can be shown to be linear and positive, but not completely positive [13].

Let us call d_A the dimension of the Hilbert space $\mathcal{H}_A$, and assume complete positivity for the quantum map $\hat{\hat{\mathcal{L}}}$. The Kraus theorem [106] states that there exists $N_K \leq d_A^2$ linear operators $\hat{K}_\ell(t)$ acting on the Hilbert space $\mathcal{H}_A$) such that

$$\rho_f = \hat{\hat{\mathcal{L}}}[\rho_{in}] = \sum_{\ell=1}^{N_K} \hat{K}_\ell \rho_{in} \hat{K}_\ell^\dagger, \tag{4.15}$$

where the operators $\hat{K}_\ell$ are called *Kraus operators*. Trace preservation imposes the extra condition that

$$\sum_{\ell=1}^{N_K} \hat{K}_\ell^\dagger \hat{K}_\ell = \hat{1}. \tag{4.16}$$

The theorem has an even stronger formulation: A linear map is completely positive if and only if it admits a Kraus representation (Stinespring's theorem). The Kraus map is indeed the general form of all physical evolutions of the density matrix.

Of course, for a Hamiltonian evolution from $t = 0$ to t, $\hat{K}_1 = \hat{U}(t)$, there is in this simple case a single, unitary Kraus operator. Reciprocally, if a Kraus operator is not unitary, then there must be more than one Kraus operator for the considered evolution.

4.2.4 Evolution Rule in the General Case

We have therefore found an expression for the quantum evolution in the general case:

(E1) *Evolution*: A quantum system initially in state ρ_{in} evolves to a final state ρ_f that has the following form:

$$\rho_f = \sum_{\ell=1}^{N_K} \hat{K}_\ell \, \rho_{in} \, \hat{K}_\ell^\dagger, \tag{4.17}$$

where the operators $\hat{K}_\ell$ are linear operators acting on the Hilbert space $\mathcal{H}$ of state vectors satisfying the condition

$$\sum_{\ell=1}^{N_K} \hat{K}_\ell^\dagger \hat{K}_\ell = \hat{1}. \tag{4.18}$$

4.2.5 Kraus Operators

By construction, the evolution ruled by Equation (4.17) is trace preserving. In addition, one has, for any state $|\phi\rangle$ and any density matrix ρ written as in Equation (2.36),

$$\langle\phi|\hat{\mathcal{L}}[\rho]|\phi\rangle = \sum_{\ell=1}^{N_K}\sum_i p_i \left|\langle\phi|\hat{K}_\ell|\psi_i\rangle\right|^2 > 0. \tag{4.19}$$

The transformed density matrix is therefore positive.

Expression (4.17) shows that the final density matrix can be interpreted as a statistical mixture of different density matrices, corresponding to different possible evolutions, each one related to a Kraus operator $\hat{K}_\ell$. Terms like $\hat{K}_\ell\rho\hat{K}_\ell^\dagger$ look like the usual Hamiltonian time evolution $\hat{U}(t)\rho\hat{U}^\dagger(t)$, but with nonunitary evolution operators, i.e. which in general have no reason to all tend to the identity when the time interval is very small. As we will discuss in more detail Section 4.5.2, they can be seen as describing instantaneous quantum jumps of the wave vector. They are reminiscent of the quantum jumps observed in experiments with single quantum objects, as we described in Chapter 1.

Let us take a unitary operator V, and build the operators

$$\hat{M}_i = \sum_j V_{ij}\hat{K}_j. \tag{4.20}$$

It is easy to show that one also has

$$\rho_f = \sum_i \hat{M}_i\,\rho_{in}\,\hat{M}_i^\dagger. \tag{4.21}$$

So the Kraus sum is not unique. One can show that Equation (4.20), with all possible unitaries V, generates all the possible Kraus decompositions.

Note also that $\hat{\hat{\mathcal{L}}}$ transforms the minimal information state $\rho_{min} = \hat{1}/d$ into $\rho_f = \frac{1}{d}\sum_\ell \hat{K}_\ell\hat{K}_\ell^\dagger$: the physical evolution creates information.

With formula (4.13), one no longer has $Tr(\hat{\hat{\mathcal{L}}}[\rho_{in}])^2 = Tr\rho_{in}^2$: The evolution does not, in general, conserve the purity of the state, so that a pure state may become mixed with time, but the opposite is also possible. In addition, the evolution given by such a map is, in general, not reversible.

4.2.6 Quantum Map in the Heisenberg Representation

We have so far adopted the so-called Schrödinger representation approach, where the quantum state of the system evolves, and not the observable $\hat{A}$. There is another approach, namely the Heisenberg representation, in which it is just the opposite.

We assume here that one performs on the system an ideal measurement of the physical quantity A, so that the usual observable $\hat{A} = \sum a_n\hat{P}_n$ can be used to determine its expectation $\langle\hat{A}\rangle_f$, or mean value:

$$\langle\hat{A}\rangle_f = Tr(\rho_f\hat{A}) = \sum_{\ell=1}^{N_K} Tr\hat{K}_\ell\,\rho_{in}\hat{K}_\ell^\dagger\hat{A} \tag{4.22}$$

$$= Tr(\rho_{in}\hat{A}_f), \tag{4.23}$$

with

$$\hat{A}_f = \sum_{\ell=1}^{N_K} \hat{K}_\ell^\dagger\hat{A}_i\,\hat{K}_\ell. \tag{4.24}$$

This operator can be used only to compute the evolution of the mean, but not that of the variance or the correlations because

$$\langle A^2\rangle_f = Tr(\rho_f\hat{A}^2) = \sum_{\ell=1}^{N_K} Tr\hat{K}_\ell\rho_{in}\hat{K}_\ell^\dagger\hat{A}^2 = Tr(\rho_{in}\hat{A}_2), \tag{4.25}$$

with

$$\hat{A}_2 = \sum_{\ell=1}^{N_K} \hat{K}_\ell^\dagger\hat{A}^2\hat{K}_\ell \neq \langle\hat{A}_2^2\rangle, \tag{4.26}$$

except when there is a single term in the sum. The difference between $\langle A_2\rangle$ and $\langle\hat{A}_2^2\rangle$ can be shown to be positive. This implies that quantum maps that are not unitary evolutions introduce an *extra fluctuating term*. It can be, for example, accounted for by introducing in the evolution equation of $\hat{A}(t)$ a random force, often called Langevin force, if one wants to use the equation not only for calculating means, but also higher moments.

4.2.7 "Unitarization" Theorem

It is the analog for quantum maps of the purification theorem for statistical mixtures (Section 2.5.2), and of the Naimark theorem for imperfect measurements (Section 5.3.4). It states that any nonunitary quantum map can be seen as a Hamiltonian evolution when the system under study in embedded in a larger Hilbert space with the help of an appropriately chosen "ancilla" space $\mathcal{H}_{anc}$. Let us assume that its dimension is N and call $|v_j\rangle$ its basis elements.

One defines a linear transformation $\hat{V}$ mapping $\mathcal{H}\otimes\mathcal{H}_{anc}$ onto $\mathcal{H}\otimes\mathcal{H}_{anc}$ satisfying the relation (omitting t for simplicity),

$$\hat{V}|\psi, v_1\rangle = \sum_{\ell=1}^{N} \left(\hat{K}_\ell|\psi\rangle\right)\otimes|v_\ell\rangle, \tag{4.27}$$

for any $|\psi\rangle$ belonging to $\mathcal{H}$; $|v_1\rangle$ is a specific basis element, which can be any vector of $\mathcal{H}_{anc}$.

It is easy to show that, because of Equation (4.16), $(\hat{V}|\psi, v_1\rangle)^\dagger\hat{V}|\psi', v_1\rangle = \langle\psi|\psi'\rangle$: The transformation, restricted to input states $|\psi, v_1\rangle$ is unitary. Mathematicians show that it is always possible to extend $\hat{V}$ to all input states $|\psi, v_j\rangle$ $(j > 1)$ of $\mathcal{H}\otimes\mathcal{H}_{anc}$ in such a way that the transformation is now unitary in the whole tensor product space. Thus, $\hat{V}$ can be considered as a Hamiltonian evolution in the whole space, transforming an input density matrix $\rho_{tot} = |\psi, v_1\rangle\langle\psi, v_1|$ into $\hat{\hat{\mathcal{L}}}[\rho_{tot}] = \hat{V}|\psi, v_1\rangle\langle\psi, v_1|\hat{V}^\dagger$. The reduced density matrix in space $\mathcal{H}$ is obtained by a partial trace over the ancilla,

$$\begin{aligned}\hat{\hat{\mathcal{L}}}[|\psi\rangle\langle\psi|] &= Tr_{\mathcal{H}_{anc}}\left(\hat{V}|\psi, v_1\rangle\langle\psi, v_1|\hat{V}^\dagger\right)\\ &= \sum_{\ell,\ell'} Tr_{\mathcal{H}_{anc}}\left(\hat{K}_\ell|\psi\rangle\right) \otimes |v_\ell\rangle\langle v_{\ell'}| \otimes \left(\langle\psi|\hat{K}_{\ell'}^\dagger\right)\\ &= \sum_{\ell,\ell'}\langle v_{\ell'}|v_\ell\rangle\hat{K}_\ell|\psi\rangle\langle\psi|\hat{K}_{\ell'}^\dagger = \sum_\ell \hat{K}_\ell|\psi\rangle\langle\psi|\hat{K}_\ell^\dagger. \end{aligned} \tag{4.28}$$

The extension of Equation (4.28) to mixed input states is then straightforward, which concludes the demonstration.

We see here that, knowing the Kraus operators, one can guess (at least partially) the expression of the unitary evolution in the whole space. Reciprocally, from a known unitary evolution $\hat{V}$ one can derive the expression of the Kraus operators by making explicit the partial trace in Equation (4.28):

$$\hat{\hat{\mathcal{L}}}[|\psi\rangle\langle\psi|] = \sum_\ell \langle v_\ell|\hat{V}|\psi, v_1\rangle\langle\psi, v_1|\hat{V}^\dagger|v_\ell\rangle = \sum_\ell \hat{K}_\ell|\psi\rangle\langle\psi|\hat{K}_\ell^\dagger, \tag{4.29}$$

with $\hat{K}_\ell = \langle v_\ell|\hat{V}|v_1\rangle$. Actually, one could have chosen another basis $\{|v'_\ell\rangle\}$ of the ancilla space and ended up with different Kraus operators $\hat{K}'_\ell = \langle v'_\ell|\hat{V}|v_1\rangle$. As the two bases being linked by a unitary transformation U of matrix elements $U_{\ell\ell'}$, the different sets of Kraus operators are linked by

$$\hat{K}_\ell = \sum_{\ell'} U_{\ell\ell'}\hat{K}_{\ell'}. \tag{4.30}$$

This a particular case of Equation (4.20).

4.2.8 Examples of Quantum Maps

Propagation of Single Photons through a Beamsplitter

The quantum description of a beamsplitter is derived in Appendix H. We consider the following situation: We send at most one photon onto a beamsplitter of amplitude transmission and reflection coefficients t and r (with $r^2 + t^2 = 1$), and we want to know the quantum state of the light that is transmitted through it, discarding the reflected part. We show in Section 5.5 that the transmitted state ρ_T is given as a function of the state ρ at the input by Equation (5.28):

$$\rho_T = \begin{pmatrix} \rho_{00} + r^2\rho_{11} & t\rho_{01} \\ t\rho_{10} & t^2\rho_{11} \end{pmatrix}. \tag{4.31}$$

It is easy to see that this map is generated by two Kraus operators: $\hat{K}_1 = |0\rangle\langle 0| + t|1\rangle\langle 1|$ and $\hat{K}_2 = r|0\rangle\langle 1|$. One checks that $\hat{K}_1^\dagger\hat{K}_1 + \hat{K}_2^\dagger\hat{K}_2 = \hat{1}$.

Evolution of Qubits

Let us now consider the evolution of qubits, where the Kraus operators are 2×2 matrices. Two extreme cases of evolution can occur, which we characterize here without reference to the detailed physics of the interaction leading to such an evolution.

- *Coherence damping*

  It corresponds to the map:

$$\hat{\hat{\mathcal{L}}}\left[\begin{pmatrix} \rho_{++} & \rho_{+-} \\ \rho_{-+} & \rho_{--} \end{pmatrix}\right] = \begin{pmatrix} \rho_{++} & D\rho_{+-} \\ D\rho_{-+} & \rho_{--} \end{pmatrix}, \tag{4.32}$$

  with $D < 1$, so that the populations are unchanged and the coherence reduced. Such an evolution is ruled by three Kraus operators: $\hat{K}_1 = \sqrt{D}\hat{1}$, $\hat{K}_2 = \sqrt{1-D}|+\rangle\langle+|$, $\hat{K}_3 = \sqrt{1-D}|-\rangle\langle-|$.

- *Population damping*

  Because of the trace preservation, the population of the two levels cannot be damped at the same time, but, for example, ρ_{--} can evolve into $D\rho_{--}$ with $D < 1$. This can be described by the map

$$\hat{\hat{L}}\left[\begin{pmatrix} \rho_{++} & \rho_{+-} \\ \rho_{-+} & \rho_{--} \end{pmatrix}\right] = \begin{pmatrix} 1 - D\rho_{--} & \sqrt{D}\rho_{+-} \\ \sqrt{D}\rho_{-+} & D\rho_{--} \end{pmatrix}, \tag{4.33}$$

  corresponding to the two Kraus operators $\hat{K}_1 = |+\rangle\langle+| + \sqrt{D}|-\rangle\langle-|$, $\hat{K}_2 = \sqrt{1-D}|+\rangle\langle-|$.

  We will see later that the beamsplitter and spontaneous emission are particular cases of population damping maps.

When the same interaction is repeated n times, in the first case the coherence evolves into $T^n\rho_{+-}$ and in the second case the population in $|-\rangle$ state evolves into $D^n\rho_{--}$. Both vanish when n goes to infinity.

4.3 Evolution in a Simple Physical Situation: Two "Colliding" Systems

In this section, we will develop as an example a simple physical coupling between two systems that leads to an evolution of the reduced density matrix where coherences are gradually destroyed, a situation named *decoherence*.

Let us consider two quantum systems of any dimension, described in Hilbert spaces $\mathcal{H}_1$ and $\mathcal{H}_2$, that are initially noninteracting. We call $\{|u_n\rangle\}$ and $\{|v_j\rangle\}$ bases of spaces $\mathcal{H}_1$ and $\mathcal{H}_2$, respectively. The two systems "collide" from time $t = 0$ for a short duration and then are again separated (see Figure 4.1). Let $\hat{U}$ be the Hamiltonian evolution operator of the total system after the end of the interaction, i.e. for a time t longer than the collision time. We assume that the whole system in $\mathcal{H}_1 \otimes \mathcal{H}_2$ is initially

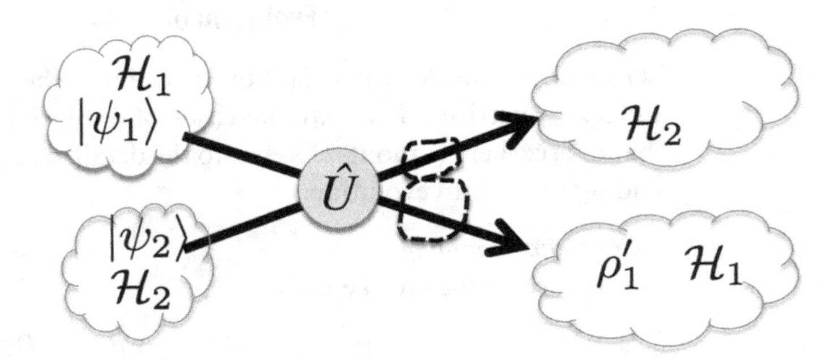

Fig. 4.1 "Collisional" interaction between two systems, 1 and 2, that interact during a short time before being separated again. This collisional interaction is then repeated many times.

a factorized state $\rho_{initial,1} \otimes \rho_{initial,2}$. Under the effect of the interaction, correlations between the two systems will appear: the final state will, in general, no be longer factorizable. It can be only described by a global density matrix ρ belonging to the total Hilbert space $\mathcal{H}_1 \otimes \mathcal{H}_2$. We are interested only in the evolution of the system in $\mathcal{H}_1$, i.e. for the reduced density matrix ρ_{1f} obtained by a partial trace over space $\mathcal{H}_2$ of ρ.

Let us first assume that the initial state is a tensor product of two pure states $\rho = |\psi_1\rangle\langle\psi_1| \otimes |\psi_2\rangle\langle\psi_2|$. The reduced density matrix after the collision will be

$$\rho_{1f} = \sum_j \langle v_j|\hat{U}|\psi_1\rangle \otimes |\psi_2\rangle\langle\psi_1| \otimes \langle\psi_2|\hat{U}^\dagger|v_j\rangle = \sum_j \hat{K}_j|\psi_1\rangle\langle\psi_1|\hat{K}_j^\dagger, \quad (4.34)$$

where $\hat{K}_j$ is an operator acting on space $\mathcal{H}_1$ and

$$\hat{K}_j|u_{n'}\rangle = \langle v_j|\hat{U}(|u_{n'}\rangle \otimes |\psi_2\rangle). \quad (4.35)$$

The demonstration can be easily extended to the case of non-pure initial states ρ_1 and ρ_2, so that one retrieves in this particular case the Kraus sum,

$$\rho_{1f} = \sum_j \hat{K}_j \rho_1 \hat{K}_j^\dagger. \quad (4.36)$$

We take as an example a simple interaction Hamiltonian:

$$\hat{H}_{int} = \hbar g(t) \hat{A}_1 \otimes \hat{B}_2, \quad (4.37)$$

in which $\hat{A}_1$ is an observable on $\mathcal{H}_1$, and $\hat{B}_2$ is an observable on $\mathcal{H}_2$. Here, $g(t)$ is the strength of the interaction, which is zero before and after the collision. The solution of the Schrödinger equation (4.2) is, in this simple case,

$$\hat{U}(t) = e^{-i\int_0^t g(t')dt' \hat{A}_1 \otimes \hat{B}_2}, \quad (4.38)$$

so that the evolution during the whole collision is

$$\hat{U} = e^{-ig\hat{A}_1 \otimes \hat{B}_2}, \tag{4.39}$$

with $g = \int_0^\infty g(t')dt'$.

For the basis $\{|u_n\rangle\}$ of $\mathcal{H}_1$, we will choose the eigenstates of $\hat{A}_1$, with corresponding eigenvalues a_n; $|\psi_1\rangle$ can then be written as $\sum_n c_n|u_n\rangle$. The global state after the collision will be

$$|\psi_f\rangle = \sum_n c_n|u_n\rangle \otimes |\psi_{2,n}\rangle, \tag{4.40}$$

with $|\psi_{2,n}\rangle = e^{-iga_n\hat{B}_2}|\psi_2\rangle$. We see that the $\mathcal{H}_2$ components $|\psi_{2,n}\rangle$ of the entangled state have received the "footprint" of their associated $\mathcal{H}_1$ component $|u_n\rangle$.

The Kraus operators are diagonal in the basis $\{|u_n\rangle\}$, with

$$\hat{K}_j|u_n\rangle = \langle v_j|\psi_{2,n}\rangle|u_n\rangle, \tag{4.41}$$

so that the matrix elements of the reduced density matrix are[1]

$$\rho^f_{1,n,n'} = \sum_j \langle v_j|\psi_{2,n}\rangle\rho_{1,n,n'}\langle\psi_{2,n'}|v_j\rangle = \langle\psi_{2,n'}|\psi_{2,n}\rangle\rho_{1,n,n'}. \tag{4.42}$$

Populations $\rho^f_{1,n,n}$ are unchanged, whereas coherences are multiplied by a number $\alpha_{n,n'} = \langle\psi_2|e^{ig(a_n-a_{n'})\hat{B}_2}|\psi_2\rangle$, which has, like any inner product of two normalized states, a modulus smaller than 1: Here, we are in the *coherence damping* situation. When the two eigenvalues are significantly different, $\alpha_{n,n'}$, as the mean of an oscillating term, is, in general, very small and the corresponding coherence will be reduced more efficiently than the one between levels of almost equal eigenvalues. The coherences between two levels of equal eigenvalues are not affected.

The conclusion of this analysis is that when two systems interact with a Hamiltonian of the form of Equation (4.37), they become entangled, as expected. The effect of this interaction (which does not affect the state vectors $|u_n\rangle$) on one of the two systems is a strong decoherence *in the basis that diagonalizes the interaction* between states of different eigenvalues. It leaves the populations unaffected.

4.4 Evolution of a Small System Coupled to a Reservoir: A Simple Model of Decoherence

Let us now consider a quantum system S, characterized by a Hilbert space $\mathcal{H}_S$, that interacts with another system, called a "reservoir," or "bath,"

[1] Note that this conclusion can be extended to the more general case where the basis states $|u_n\rangle$ are eigenstates of the evolution operator, whatever its precise expression, more precisely when we can write $\hat{U}(|u_n\rangle \otimes |\psi_2\rangle) = |u_n\rangle \otimes |\psi_{2n}\rangle$.

characterized by a Hilbert space $\mathcal{H}_R$ of very large size. The situation is analogous to the canonical ensemble in statistical physics, in which one defines the thermostat, or bath, as a large system that interacts with the small system of interest, but which is not affected much by this interaction and always stays in the same state.

In this section we give an illustrative example of such a reservoir that allows us to make extensive calculations of the decoherence process. It will allow us to clarify the concept of short memory, or Markov, approximation. We will then adopt a more formal and general approach, leading to the very important "Lindblad equation."

4.4.1 Case of Repeated Interactions

As an example, we now consider a model case for which we can directly use our previous results: The system of interest, labeled as S_1, is experiencing "collisions" (i.e. interactions during a limited amount of time) with many identical systems, labeled as S_2, the totality of which is described by a large Hilbert space $\mathcal{H}_R$ of the reservoir.

In such a situation, it is reasonable to assume that system S_1 will interact with a series of systems S_2 in $\mathcal{H}_R$ that are all in a "fresh" initial state ρ_2, i.e. not entangled with the reservoir because of a possible previous collision with it, not with anything else.

Using the results of the previous section repeatedly, one finds that the decoherence effect accumulates, and that the reduced density matrix of system S_1 becomes completely diagonal after many collisions, except between levels of degenerate eigenvalues. The final density matrix $\rho_{1,f}$ is therefore the statistical mixture

$$\rho_{1,f} = \sum_n \hat{P}_n \rho_{in} \hat{P}_n. \tag{4.43}$$

The projectors $\hat{P}_n$ on the eigenspaces of $\hat{A}_1$ are therefore the Kraus operators of the evolution under the effect of the reservoir.

4.4.2 Markov Approximation

We now want to determine the temporal evolution of the small system S_1 from time t to a later, but nearby, time $t + \Delta t$. The evolution of the whole system, being an isolated system, is Hamiltonian and governed by the unitary evolution operator $\hat{U}_{SR}(t, t + \Delta t)$. We make the important assumption that, because of decoherence, the state of the whole system at time t is factorized, the reservoir being in state ρ_R during the evolution. We therefore write the state of the total system as $\rho_1(t) \otimes \rho_R$. Let us write ρ_R as $\sum_i p_i |\phi_i\rangle\langle\phi_i|$, which is always possible. The reduced density matrix $\rho_1(t+\Delta t)$ of the small system is now given by

$$\rho_1(t + \Delta t) = Tr_R \hat{U}_{SR}(t, t + \Delta t) \rho_1(t) \otimes \rho_R \hat{U}^\dagger_{SR}(t, t + \Delta t). \tag{4.44}$$

Introducing a basis $\{|v_j\rangle\}$ of the Hilbert space $\mathcal{H}_R$, one obtains, as in Section 4.3:

$$\begin{aligned}\rho_1(t+\Delta t) &= \sum_{i,j} p_i \langle v_j|\hat{U}_{SR}(t,t+\Delta t)\rho_1(t)|\phi_i\rangle\langle\phi_i|\hat{U}^{\dagger}_{SR}(t,t+\Delta t)|v_j\rangle \\ &= \sum_{i,j} \hat{K}_{i,j}(t,\Delta t)\rho_S(t)\hat{K}^{\dagger}_{i,j}(t,\Delta t),\end{aligned} \quad (4.45)$$

where the Kraus operators $\hat{K}_{i,j}$ are the linear operators, acting only in the Hilbert state of the small system $\mathcal{H}_S$, given by

$$\hat{K}_{i,j} = \sqrt{p_i}\langle v_j|\hat{U}_{SR}(t,t+\Delta t)|\phi_i\rangle. \quad (4.46)$$

Let us now consider the state of the system at time $t + 2\Delta t$. Generally speaking, although the reservoir is stationary, we cannot calculate $\rho_1(t+2\Delta t)$ from $\rho_1(t+\Delta t)$ using formula (4.45) giving $\rho_1(t+\Delta t)$ from $\rho_1(t)$ because the new initial state is not factorized: the different "time slices" of duration Δt are not equivalent. The expressions for the Kraus operators depend on t and are, in general, difficult to determine.

The problem is drastically simplified if one makes the approximation, called the "short memory" or Markov approximation. Our aim is to describe only the "coarse-grained" temporal evolution, i.e. the evolution by time steps Δt that are not infinitely small. We assume that these time steps are long enough compared to the "collision time" so that we can use Equation (4.44). In addition, the reservoir is so big and the considered system so small that the reservoir state is insensitive to changes occurring in the system S. It remains always in state ρ_R, and therefore the total state of the system is in the factorized state $\rho_1(t)\otimes\rho_R$ at all times. The reservoir in some way "forgets" that it has interacted with the small system.[2] As the global state $\rho(t+\Delta t)$ is factorized, we can use the previous derivation, and we have, for any t, that

$$\rho_1(t+\Delta t) = \sum_j \hat{K}_j(\Delta t)\rho_1(t)\hat{K}^{\dagger}_j(\Delta t). \quad (4.47)$$

Note that now, the Kraus operators $\hat{K}_j(\Delta t)$ do not depend on t: In the Markov approximation the "time slices" of large enough duration Δt are equivalent.

4.5 Quantum Map at the Markov Approximation: Lindblad Equation

Let us now leave aside a given model of evolution, as we did in the previous sections, and consider the general problem of a small system S coupled by one way or another to a large one, the reservoir or bath, that is stationary,

[2] This behavior must of course be checked for each particular form of the interaction between the system and the bath.

and at the short memory, or Markov approximation. We take a time interval Δt that is much shorter than the characteristic evolution time of the density matrix $\rho_S(t)$, but longer than the decoherence time, so that the state of the whole system remains factorized.

The Markov approximation implies that $\rho_S(t)$ is a differentiable function of time at this scale, with derivatives that do not depend on time. This allows us to write the state at time $t + \Delta t$ in form of a Taylor expansion up to first order in Δt:

$$\rho_S(t + \Delta t) \simeq \rho_S(t) + \Delta t \frac{d\rho_S}{dt}, \tag{4.48}$$

where $\rho_S(t + \Delta t)$ is related to $\rho_S(t)$ by a quantum map, which can always be written as a Kraus sum (4.47). A comparison of Equation (4.48) with Equation (4.47) shows that, in order to get a time dependence of the form (4.48), i.e. linear in Δt, for the density matrix $\rho_S(t + \Delta t)$ deduced from the Kraus sum, the simplest possibility is that one of the Kraus operators is close to the identity and linear in Δt, while the other Kraus operators are proportional to $\sqrt{\Delta t}$, so that the terms $\hat{K}_j(\Delta t)\rho_S(t)\hat{K}_j^\dagger(\Delta t)$ are linear in Δt.

Let $\hat{K}_0$ be the Kraus operator close to the identity. We assume the following expressions for the Kraus operators in terms of linear operators $\hat{B}_0$ and $\hat{A}_j$:

$$\hat{K}_0(\Delta t) = \hat{1} + \Delta t \hat{B}_0, \quad \hat{K}_j(\Delta t) = \sqrt{\Delta t}\hat{A}_j, \tag{4.49}$$

so that

$$\sum_{j=0}^{N} \hat{K}_j(\Delta t)\rho_S(t)\hat{K}_j^\dagger(\Delta t) \simeq \rho_S(t) + \Delta t\left(\hat{B}_0\rho_S + \rho_S\hat{B}_0^\dagger + \sum_{j=1}^{N}\hat{A}_j\rho_S\hat{A}_j^\dagger\right). \tag{4.50}$$

Comparing with Equation (4.48) we get a simple differential equation for the evolution of the reduced density matrix at the Markov approximation

$$\frac{d\rho_S}{dt} = \hat{B}_0\rho_S + \rho_S\hat{B}_0^\dagger + \sum_{j=1}^{N}\hat{A}_j\rho_S\hat{A}_j^\dagger, \tag{4.51}$$

where the operators $\hat{B}_0$ and $\hat{A}_n$ are independent of time. The constraint $\sum_{j=0}^{N}\hat{K}_j^\dagger\hat{K}_j = \hat{1}$ up to second order in Δt transfers into:

$$\hat{B}_0 + \hat{B}_0^\dagger + \sum_{j=1}^{N}\hat{A}_j^\dagger\hat{A}_j = 0. \tag{4.52}$$

$\hat{B}_0$, as any linear operator, can be written as the sum of commuting anti-Hermitian and Hermitian operators, so that

$$\hat{B}_0 = -i\hat{H}/\hbar - \hat{\Gamma}, \tag{4.53}$$

where $\hat{H}$, $\hat{\Gamma}$ are Hermitian. Finally, we obtain the equation governing the "coarse-grained" evolution of the reduced density matrix, called the *master equation*, in the form of a first-order differential equation:

$$\frac{d\rho_S}{dt} = -\frac{i}{\hbar}[\hat{H}, \rho_S] - (\rho_S\hat{\Gamma} + \hat{\Gamma}\rho_S) + \sum_{n=1}^{N} \hat{A}_j\rho_S\hat{A}_j^\dagger, \tag{4.54}$$

with

$$\hat{\Gamma} = \frac{1}{2}\sum_{j=1}^{N} \hat{A}_j^\dagger \hat{A}_j, \tag{4.55}$$

which is equivalent to

$$\frac{d\rho_S}{dt} = -\frac{i}{\hbar}[\hat{H}, \rho_S] - \frac{1}{2}\sum_{n=1}^{N} \left(\rho_S\hat{A}_j^\dagger\hat{A}_j + \hat{A}_j^\dagger\hat{A}_j\rho_S - 2\hat{A}_j\rho_S\hat{A}_j^\dagger\right). \tag{4.56}$$

The equation in the form (4.54) or (4.56) is a particular form of *master equation*. It is known as the *Lindblad equation*.

4.5.1 Discussion

Let us consider successively the different contributions to the evolution of the density matrix in Equation (4.56):

1. The first term in the master equation (4.56) obviously corresponds to the Hamiltonian, von Neumann, evolution of the density matrix (Equation (4.6));
2. The second and third terms are in the form of an anti-commutator. They can be grouped with the first term, so that this part of the evolution can also be written as

$$\frac{d\rho_S}{dt} = -\frac{i}{\hbar}\left(\hat{H}_{\text{eff}}\,\rho_S - \rho_S\,\hat{H}_{eff}^\dagger\right), \tag{4.57}$$

with $\hat{H}_{eff} = i\hbar\hat{B}_0 = \hat{H} - i\hbar\hat{\Gamma}$. One has a kind of pseudo-Hamiltonian evolution with an effective, non-Hermitian, Hamiltonian $\hat{H}_{eff}$. Let us assume for simplicity that the operators $\hat{H}$ and $\hat{\Gamma}$ commute. They diagonalize in the same basis, and the diagonal elements in this basis are of the form $E_n - i\hbar\gamma_n$: The solution for the state vector of the "partial" time evolution (4.57) consists of a combination of terms $e^{-iE_nt}e^{-\gamma_nt}|\psi_n\rangle$. We retrieve the usual oscillatory behaviour of energy eigenstates, here multiplied by decaying exponentials $e^{-\gamma_nt}$, characterizing a non-unitary, irreversible damping process.
3. The remaining terms are of the form $\hat{A}_j\rho_S\hat{A}_j^\dagger$. Because of relation (4.55), they are necessarily present when there is a dissipative term $-i\hbar\hat{\Gamma}$ in the effective Hamiltonian. They require a more detailed discussion, which is the subject of Section 4.5.2.

4.5.2 Quantum Jumps

As we already noticed, the Kraus form of a quantum map implies that the evolution of the density matrix ρ_1 between times t and $t+\Delta t$ is a statistical sum of different nonunitary evolutions, each one associated with one term $\hat{K}_j \rho_1 \hat{K}_j^\dagger$ of the Kraus sum. The evolution of the state vector $|\psi_1\rangle$ that is compatible with the Lindblad equation for the density matrix ρ_1 is that for each term, $|\psi_1(t)\rangle \to |\psi_1(t+\Delta t)\rangle = \hat{K}_j|\psi_1(t)\rangle$. In Section 4.5.1, because the smooth evolution of the density matrix can be written as a Taylor expansion, we have been led to introduce two forms of evolutions, one related to $\hat{K}_0(\Delta t) = \hat{1} + \Delta t \hat{B}_0$, and the others to $\hat{K}_j(\Delta t) = \sqrt{\Delta t}\hat{A}_j$, up to second order in Δt.

$\hat{K}_0$ gives a "smooth" evolution of the state vector, as it tends to the identity when Δt goes to zero. It is easy to show that this evolution obeys a Schrödinger equation with a non-Hermitian, dissipative Hamiltonian $\hat{H}_{eff} = i\hbar\hat{B}_0$. The rate of change of the state vector, $(|\psi_1(t+\Delta t)\rangle - |\psi_1(t)\rangle)/\Delta t = \hat{B}_0|\psi_1(t)\rangle = -i\hat{H}_{eff}|\psi_1(t)\rangle/\hbar$ is finite.

In contrast, $\hat{K}_j$ does not lead to a smooth evolution of the state vector. The rate of change of the state vector, $(|\psi_1(t+\Delta t)\rangle - |\psi_1(t)\rangle)/\Delta t = (\hat{1} - \sqrt{\Delta t}\hat{A}_j)/\Delta t$ goes to infinity when $\Delta t \to 0$; the evolution is not differentiable, and suddenly generates the state vector $\hat{K}_j|\psi_1\rangle$ by a *quantum jump*. $\hat{K}_j$ is often called a *jump operator*. The total evolution is a statistical mixture of the different evolutions. These quantum jumps occur in a random way and are intermixed with occurrences of Hamiltonian evolutions related to $\hat{K}_0$. One can say that an open quantum system *is governed by chance and by law*: chance because of the random quantum jumps, and law because of the Hamiltonian evolution between the jumps.

In Section 4.6 we will see more detailed examples where the jump operators are, respectively, the lowering operator $\hat{a}$ for the damped atomic oscillator and the atomic lowering operator $\hat{\sigma}_-$ for spontaneous emission. As we saw in the Chapter 1, such sudden changes in the energy of the state are observable and occur randomly. When one repeats the process, the quantum jumps occur at different randomly distributed times: When averaged, the evolution due to the quantum jumps is smoothed out, and one ends up with a smooth, continuous evolution given by the master equation (Figure 8.1 gives an example of this behavior in the case of photon detection). We will come back to these quantum jumps in Chapter 5.

A very efficient numerical method, called the "quantum Monte-Carlo method" [33, 48] is based on these considerations. The calculated evolution of the wave vector consists of "trajectories" that randomly alternate between periods of smooth exponential decays and sudden quantum jumps occurring at randomly chosen times. The averaged evolution over many samples gives the smooth evolution ruled by the master equation. As the two effects necessarily come together because of relation (4.55), exponential decay is always accompanied by quantum jumps, whereas Hamiltonian evolution does not involve quantum jumps.

Note that the set of Kraus operators, and therefore the set of possible nonunitary evolutions, is not unique for a given quantum map (Equation (4.21)); there can be different physical interpretations of these evolutions according to the chosen set.

4.6 Examples

We now consider two physical situations that we have already encountered: a damped harmonic oscillator and the spontaneous emission of a two-level system. They will allow us to find from first principles the quantum maps for each situation and to relate them to the general approach that we have just presented.

4.6.1 The Damped Harmonic Oscillator

Let us consider the trapped ion that was described in Chapter 1. It is a quantum harmonic oscillator, with annihilation operators $\hat{a}$ and energies $E_n = \hbar\omega(n + 1/2)$. We suppose that it is not isolated, but coupled to a big set of other harmonic oscillators (phonons, photons, ...) described by annihilation operators $\hat{b}_n$, called the reservoir. In addition, we assume that the reservoir is not affected by the presence of the oscillator and always stays in the same state that we assume to be the vacuum state, to simplify the analysis. The Hamiltonian of the whole system can be shown to be

$$\hat{H} = \hat{H}_0 + \hat{H}_{int},$$
$$\hat{H}_0 = \hbar\omega\hat{a}^\dagger\hat{a} + \sum_n \hbar\omega_n\hat{b}_n^\dagger\hat{b}_n; \quad \hat{H}_{int} = \sum_n \hbar g_n\left(\hat{a}^\dagger\hat{b}_n + \hat{a}\hat{b}_n^\dagger\right), \tag{4.58}$$

where the coupling term $\hat{H}_{int}$ corresponds to an energy-preserving transfer of a single quantum from the ionic harmonic oscillator to the others, and the reverse, and g_n is assumed to be real.

The total system is isolated and obeys the usual Schrödinger equation, which we can write in the Heisenberg representation for the operators $\hat{a}(t)$ and $\hat{b}_n(t)$:

$$\frac{d\hat{a}}{dt} = \frac{1}{i\hbar}[\hat{a}, \hat{H}] = -i\omega\hat{a} - i\sum_n g_n\hat{b}_n, \tag{4.59}$$

$$\frac{d\hat{b}_n}{dt} = \frac{1}{i\hbar}\left[\hat{b}_n, \hat{H}\right] = -i\omega_n\hat{b}_n - ig_n\hat{a} \tag{4.60}$$

Formally solving the second equation and inserting the result into the first equation one obtains

$$\frac{d\hat{a}}{dt} = -i\omega\hat{a} - \int_0^t dt'\hat{a}(t')\mathcal{K}(t-t') + \hat{f}(t), \tag{4.61}$$

where the Kernel $\mathcal{K}$ and additional noise term $\hat{f}(t)$ are given by

$$\mathcal{K}(t) = \sum_n g_n^2 e^{-i\omega_n t}; \quad \hat{f}(t) = -i\sum_n g_n\hat{b}_n(0)e^{-i\omega_n t}. \tag{4.62}$$

Being the sum of a great number of terms oscillating at different frequencies that are in phase only at $t = 0$, $\mathcal{K}(t)$ is nonnegligible only for very small times, and we will approximate it by $\kappa\delta(t)$; this is an example of the Markov approximation introduced in the previous section. We then have that:

$$\frac{d\hat{a}}{dt} = -i\omega\hat{a} - \kappa\hat{a} + \hat{f}(t), \tag{4.63}$$

where $\hat{f}(t)$ satisfies $\langle\hat{f}(t)\rangle = 0$ and $\langle\hat{f}^{\dagger}(t)\hat{f}(t')\rangle = \kappa\delta(t - t')$.

This noise term is similar to the one appearing in the classical equation of the Brownian motion of a particle submitted to the collision of a surrounding liquid:

$$\frac{dv}{dt} = -\kappa v + f(t), \tag{4.64}$$

where v is the velocity of the particle, κ the damping constant, and f the fluctuating "Langevin" force responsible for the random path followed by the particle and satisfying $\overline{f(t)} = 0, \overline{f(t)f(t')} \propto \delta(t - t')$, where the overline indicates a classical ensemble average.

Going back to the quantum equation, the two first terms of Equation (4.63) correspond to the pseudo-Hamiltonian evolution with a complex frequency leading to exponential decay. They are the consequence in the evolution of observable $\hat{a}$ of the pseudo-Hamiltonian part of the master equation (4.54) governing the evolution of the density matrix. The effective non-Hermitian operator is $\hat{H}_{eff} = \hbar(\omega - i\kappa)\hat{a}^{\dagger}\hat{a}$, so that the associated damping operator of Equation (4.54) is $\hat{\Gamma} = \kappa\hat{a}^{\dagger}\hat{a}$. Relation (4.52) implies that the Kraus operator $\hat{A}_1$ can be taken as $\hat{A}_1 = \sqrt{2\kappa}\hat{a}$. The Lindblad master equation (4.56) for ρ must necessarily have the form

$$\frac{d\rho}{dt} = -i\omega(\hat{a}^{\dagger}\hat{a}\rho - \rho\hat{a}^{\dagger}\hat{a}) - \kappa(\hat{a}^{\dagger}\hat{a}\rho + \rho\hat{a}^{\dagger}\hat{a}) + 2\kappa\hat{a}\rho\hat{a}^{\dagger}. \tag{4.65}$$

According to the discussion of Section 4.5.2, the quantum jumps are related to the last term in Equation (4.64), associated with the jump operator $\hat{A}_1 = \hat{a}$. This operator indeed suddenly annihilates one quantum excitation of the quantum harmonic oscillator.

Here, we have derived the master equation in a kind of indirect way through the evolution of operators, and not of the density matrix. The Lindblad master equation (4.56) for the density matrix evolution in the Schrödinger representation could have been derived more directly by calculating the Hamiltonian evolution of the total system and making the partial trace over the reservoir, but this calculation is far more complex than the one we present here concerning the evolution of observables in the Heisenberg representation. This shows that there are different ways of tackling the problem of the time evolution at the Markov approximation: the Lindblad master equation for the density matrix and the quantum Langevin evolution equation like Equation (4.63) for observables that contain stochastic, fluctuating terms called Langevin forces, in order to derive the evolution not only of mean values but also of second moments.[3]

[3] Due to the quantum regression theorem [75], the quantum Langevin equations are also valid for calculating the evolution of higher order moments.

The evolution of the mean values $\langle\hat{a}\rangle = Tr\rho\hat{a}$ and $\langle\hat{a}^\dagger\hat{a}\rangle = Tr\rho\hat{a}^\dagger\hat{a}$ can be derived from Equations (4.59) and (4.60). One obtains

$$\frac{d}{dt}\langle\hat{a}\rangle = (-i\omega - \kappa)\langle\hat{a}\rangle; \quad \frac{d}{dt}\langle\hat{a}^\dagger\hat{a}\rangle = -2\kappa\langle\hat{a}^\dagger\hat{a}\rangle. \tag{4.66}$$

The evolution of these mean values is in the form of a smooth exponential decay, with or without oscillations, because the noise term $\hat{f}(t)$ has a zero mean value. However, the evolution of a given realization of the system is governed by the random value of the noise term, as in Brownian motion. Seen from the master equation, this randomness is related to the presence of the quantum jump operator, which suddenly lowers the energy of the oscillator by one quantum.

4.6.2 Spontaneous Emission

Quantum Map for Spontaneous Emission

We consider a two-level system $\{|+\rangle, |-\rangle\}$, like a qubit or a two-level atom, interacting with the quantized electric field in the vacuum state. The system atom+field is isolated and therefore described by a state vector $|\Psi(t)\rangle$, as calculated in Section G.2, (Equations (G.18) and (G.19)) in the case where the atom is initially in the excited state $|+\rangle$. The probability of finding the atom in the excited state at later times is then given by the usual decaying exponential $P_+(t) = |\gamma_0(t)|^2 = e^{-\Gamma t}$, while the probability to find the photon in mode ℓ is given by $P_\ell(t) = |\gamma_\ell(t)|^2$. The vector $|\Psi(t)\rangle$ is not factorizable, meaning that there are strong correlations between the atomic and field parts.

It is easy to show that, when the atomic initial state has the general form $|\psi_{at}\rangle = \alpha|+\rangle + \beta|-\rangle$ (the field being still in the vacuum state), the $\beta|-\rangle$ dependent part of the state vector does not evolve because the atom and the field are in their ground states, so that the whole system at any time t is now given by

$$|\psi(t)\rangle = \alpha|\Psi(t)\rangle + \beta|-;0\rangle, \tag{4.67}$$

where $|\Psi(t)\rangle$ is given by Equations (G.18) and (G.19).

We now know the exact expression of the total state vector for any initial atomic state within the short memory, or Markov, approximation. We can deduce from it the atomic state density matrix in the general case by making a partial trace over the field variables:

$$\rho_{ii'} = \sum_R \langle i; R|\,\Psi(t)\rangle\,\langle\Psi(t)\,|i'; R\rangle\,, \tag{4.68}$$

where R is a radiation state basis vector ($|0\rangle$ or $|1:\ell\rangle$), and i, i' the atomic levels.

One finds for the populations that

$$\rho_{++} = |\alpha|^2|\gamma_0|^2 = \rho_{++}(0)e^{-\Gamma t}, \tag{4.69}$$

$$\rho_{--} = 1 - \rho_{++}(0)e^{-\Gamma t} = 1 - e^{-\Gamma t} + \rho_{--}(0)e^{-\Gamma t}, \tag{4.70}$$

and for the coherence ρ_{+-},

$$\rho_{+-} = \alpha\beta^*\gamma_0 = \rho_{+-}(0)e^{-\Gamma t/2}e^{-i(\omega_0+\delta\omega)t}. \tag{4.71}$$

The coupling of the excited atom to the vacuum field leads to an exponential decay of the excited state population and of the coherence. One notes that *for spontaneous emission, the relaxation rate of the coherence is the half of the relaxation rate of the population.* To simplify the notation, one includes the energy shift (or "Lamb shift") $\delta\omega$ in the energy of the excited state $|+\rangle$ ($\omega_0' = \omega_0 + \delta\omega$) and therefore in the atomic Hamiltonian, which we note as $\hat{H}_{at}$. The spontaneous emission quantum map is then equal to

$$\rho_{at}(t) = \begin{bmatrix} \rho_{++}(0)e^{-\Gamma t} & \rho_{+-}(0)e^{-\Gamma t/2}e^{i\omega_0' t} \\ \rho_{-+}(0)e^{-\Gamma t/2}e^{-i\omega_0' t} & 1-\rho_{++}(0)e^{-\Gamma t} \end{bmatrix}. \tag{4.72}$$

It is a special case of a population damping map (Equation (4.33)) introduced above as an example. It can be written as a Kraus sum with two terms:

$$\rho_{at}(t) = \hat{K}_1(t)\,\rho_{at}(0)\,\hat{K}_1^\dagger(t) + \hat{K}_2(t)\,\rho_{at}(0)\,\hat{K}_2^\dagger(t), \tag{4.73}$$

with two Kraus operators equal to

$$\hat{K}_1(t) = |+\rangle\langle+|e^{-\Gamma t/2} + e^{i\omega_0' t}|-\rangle\langle-|; \quad \hat{K}_2(t) = \sqrt{1-e^{-\Gamma t}}|-\rangle\langle+|. \tag{4.74}$$

It is also interesting to extract from the quantum map (4.73) the first order differential equation for the atomic density matrix, so as to compare it with the Lindblad equation demonstrated in the last section. Differentiating Equation (4.72), one obtains

$$\frac{d}{dt}\rho_{at}(t) = \begin{bmatrix} -\Gamma\rho_{++}(0)e^{-\Gamma t} & (-\Gamma/2 + i\omega_0')\rho_{+-}(0)e^{-\Gamma t/2}e^{i\omega_0' t} \\ (-\Gamma/2 - i\omega_0')\rho_{-+}(0)e^{-\Gamma t/2}e^{-i\omega_0' t} & \Gamma\rho_{++}(0)e^{-\Gamma t} \end{bmatrix}. \tag{4.75}$$

A straightforward calculation shows that $\frac{d}{dt}\rho_{at}(t)$ can also be more simply written as

$$\frac{d}{dt}\rho_{at} = \frac{1}{i\hbar}\left[\widehat{H}_{at}, \rho_{at}\right] - \frac{\Gamma}{2}(\rho_{at}|+\rangle\langle+| + |+\rangle\langle+|\rho_{at}) + \Gamma|-\rangle\langle+|\rho_{at}|+\rangle\langle-|, \tag{4.76}$$

where $\widehat{H}_{at}$ is the atomic Hamiltonian with the Lamb shift $\delta\omega$ included.

The last term can also be written with the help of the raising and lowering operators $\hat{\sigma}^+$ and $\hat{\sigma}^-$ (relations (D.3) and (D.4)) as $\Gamma\hat{\sigma}_-\rho_{at}\hat{\sigma}_+$. We have therefore been able to write the evolution equation as a Lindblad equation (4.54), with two Kraus operators and a single quantum jump operator, $\hat{A}_1$, equal to the atomic lowering operator $\hat{\sigma}_- = |-\rangle\langle+|$.

Up-Jumps and Down-Jumps in the Shelving Technique

Let us return to the shelving situation, outlined in Section 1.3, that of a single three-level atom subject to a weak excitation on transition $|G\rangle \rightarrow |D\rangle$, a strong interaction on transition $|G\rangle \rightarrow |B\rangle$, and spontaneous decay on transitions $|D\rangle \rightarrow |G\rangle$ and $|B\rangle \rightarrow |G\rangle$ (Figure 1.4).

The fluorescent light on transition $|B\rangle \rightarrow |G\rangle$ (Figure 1.5) features up- and down-jumps, which look at first sight rather similar, but are not: the down-jump is generated by a non-Hermitian Hamiltonian (the commutator term in Equation (4.56)), whereas the up-jump is generated by a spontaneous decay due to the coupling with the environment and the full Lindblad equation (4.56) including a jump operator.

The up-jumps are fundamentally discontinuous since they are heralded by the emission of a photon. If at time $t + \Delta t$ the detector "clicks," conservation of energy demands the atom to be in the ground state $|\psi(t + \Delta t)\rangle = |G\rangle$, even if previously its state $|\psi(t)\rangle$ was not close to the ground state ($\langle\psi(t)|G\rangle \sim 0$). On the other hand, the experimental signal of "no-click" after a down-jump is much more subtle: it produces only a smooth Bayesian update. A much more detailed discussion of the present situation is given in reference [88, p. 192].

4.7 Exercises for Chapter 4

Exercise 4.1 Show that the transpose operation (T) is positive, but not completely positive and thus not a physical quantum map.

(a) Prove that any valid quantum map M_A over a system A needs to be a positive map. That is, given a nonnegative operator ρ_A as an input, the output is also nonnegative.

(b) Show that any valid quantum map M_A over a system A needs to be completely positive. That is, when a map M_A is performed over a system A, which is understood as a subsystem of a larger system AB, then the map over the full system $M_{AB} = M_A \bigotimes \hat{1}_B$ is positive. If so, M_A is said to be completely positive.

(c) Show that taking the transpose of a density operator is not a physically realizable quantum map. Show that the map is positive but not completely positive. Do this by studying the effects of transposing system A over $|\psi^-_{AB}\rangle = \frac{1}{\sqrt{2}}(|0_A, 1_B\rangle - |1_A, 0_B\rangle)$. What about the effect of the transformation over the state $|\Psi\rangle = \frac{1}{\sqrt{N}} \Sigma_k^N |k_A, k_B\rangle$?

(d) What is the relation of the complete positivity of a map over a system A and the entanglement of A with a system B?

(e) For what values of p is the (Werner) state $\rho_{AB} = p|\psi^-_{AB}\rangle\langle\psi^-_{AB}| + (1-p)\frac{1}{4}$ entangled? Is the positive partial transpose criterion necessary and sufficient to prove entanglement?

Exercise 4.2 Show that if the Kraus map contains a single operator then it must be unitary.

Exercise 4.3 Consider a whole set of smooth Kraus operators ("smooth" in the sense that they allow for a continuous and differentiable Taylor expansion without nonanalytical "jumps": $\hat{K}_i(dt) = \sqrt{p_i}(1 + \hat{A}_i dt + \cdots)$).

(a) Show that the evolution for the density operator can only be unitary.
(b) Show that the evolution for the state vector must also be unitary.
(c) Can one have a nonunitary evolution driven by a continuous trajectories of the state vector (this is by analytical $\hat{K}_i(dt) = \sqrt{p_i}(1 + \hat{A}_i dt + \cdots)$)?
(d) Give an example (the explicit operators) for such a unitary evolution arising from underlying randomness. Consider the Rabi drive of a qubit.

Exercise 4.4 Show that the Kraus map is an average over the action of the individual Kraus operators.

Exercise 4.5 Derive the Lindblad equation in the Heisenberg representation for an operator $\hat{A}(t)$.

Exercise 4.6 Discuss the relation between the dissipation and the fluctuations by focusing on the Lindblad equation for observables.

Exercise 4.7 Derive the differential equations for the state vector that are equivalent to the Lindblad equation for the density operator.

(a) Derive a Schrödinger-like equation for the state vector from the expression in Equation (4.54).
(b) Show that to include the jump term $\left(\hat{A}_i \rho \hat{A}_i^\dagger\right)$ one requires a stochastic differential formulation.
(c) Derive the properties of the stochastic differential element by comparison to the Lindblad equation.
(d) Discuss the quantum Monte Carlo method to simulate open systems in this context.

Exercise 4.8 Partial trace.

(a) Discuss the relationship between classical probability marginalization and the partial trace in quantum mechanics.
(b) Consider the two qubit state $|\phi\rangle = \alpha|+, a\rangle + \beta|-, b\rangle$. Obtain the state of the first particle ($\hat{\rho}_1$) by tracing out the second one. Find the relation between entanglement and decoherence D in Equation (4.33).

(c) Discuss how the calculation for ρ_1 is modified if $|\phi\rangle$ represents the state of a spin 1/2 and its own position wave function ($\langle x|a\rangle = \psi_a(x)$ and $\langle x|b\rangle = \psi_b(x)$).

Exercise 4.9 Consider a two-level system evolving under $\hat{H}_i = \hbar\omega_i/2\hat{\sigma}_z$, where ω_i fluctuates slowly with respect to the experimental run (typically μs) but fast with respect to the repetition rate (seconds).

During the ith realization of the experiment the pure state evolves as $|\psi_i(t)\rangle = \alpha_0|0\rangle + e^{i\omega_i t}\beta_0|1\rangle$.

(a) Give an example of a physical system that will fit this description.
(b) Naming $Proba(\omega_i)$ the probability of a given ω_i write down the density matrix that describes the system and derive its evolution equation. Discuss the evolution of the coherences and of the populations. Compare with the result of Exercise 4.8.
(c) Discuss the particular case where $Proba \sim$ Normal(0, σ_ω) (a normal distribution with zero mean and finite variance). Show that by the time the variance of the random variable $\sqrt{\langle\omega_i^2\rangle_i}t = \sigma_\omega t$ is of the order of 2π, decoherence has taken its toll.

Exercise 4.10 Consider the "phase flip decoherence channel" with probability p_1 the quits will be flipped: $|0\rangle \rightarrow |1\rangle$ and $|1\rangle \rightarrow |0\rangle$. This corresponds to the application of the jump operator $\hat{\sigma}_x$. Otherwise, nothing happens.

(a) Write down the Kraus map.
(b) Derive the Lindblad equation assuming a memoryless process: $p_1 = \gamma dt$.
(c) Discuss the difference in the trajectories with respect to the noise model considered in Exercise 4.9.

Exercise 4.11 Consider the "population damping channel" for a qubit (jump operator is $\hat{\sigma}_-$ (Equation (4.33)). Explicitly verify that the Lindblad equation is invariant under a unitary transformation for the Kraus map.

5 Measurement

Measurements play a central and specific role in quantum physics, as formalized in the early days of quantum mechanics by von Neumann [173]. In Chapter 2 we recalled the quantum rules concerning the probability of measurement and the expression of the quantum state after measurement. These rules concern perfect measurements of parameters that are associated with projection operators. Actual measurements may simultaneously deliver a list of several numbers as a result. They may present imperfections and limited efficiency. In a way analogous to the previous chapters, in which we have shown that it is possible to describe more complex situations and evolutions than the ones covered by the basic rules of quantum mechanics, we present in this chapter a generalized approach of the measurement process. This approach will allow us to treat imperfect measurement devices, but also extended measurement strategies, such as the simultaneous measurement of two noncommuting observables.

5.1 Rules for Generalized Measurements

We will now introduce modified versions of the rules **M1, M2, M3** from Appendix A (Equations (A.3) and (A.4)) for the measurement so as to be able to describe extended categories of measurements.

A generic measurement is sketched in Figures 5.1 and 5.2: A quantum system described by a density matrix ρ acting in a Hilbert space $\mathcal{H}$ interacts with a measurement device, or detector. Let a_n be the result of the measurement. This result varies randomly from one measurement to the next. Many measurements are performed on successive identical copies of ρ. The successive measurement results are read and either

- stored in a computer memory (Figure 5.1) that records the whole set of measured values. The conditional density matrix $\rho^{after|a_n}$ gives the quantum state of the system just after the measurement and after a post-selection process has been performed on these data, i.e. when one has got rid of all the recorded data that are associated with a result other than a_n. One can just as well ignore the data corresponding to measurement results different from a_n and not store them, a procedure called *heralding*.
- used to perform a post-selection on the quantum system itself, i.e. not destroy it when the measurement has given the result a_n and block it

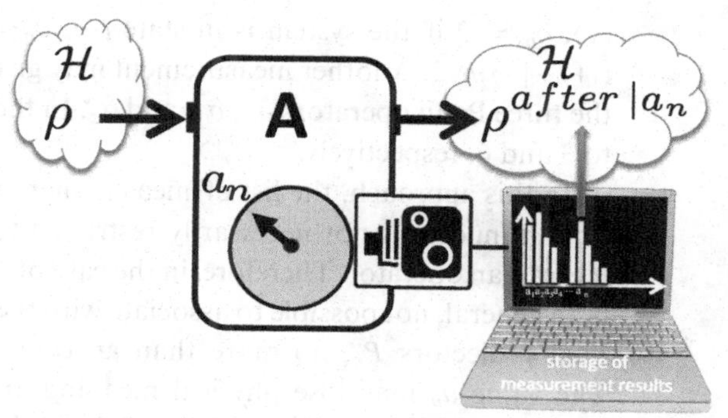

Fig. 5.1 Diagram of a generic measurement process. The results of the measurement are read and stored in computer memory, which can display the histogram of results. Rule M3 gives the conditional quantum state after the measurement, obtained by post-selecting the events when the measurement has given the result a_n.

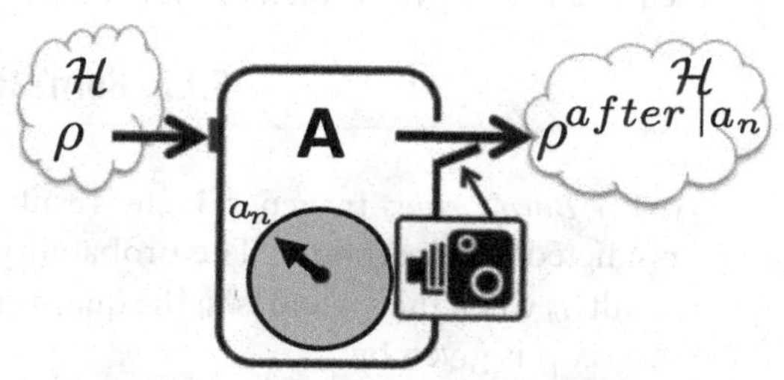

Fig. 5.2 Diagram of a generic measurement process. The results of the measurement are used to either open (when the result is a_n) or close (when it is different from a_n) a shutter. Rule M3 gives the quantum state actually exiting the device.

otherwise (for example, in optics by opening or closing a shutter on the light beam carrying the state ρ (Figure 5.2). The quantum state of the system after the measurement is described by the density matrix $\rho^{after|a_n}$.

These two configurations involve a significant loss of information, either by deliberately ignoring a part of the data, or by destroying the quantum system itself when the measurement has not given the expected result.

The three general rules describing the measurement process are given below and recalled in Appendix B, which gathers together all the generalized postulates of quantum mechanics. They introduce new and important physical quantities that we will study in the following sections.

5.1.1 Quantization Rule

(M1) *Quantization*: A measurement can only give as a result a value a_n belonging to a list $\{a_1, a_2, ..., a_{N_A}\}$. The number N_A of these possible values is not limited to the dimension of the Hilbert space $\mathcal{H}$.

For example, for a measurement determining the quantum state of a qubit, the list can include the following results: $a_n = 1$ if the system is in state

$|+\rangle$, $a_n = 2$ if the system is in state $|-\rangle$, $a_n = 3$ if the system is in state: $(|+\rangle + |-\rangle)/\sqrt{2}$. Another measurement may give as results the eigenvalues of the three Pauli operators $\hat{\sigma}_x$, $\hat{\sigma}_y$, and $\hat{\sigma}_z$. In these two examples, N_A is equal to 3 and 6, respectively.

In this approach, the list of measurement results, which can be discrete or continuous, is not necessarily restricted to the list of eigenvalues of a Hermitian operator. Therefore, in the case of a generalized measurement, it is, in general, not possible to associate with the measurement device orthogonal projectors $\hat{P}_n$, no more than an observable operator $\hat{A} = \sum_n a_n \hat{P}_n$. The value a_n may lose physical meaning and can be just the number n labeling the nth possible result, like in the examples we have just given.

This first rule opens up a lot of new possibilities for measurements on quantum states: for example, the list of possible results can include results that are eigenvalues of non commuting operators, like the simultaneous measurement of position *and* momentum of a particle.

5.1.2 Born's Rule

(**M2**) *Born's rule*: In general, the result of a measurement cannot be predicted with certainty. The probability $Proba(a_n|\rho)$ of obtaining the result a_n when the system is in the quantum state described by the density matrix ρ is given by

$$Proba(a_n|\rho) = Tr\rho\hat{\Pi}_n, \tag{5.1}$$

where the $\hat{\Pi}_n$ are N_A positive linear operators associated with the measurement device and the value a_n of the property satisfying the condition

$$\sum_{n=1}^{N_A} \hat{\Pi}_n = \hat{1}. \tag{5.2}$$

The set of N_A operators $\hat{\Pi}_n$ completely describes a given generalized measurement device. It is usually called a POVM, for positive operator valued measure. Following [13], we will call each $\hat{\Pi}_n$ a probability operator (PO).

5.1.3 State Collapse Rule

(**M3**) *State collapse*: The conditional, or post-selected, state of the system just after the measurement *and in the subset of measurements that have given the result* a_n, called $\rho^{after|a_n}$, is

$$\rho^{after|a_n} = \frac{1}{Tr\rho\hat{\Pi}_n} \hat{M}_n \rho \hat{M}_n^\dagger, \tag{5.3}$$

where the $\hat{M}_n$ are N_A linear operators (but not necessarily positive) acting on the Hilbert space $\mathcal{H}$ and satisfying $\hat{\Pi}_n = \hat{M}_n^\dagger \hat{M}_n$.

The state described by $\rho^{after|a_n}$ is called a *conditional state* because it concerns the fundamentally random and uncontrolled opportunities when the measurement gives the precise result a_n, discarding all the other measurement results. We will call the post-measurement operator (PMO) the operator $\hat{M}_n$. Note the similarity between the present state collapse formula (Equation (5.3)) and the Kraus sum formula (Equation (4.17)) describing a quantum map, the post-measurement operators $\hat{M}_n$ replacing the Kraus operators $\hat{K}_\ell$: The quantum state after the measurement has given the result a_n can be considered as a statistical mixture of different evolutions. Note also a difference of paramount importance between the two formulas: a normalizing factor is present in relation (5.3) but not in relation (4.17). This implies that *the relation* (5.3) *is not linear in* ρ. We will come back to this key point in Section 5.7.

5.1.4 Link with the Previous Rules

It is simple to check that the new rules include the old ones. In order to retrieve the usual formulation, one must simply take the projector $\hat{P}_n$ (which is a positive operator) as both the probability operator $\hat{\Pi}_n$ and the PMO $\hat{M}_n$, in which case there is only one term in the sum (5.2) ($N_A = 1$). Hence von Neumann measurements are associated with probability operators that are projectors, whereas generalized measurements are associated with probability operators that are just positive operators, hence the name of projective measurement often given to the von Neumann measurements.

5.2 Properties of the Probability Operator-Valued Measure

5.2.1 General Properties

The positivity of operators $\hat{\Pi}_n$ is a direct consequence of Equation (5.1) and of the fact that a probability is a positive quantity.

We assume that the *list of measurement results* $\{a_1, a_2, ..., a_{N_A}\}$ *is complete*: A measurement performed on state ρ must necessarily give one of the results a_n, and therefore $\sum_n Proba(a_n|\rho) = 1$ for any ρ. One has the following completeness relation:

$$\sum_{n=1}^{N_A} \hat{\Pi}_n = \hat{1}. \tag{5.4}$$

The POs $\hat{\Pi}_n$ are not necessarily orthogonal, nor projectors, i.e. the product $\hat{\Pi}_n \hat{\Pi}_{n'}$ is not necessarily equal to $\delta_{n,n'} \hat{\Pi}_n$.

5.2.2 Gleason's Theorem

Equation (5.1), the generalized version of Born's rule can be derived from very general arguments, as shown by Gleason's theorem [107], which we will state here without demonstration.

Let $P(a_n|\rho)$ be functions of a_n and ρ that are positive and smaller than one. We also assume that they have the properties

$$\sum_{n=1}^{N_A} P(a_n|\rho) = 1; \quad P(a_n \text{ or } a_{n'}|\rho) = P(a_n|\rho) + P(a_{n'}|\rho), \tag{5.5}$$

and if the mathematical object describing the quantum state belongs to a Hilbert space $\mathcal{H}$ of dimension larger than 2, then there exists a set of operators $\hat{\Pi}_n$, independent of the state ρ of the system, such that

$$P(a_n, \rho) = Tr\left(\rho\hat{\Pi}_n\right). \tag{5.6}$$

One therefore retrieves relation (5.1). Conditions (5.5) are necessarily fulfilled by probabilities, so that this theorem applies to the determination of the probability of measuring a_n on ρ.

The important point to keep in mind is that the generalized Born's rule *is intrinsically linked to the linear character of the space in which the quantum states are defined*, i.e. ultimately to the superposition principle.

5.2.3 Mean Values, Variances

Expression (5.1) allows us to calculate any statistical moment of the physical quantity A:

$$\langle A\rangle = \sum_{n=1}^{N_A} a_n Proba(a_n|\rho) = Tr\rho\hat{A}_1, \tag{5.7}$$

$$\langle A^k\rangle = \sum_{n=1}^{N_A} a_n^k Proba(a_n|\rho) = Tr\rho\hat{A}_k, \tag{5.8}$$

where

$$\hat{A}_1 = \sum_{n=1}^{N_A} a_n\hat{\Pi}_n; \quad \hat{A}_k = \sum_{n=1}^{N_A}(a_n)^k\hat{\Pi}_n. \tag{5.9}$$

Here, $\hat{A}_1 = \sum_{n=1}^{N_A} a_n\hat{\Pi}_n$ can be considered as the generalization of the observable operator $\hat{A} = \sum_n a_n\hat{P}_n$ used in the case of a von Neumann measurement. In this case $\hat{A}_1 = \hat{A}$ and $\hat{A}_k = \hat{A}^k$ because $\hat{P}_n\hat{P}_{n'} = \delta_{n,n'}$. For a generalized measurement there is no simple relation between $(\hat{A}_1)^k$ and $\hat{A}_k$. Therefore, *the observable $\hat{A}_1 = \sum a_n\hat{\Pi}_n$ is not, in general, sufficient to determine all the statistical properties of the measurement results.*

An interesting question then arises: How do we characterize the difference in terms of quantum fluctuations between a generalized and a von Neumann measurement, and how does it affect the usual Heisenberg inequality that is demonstrated for usual variances of Hermitian operators? One has, in the case of a nonprojective measurement,

$$Var(A) = \langle \hat{A}_2 \rangle - \langle \hat{A}_1 \rangle^2 = Var(A_1) + \Delta, \qquad (5.10)$$

where Δ is the additional term that arises because one is performing a generalized measurement of quantity A instead of a projective measurement of the observable $\hat{A}_1$. Δ can be written as

$$\Delta = \sum_n a_n^2 \langle \hat{\Pi}_n \rangle - \langle \hat{A}_1^2 \rangle = \sum_n \langle (\hat{A}_1 - a_n) \hat{\Pi}_n (\hat{A}_1 - a_n) \rangle, \qquad (5.11)$$

remembering that $\hat{A}_1 = \sum a_n \hat{\Pi}_n$ and $\sum \hat{\Pi}_n = 1$. Δ is manifestly a nonnegative operator: As expected, the nonideality of the measurement implies an additional uncertainty (equal to zero for a projective measurement) on top of the basic quantum one. As a consequence, *the Heisenberg inequality also holds for generalized measurements.*

5.3 Properties of the Post-measurement Operators $\hat{M}_n$

5.3.1 General Properties

The normalization of the density matrix $Tr\rho^{after|a_n} = 1$ implies that

$$\hat{\Pi}_n = \hat{M}_n^\dagger \hat{M}_n. \qquad (5.12)$$

A possible solution of Equation (5.12) for $\hat{M}_n$ is $\hat{U}\sqrt{\hat{\Pi}_n}$, where $\hat{U}$ is an arbitrary unitary operator (the operator $\sqrt{\hat{\Pi}_n}$ exists because $\hat{\Pi}_n$ is a positive operator). PMOs satisfying relation (5.12) for a given $\hat{\Pi}_n$ are not at all unique. This relates to the fact that several experimental measurement set-ups can lead to the same probabilities of results, but to different post-measurement histories. To determine the operators $\hat{M}_n$, one needs to know the exact mechanism of the measurement apparatus.

5.3.2 Conditional State after a Measurement Made on a Pure State

Let us now take a pure state $\rho = |\psi\rangle\langle\psi|$ as the input state. The conditional state after a measurement giving result a_n is

$$\rho^{after|a_n} = \frac{\hat{M}_n |\psi\rangle\langle\psi| \hat{M}_n^\dagger}{Proba(a_n|\psi)} = \left|\psi^{after|a_n}\right\rangle \left\langle \psi^{after|a_n}\right|, \qquad (5.13)$$

with

$$\left|\psi^{after|a_n}\right\rangle = \frac{\hat{M}_n|\psi\rangle}{\sqrt{Proba(a_n|\psi)}}. \qquad (5.14)$$

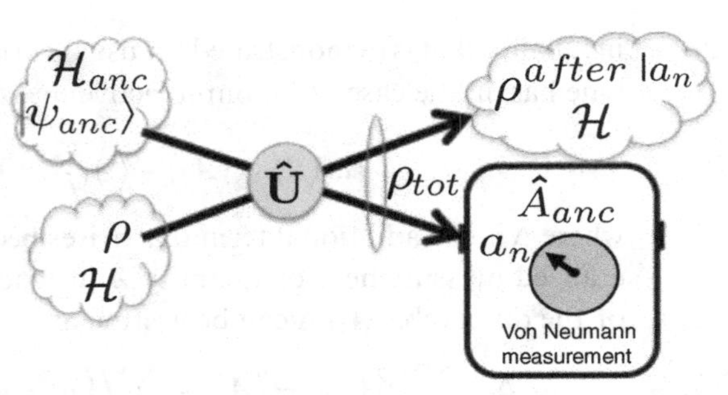

Fig. 5.3 Naimark theorem: A generalized measurement in a given Hilbert space appears as a von Neumann measurement performed on an ancilla system that has interacted with the system under study.

This means that the conditional state after a generalized measurement implemented on a pure state is always a pure state. Compared to the von Neumann case, one just has to replace the projection operator $\hat{P}_n$ by the PMO $\hat{M}_n$.

5.3.3 Not Post-selected Measurement

Formula (5.3) gives the expression of the conditional state that results from a post-selection process, among the record of all performed measurements, of those having given the result a_n, discarding all the other ones. If one does not perform this post-selection, one gets (after the measurement) a density matrix ρ^{after} that is the statistical mixture of all the post-selected states weighted by the corresponding probabilities, i.e.,

$$\rho^{after} = \sum_{n=1}^{N_A} \rho^{after|a_n} Proba(a_n|\rho) = \sum_{n=1}^{N_A} \hat{M}_n \rho \hat{M}_n^{\dagger}. \tag{5.15}$$

This expression is also known as the result of an "unread measurement." Such a name can be misleading as it seems to imply a mysterious effect of the observer, which obtains different after-measurement states depending on whether he chooses to look or not at the dial of the measurement device. Actually, the fact that the observer chooses to read or not to read the dial is not relevant. What is important is whether the observer chooses to consider the whole list of measurement results stored in the computer, or only the smaller part of it corresponding to a given value a_n.

We note also that Equation (5.13) relating the conditional collapsed output state $\rho^{after|a_n}$ to the input state ρ *is not linear in* ρ, as already noted, whereas Equation (5.15) relating the not post-selected output state ρ^{after} to the input state ρ is linear in ρ.

5.3.4 Naimark Theorem

The Naimark theorem [13, 142], which we will not demonstrate here, is the equivalent for nonprojective measurements of the purification theorem for mixed states. Any generalized measurement in a given Hilbert space $\mathcal{H}$ can be considered as a von Neumann measurement in a larger Hilbert space.

More precisely, let us introduce an ancilla system S_{anc}, described by the Hilbert space $\mathcal{H}_{anc}$, that interacts during some time with the system S under study, described by states in Hilbert space $\mathcal{H}$ (see Fig. 5.3). The initial state is the factorized state $\rho \otimes |\psi_{anc}\rangle\langle\psi_{anc}|$, where $|\psi_{anc}\rangle$ is the initial pure state of the ancilla system. The evolution due to this coupling, ruled by the Schrödinger equation, is described by the unitary evolution operator $\hat{U}$ on the whole space (see Chapter 4). It creates the entangled state $\rho_{tot} = \hat{U}\rho \otimes |\psi_{anc}\rangle\langle\psi_{anc}|\,\hat{U}^\dagger$. After the interaction, one performs a von Neumann measurement within the ancilla space. It concerns an observable that we call $\hat{A}_{anc}$, having the list of results $\{a_n\}$ as eigenvalues and characterized by projectors $\hat{P}_{anc,n}$ (which implies that the ancilla space has a dimension which can be larger than the dimension of $\mathcal{H}$). The theorem shows that there is an initial ancilla state $|\psi_{anc}\rangle$ and a unitary coupled evolution operator $\hat{U}$ such that the probability $Proba(a_n|\rho_{tot})$ of the von Neumann measurement, calculated by Equation (2.7), is equal to $Proba(a_n|\rho)$ given by Equation (5.1). In addition, when the measurement has given the result a_n, the reduced density matrix in space $\mathcal{H}$ resulting from the generalized state collapse postulate (2.8) is precisely given by Equation (5.3).

There are actually many different ways to perform such a generalized measurement in a larger space. Note also that the theorem implies that, if one considers only a part of a composite system on which a von Neumann measurement is performed, the statistics of the measurement and the conditional state obey the postulates (5.1) and (5.3) of generalized measurements.

5.4 Imperfect Measurements

We quickly described in Chapter 1 a few examples of real measurements that are currently used by experimentalists and are so sensitive that they give information about single quantum objects. For example, single-photon and single-particle detectors, photon number resolving photon detectors, devices able to measure the energy level of a single trapped ion or the presence of a single photon inside an optical cavity, photodiodes, and superconducting devices sensitive enough to record the instantaneous fluctuations of quantum systems. Some measurement devices destroy the system under study, like the photomultiplier, most particle detectors, the photodiode, the bolometer, and the selective ionization by an electrostatic field. In these measurement set-ups, the quantum system under study cannot be submitted to a second measurement. Others are nondestructive, like the "shelving" technique used to measure the energy level of an ion, or the nondestructive detection of a photon in a cavity. These techniques allow experimentalists to monitor in real time the evolution of a single observable or of a single quantum object. These are called *quantum nondemolition* measurements. We will give more examples of such measurements in the following sections.

In addition, detectors may have different kinds of imperfections:

1. They may "miss" events. More precisely, they may be sensitive enough to detect individual quantum objects, but they do not detect all of them. The ratio of actually recorded particles over the incident ones is called the *quantum efficiency* of the detector.
2. They can give wrong information, indicating, for example, that an atom is in a given state when it is actually in another state. We have examples of wrong information about the Rydberg states $|+\rangle$ and $|-\rangle$ in the recording of Figure 3.2, for example, the rare presence of red clicks inside the series of blue clicks witnessing the absence of photons in the cavity. As the two levels can be seen as forming a qubit, this defect is characterized by what is called the *bit error rate*, or *bit flip rate*.
3. Some detectors can give a nonzero signal, i.e. a "click," even when no particle is incident on it. This defect is characterized by the *dark count rate*. Others have a small signal-to-noise ratio and hide very small signals in the detector noise.
4. Most single particle detectors are able to detect the absence or presence of particles, but not to count their exact number. Others have a *particle number resolving capacity*, up to a maximum number that is of the order of a few units. On the other hand, a high quantum efficiency photodiode delivers a photocurrent proportional to the mean number of photons $\langle\hat{N}\rangle$ for very large values of $\langle\hat{N}\rangle$ (of the order of 10^{16} photons/s), but is not able to distinguish between states of light $|n\rangle$ and $|n+1\rangle$.

5.5 Examples

5.5.1 Probability Operators

Bit Error Rate

Let us consider an imperfect measurement of the eigenstate of a qubit that sometimes gives wrong answers: When the input state is $|+\rangle$, it has a probability p_+ of giving the answer "plus," and a probability $1-p_+$ of giving the answer "minus." When the input state is $|-\rangle$, it has a probability p_- of giving the answer "minus," and a probability $1-p_-$ of giving the answer "plus," The POVM in this case is

$$\hat{\Pi}_{plus} = p_+|+\rangle\langle+| + (1-p_-)|-\rangle\langle-|; \quad \hat{\Pi}_{minus} = (1-p_+)|+\rangle\langle+| + p_-|-\rangle\langle-|. \tag{5.16}$$

Note that, a priori, nondiagonal terms like $\alpha|+\rangle\langle-| + \alpha^*|-\rangle\langle+|$ could be added in the expressions (5.16) of the POs without changing the value of the probabilities given above. But then these operators are no longer positive: $Tr\rho\hat{\Pi}_{plus}$ and $Tr\rho\hat{\Pi}_{min}$ take negative values for some states unless $\alpha = 0$.

One obviously has $\hat{\Pi}_{plus} + \hat{\Pi}_{minus} = \hat{1}$, but now $\hat{\Pi}^2_{plus} \neq \hat{\Pi}_{plus}$: The POs are no longer projectors.

Quantum Efficiency η and Dark Count Probability d

Let us now consider an imperfect single photon detector. When the input state is a single photon state $|1\rangle$, it has a probability η of giving a click (and a probability $1-\eta$ of giving no click). When there is no input light, i.e. when the input state in the vacuum is $|0\rangle$, it has a probability d of giving a click (and a probability $1 - d$ of giving no click). In this case, the POVM is

$$\hat{\Pi}_{click} = d|0\rangle\langle 0| + \eta|1\rangle\langle 1|, \tag{5.17}$$

$$\hat{\Pi}_{no\ click} = (1-d)|0\rangle\langle 0| + (1-\eta)|1\rangle\langle 1|. \tag{5.18}$$

If one associates the measurement value $n = 0$ with the no click situation and $n = 1$ to the click occurrence, then the operator describing the number observable, defined by $\hat{N}_1 = \sum n\hat{\Pi}_n$ (Equation (5.9)), is equal to:

$$\hat{N}_1 = d|0\rangle\langle 0| + \eta|1\rangle\langle 1|. \tag{5.19}$$

It has the same eigenstates as the ideal counter, but different eigenvalues. Let us recall that this operator is only useful when one is interested in determining the mean value of the photon number, not the variance.

5.5.2 Post-measurement Operators

Case of a Projective and Destructive Measurement

Let us show that expression (5.1) includes the case of a measurement that fulfils the usual Born's rule for the probabilities of measurements (2.5), but which is destructive and does not follow the von Neumann state collapse postulate (A.2).

This can be seen in the following simple case of an observable $\hat{A}$ having nondegenerate eigenvalues a_n and eigenstates $|u_n\rangle$. Then the POVM is the set of projectors $\hat{P}_n = |u_n\rangle\langle u_n|$, and expression (5.1) reduces to the usual Born's rule (A.2). In the destructive case, the measurement device transforms the input state into a kind of "destroyed state," $|\phi_D\rangle$, which, for example, in the case of an atomic Rydberg state energy measurement, is the quantum state describing the ionized atom and the ejected electron (Figure 3.2). The corresponding PMO is $\hat{M}_n = |\phi_D\rangle\langle u_n|$. In such a case,

$$\rho^{after|a_n} = \frac{|\phi_D\rangle\langle u_n|\rho|u_n\rangle\langle\phi_D|}{Tr\rho|u_n\rangle\langle u_n|} = |\phi_D\rangle\langle\phi_D|, \tag{5.20}$$

whatever the result of the measurement.

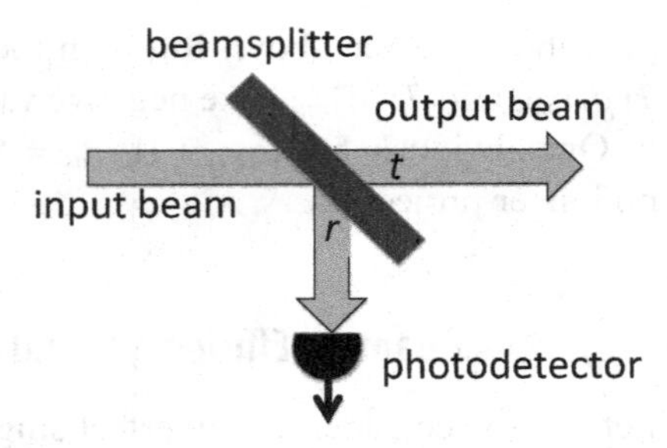

Fig. 5.4 Partially nondestructive detection of a photon, with the help of a beamsplitter that reflects part of the input light on a single photon detector and destroys it, while transmitting the other part in a nondestructive way.

Photodetection with a Beamsplitter

A simple but imperfect nondestructive photodetector is sketched in Figure 5.4. The input beam to be measured crosses a beamsplitter of amplitude transmission and reflection coefficients t and r (both real, with $t^2 + r^2 = 1$). To simplify, we assume that the input state contains only 0 or 1 photon, so that we can restrict the Hilbert space to the one spanned by states $|0\rangle$ and $|1\rangle$. The partially reflected light is measured by a photodetector, which we assume for simplicity to be perfect: no dark count and quantum efficiency 1. Note the similarity of this scheme (Figure 5.4) with the one considered by the Naimark theorem (Figure 5.3).

The system is simple enough to be the object of an exact calculation with the help of the input-output relations of quantum optics recalled in Section H.3. If the input beam state is $|0\rangle$, then the output two-beam state is $|0, 0\rangle$, the two numbers giving the number of photons in the transmitted and reflected beams, respectively; if the input beam state is $|1\rangle$, then the output state is $r|0, 1\rangle + t|1, 0\rangle$. So the quantum map giving the density matrix describing the two output beams for any input density matrix ρ is

$$\rho^{out} = \begin{pmatrix} \rho_{00} & r\rho_{01} & t\rho_{10} & 0 \\ r\rho_{10} & r^2\rho_{11} & rt\rho_{11} & 0 \\ t\rho_{01} & rt\rho_{11} & t^2\rho_{11} & 0 \\ 0 & 0 & 0 & 0 \end{pmatrix}, \tag{5.21}$$

the order of the basis elements being $|0, 0\rangle, |0, 1\rangle, |1, 0\rangle, |1, 1\rangle$.

As the measurement is of the von Neumann type, the PO associated with a click is the projector $|1\rangle\langle 1|$ acting on the reflected beam, i.e. the operator $\hat{P}_{click} = |0, 1\rangle\langle 0, 1| + |1, 1\rangle\langle 1, 1|$ in the total space. So the corresponding probability is $Tr\rho\hat{P}_{click} = r^2\rho_{11}$, and the total state after a click is:

$$\hat{\rho}_{tot}^{after|click} = \frac{\hat{P}_{click}\rho\hat{P}_{click}}{Tr\rho\hat{P}_{click}} = \frac{1}{r^2\rho_{11}} \begin{pmatrix} 0 & 0 & 0 & 0 \\ 0 & r^2\rho_{11} & 0 & 0 \\ 0 & 0 & 0 & 0 \\ 0 & 0 & 0 & 0 \end{pmatrix} = |01\rangle\langle 01|. \tag{5.22}$$

The output beam state is obtained by a partial trace over the reflected beam space,

$$\hat{\rho}^{after|click} = \begin{pmatrix} 1 & 0 \\ 0 & 0 \end{pmatrix} = |0\rangle\langle 0|, \tag{5.23}$$

which is rather obvious since the single photon, *counted on the reflected beam*, cannot be measured at the same time on the transmitted beam. One notes that, considered as acting on an initial state belonging to the Hilbert space of the beam going straight through the beamsplitter, the measurement is not ideal and corresponds to a PO and a PMO given, respectively, by

$$\hat{\Pi}_{click} = r^2|1\rangle\langle 1|; \; \hat{M}_{click} = r|0\rangle\langle 1|. \tag{5.24}$$

One can do the same for the no click case, with $\hat{P}_{no\,click} = |0,0\rangle\langle 0,0| + |1,0\rangle\langle 1,0|$. The corresponding probability is $Tr\rho\hat{P}_{no\,click} = t^2\rho_{11} + \rho_{00}$, and the total state after the measurement is

$$\hat{\rho}_{tot}^{after|no\,click} = \frac{1}{t^2\rho_{11} + \rho_{00}} \begin{pmatrix} \rho_{00} & 0 & t\rho_{10} & 0 \\ 0 & 0 & 0 & 0 \\ t\rho_{01} & 0 & t^2\rho_{11} & 0 \\ 0 & 0 & 0 & 0 \end{pmatrix}. \tag{5.25}$$

The partial trace gives the output beam state

$$\hat{\rho}^{after|no\,click} = \frac{1}{t^2\rho_{11} + \rho_{00}} \begin{pmatrix} \rho_{00} & t\rho_{01} \\ t\rho_{10} & t^2\rho_{11} \end{pmatrix}. \tag{5.26}$$

The corresponding PO and PMO in the output beam space are given, respectively, by

$$\hat{\Pi}_{no\,click} = |0\rangle\langle 0| + t^2|1\rangle\langle 1|; \quad \hat{M}_{no\,click} = |0\rangle\langle 0| + t|1\rangle\langle 1|. \tag{5.27}$$

If the input state is the pure state $\alpha|0\rangle + \beta|1\rangle$, then the output state is also pure: $|0\rangle$ in the click case and $(\alpha|0\rangle + \beta t|1\rangle)/\sqrt{|\alpha|^2 + |\beta|^2 t^2}$ in the no click case.

The beamsplitter has the same POVM as a detector with limited quantum efficiency $\eta = t^2$, but of course not the same post-measurement operator. One checks that $\hat{M}^{\dagger}_{click}\hat{M}_{click} = \hat{\Pi}_{click}$, $\hat{M}^{\dagger}_{no\,click}\hat{M}_{no\,click} = \hat{\Pi}_{no\,click}$ and $\hat{\Pi}_{click} + \hat{\Pi}_{no\,click} = \hat{1}$.

If the output state is not post-selected according to the measurement results, its expression after the crossing of the beamsplitter is

$$\hat{\rho}^{after} = \begin{pmatrix} \rho_{00} + r^2\rho_{11} & t\rho_{01} \\ t\rho_{10} & t^2\rho_{11} \end{pmatrix}. \tag{5.28}$$

This equation gives the effect of a beamsplitter on any input state. It could have been obtained more straightforwardly by taking the partial trace of ρ_{out} (Equation (5.21)). It is a linear transformation of the different components of the input density matrix, which is trace preserving, i.e. a good example of a *quantum map* that we have studied in a general way in Chapter 4. Note that, as $t < 1$, *the coherence is reduced by the physical interaction with the measurement set-up.*

5.5.3 Single Photons from Twin Beams

As we will see in Section I.2, spontaneous parametric down conversion in nonlinear optical media allows one to generate correlated signal and idler "twin beams," labeled (s) and (i), emitted in different directions. The quantum state that is generated using a low power pump beam can be written as the pure state

$$|\psi\rangle = \sqrt{n_0}|0 : s, 0 : i\rangle + \sqrt{n_1}|1 : s, 1 : i\rangle, \tag{5.29}$$

with $n_0 \simeq 1$. One inserts on beam (i) an imperfect single photon detector of dark count $d \simeq 0$ and quantum efficiency $\eta \simeq 1$, described by the POVM of Equations (5.17) and (5.18).

The probability of a click is given by $Proba_{click} = Tr(\rho_i \hat{\Pi}_{click})$, where $\rho_i = n_0|0 : i\rangle\langle 0 : i| + n_1|1 : i\rangle\langle 1 : i|$ is the reduced density matrix of the idler beam traced over the signal Hilbert space, which is not subject to measurement, and $\hat{\Pi}_{click}$ is the PO of the occurrence of a click (Equation (5.17)). The result is

$$Proba_{click} = n_0 d + n_1 \eta \simeq d + n_1. \tag{5.30}$$

The conditional state after the occurrence of a click on the idler channel is

$$\rho^{after|click} = \frac{1}{Proba_{click}} \hat{M}_{click}|\psi\rangle\langle\psi|\hat{M}^{\dagger}_{click}, \tag{5.31}$$

where $\hat{M}_{click}$, the operator on the idler Hilbert space, is given by

$$\hat{M}_{click} = \sqrt{d}|0 : i\rangle\langle 0 : i| + \sqrt{\eta}|1 : i\rangle\langle 1 : i|. \tag{5.32}$$

The post-measurement conditional state is pure and belongs to the signal beam Hilbert space. The result is

$$\left|\psi^{after|click}\right\rangle = \frac{1}{n_0 d + n_1 \eta}\left(\sqrt{dn_0}|0 : s\rangle + \sqrt{\eta n_1}|1 : s\rangle\right). \tag{5.33}$$

If the dark count is low enough ($n_0 d + n_1\eta \simeq n_1\eta$), one gets from the output *a pure, conditional, single photon state*, even if the quantum efficiency of the photodetector is far from 1. This technique is widely used by quantum optics experimentalists to generate single photon states, with a typical success rate of the order of 10^7 single photons per second.

5.5.4 Discrimination between Nonorthogonal States

The formalism we have introduced in this chapter is useful for treating the unavoidable imperfections of real measurements in a rigorous way, but it is also useful for analyzing nonprojective measurement configurations.

We will see that this implies extra fluctuations in the measurements results, as already discussed in Section 5.4.

Let us consider the problem of optimizing the discrimination between two non-orthogonal pure qubit states $|\psi_1\rangle$ and $|\psi_2\rangle$, a problem studied by Helstrom [93] and relevant for quantum communication issues. Here, we describe a simple example and restrict ourselves to a case where these (and only these) two states are generated with equal probabilities 0.5 and submitted to a measurement described by the POVM $\left(\hat{\Pi}_1, \hat{\Pi}_2\right)$. If the measurement gives result 1, then the state of the system is believed to be $|\psi_1\rangle$. If the measurement gives result 2, then the state of the system is believed to be $|\psi_2\rangle$. This statement is, of course, perfectly correct when the states are orthonormal and the POs are the corresponding projectors, and partly correct if they are not orthogonal. To optimize the discrimination, the strategy is to calculate the error probability in the POVM formalism, to minimize this error, and to determine the associated forms of the POs. We will not go into the details of this derivation, which can be found in the "Introduction to Quantum Information" chapter of reference [67], but simply provide an overview.

We assume the following simple form for the nonorthogonal states $|\psi_1\rangle$ and $|\psi_2\rangle$ [13, 67]:

$$|\psi_1\rangle = \cos\theta|+\rangle + \sin\theta|-\rangle; \quad |\psi_2\rangle = \cos\theta|+\rangle - \sin\theta|-\rangle. \tag{5.34}$$

The probability of a wrong decision $Proba_{wrong}$ is

$$Proba_{wrong} = 0.5\, Tr(\rho_1\hat{\Pi}_2) + 0.5\, Tr\rho_2\hat{\Pi}_1. \tag{5.35}$$

One can show that the minimum value of this probability is $\left(1 - \sqrt{1 - |\langle\psi_1|\psi_2\rangle|^2}\right)$ (Helstrom bound), equal to $(1 - \sin 2\theta)/2$ in the present case. The best choice for the POVM can be shown to be the von Neumann measurement associated with projection operators on states:

$$|\phi_1\rangle = (|+\rangle + |-\rangle)\sqrt{2}, \quad |\phi_2\rangle = (|+\rangle - |-\rangle)\sqrt{2}. \tag{5.36}$$

In more general cases, for example of more than two states, the optimized POVM is not a von Neumann one.

5.6 Is the Measurement Process a Physical Evolution?

5.6.1 Position of Problem

As seen in the first chapters, the rules of quantum mechanics pay a special attention to the measurement process, which follows specific rules distinct from the postulate of evolution. But a measurement set-up seems indeed to be a physical system, which, like any quantum physical system, must obey the rules of "normal" evolution.

Physicists and philosophers have considered from the beginning of the quantum age this central problem of quantum mechanics and brought to it different answers, as can be found for example in [107].

An important contribution to the understanding of the measurement process was made by Zurek [187]. The present derivation is close to the one he has introduced.

5.6.2 Physical Model of the Measurement Device

One wants to measure the physical quantity A on a quantum system described by a density matrix ρ in Hilbert space $\mathcal{H}$. For this purpose we couple it to a measurement device, the "meter," which is another quantum physical system described by states in another Hilbert space $\mathcal{M}$. The fact that it is able to "measure" A means that experimentalists are able to "read" specific states $|\chi_n\rangle$ of the meter that form the "pointer basis": each $|\chi_n\rangle$ corresponds, for example, to a definite position of a needle that points to the value a_n, one of the eigenvalues of the associated observable $\hat{A}$. A good measurement device is a one in which the needle points to a_n when the input state in $\mathcal{H}$ is the corresponding eigenstate $|u_n\rangle$. The result of the reading is then kept, for example stored in a memory, and the whole process is repeated many times on successive initial states all prepared in state ρ. From the collection of stored results, one can infer the probability of obtaining a specified pointer position a_n.

One immediately notes that the evolution resulting from the interaction between the system and the meter cannot be Hamiltonian, i.e. a combination of oscillatory terms. The needle in this case would oscillate for ever and never point to a given position, i.e. would never have a stationary final state. This is equally true for a classical measurement. Imagine a weighing scale without damping: Putting a weight on one platform of the scale induces a never-ending oscillation of the scale beam, and it will not be possible to infer the value of the weight. In contrast, if the motion of the beam is damped, the system will reach a final state associated with the value of

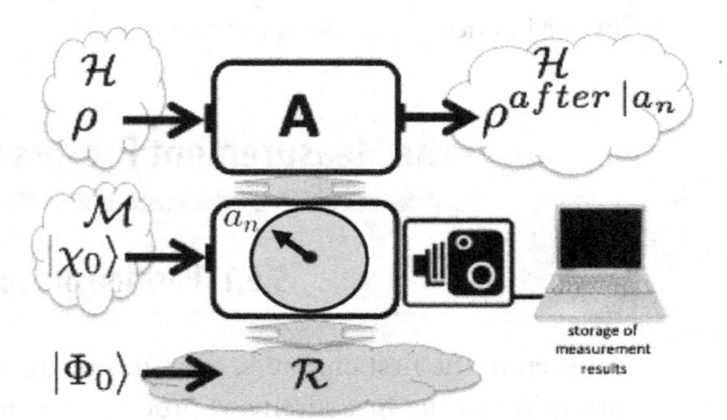

Fig. 5.5 A measurement device couples the system to measure to a physical "meter" that can be read by the experimentalist and stored by a computer. The evolution of the meter is damped by its interaction with a reservoir.

the weight. This means that the measurement set-up is coupled to a large Hilbert space, the "reservoir" $\mathcal{R}$, which induces an irreversible evolution of the meter quantum state, as described in Chapter 4. The whole device is sketched out in Figure 5.5.

5.6.3 Zurek's Model

Zurek's model is a toy model, extensively described in [185, 186]. It relies on several assumptions that simplify the resolution of the problem but keep the essence of the physics involved:

1. One assumes that the coupling between the system and the meter is of the type described in Section 4.3. The interaction Hamiltonian is
$$\hat{H}_{int} = \hbar g(t)\hat{A} \otimes \hat{M}, \tag{5.37}$$
where $\hat{M}$ is the "meter observable," which has the readable pointer states $|\chi_n\rangle$ as eigenstates. For the sake of simplicity, $\hat{H}_{int}$ is supposed to act only during a limited amount of time.
2. One assumes that the coupling of the meter with the reservoir, described by an interaction term $\hat{H}_{MR}$, takes place after the interaction between the meter and the system. The reservoir is not directly coupled to the system, so that the system state does not change during the meter-reservoir interaction.
3. One assumes that $\hat{H}_{MR}$ is *diagonal in the pointer basis* $\{|\chi_n\rangle\}$:
$$\hat{H}_{MR} = \sum_n |\chi_n\rangle\langle\chi_n| \otimes \hat{V}^n_{MR}, \tag{5.38}$$
where $\hat{V}^n_{MR}$ is an operator acting only in the reservoir Hilbert space $\mathcal{R}$. This basic assumption implies that the pointer states $|\chi_n\rangle$ are stationary and stay pointed to the appropriate value a_n even in presence of the reservoir (the beam of the scale stops its motion when it points to the right weight).

Evolution of the Quantum System

The initial state in the system+meter+reservoir Hilbert space is the factorized state $|\psi\rangle \otimes |\chi_0\rangle \otimes |\phi_0\rangle$, with $|\psi\rangle = \sum_n c_n|u_n\rangle$. After interaction (5.37), the system+meter is, according to the discussion of Section 4.3, in the entangled state $|\Psi_{SM}\rangle$ given by:
$$|\Psi_{SM}\rangle = \sum_n c_n|u_n\rangle \otimes |\chi_n\rangle. \tag{5.39}$$
After a duration t of reservoir-meter interaction the system+meter+reservoir is in state
$$|\Psi_{SMR}\rangle = \sum_n c_n|u_n\rangle \otimes |\chi_n\rangle \otimes |\phi_n(t)\rangle, \tag{5.40}$$

with $|\phi_n(t)\rangle = \exp(-i\hat{V}^n_{MR}t/\hbar)|\Phi_0\rangle$. One notices that the initial state of the system $|u_n\rangle$ has created prints both in the meter and in the reservoir. But the reservoir is a large system that one does not measure. As a result, we need to make a partial trace over it, which will give rise to the reduced system+meter density matrix

$$\rho_{SM} = \sum_{n,n'} c_n c^*_{n'} \langle \phi_{n'}(t)|\phi_n(t)\rangle |u_n, \chi_n\rangle\langle u_{n'}, \chi_{n'}|. \tag{5.41}$$

Because of the large size of the reservoir, the matrix elements $\langle \phi_{n'}(t)|\phi_n(t)\rangle$ quickly decay to zero when $n \neq n'$. The final state is therefore described by the diagonal density matrix:

$$\rho^{final}_{SM} = \sum_n |c_n|^2 |u_n, \chi_n\rangle\langle u_n, \chi_n|. \tag{5.42}$$

This state contains perfect correlations between the system and the meter that are required for the meter to properly measure the input system state. By performing partial traces, one also gets the reduced states of the meter alone and of the system alone:

$$\rho^{final}_M = \sum_n |c_n|^2 |\chi_n\rangle\langle\chi_n|; \quad \rho^{final}_S = \sum_n |c_n|^2 |u_n\rangle\langle u_n|, \tag{5.43}$$

where ρ^{final}_M is a mixed state that gives the probability of finding the needle pointing to the value a_n in agreement with Born's rule. Here, ρ^{final}_S is the state of the system after its interaction with the measurement device, made of the meter and the reservoir. It can be also written in terms of the input density matrix ρ as

$$\rho^{final}_S = \rho^{out} = \sum_n \hat{P}_n \rho \hat{P}_n. \tag{5.44}$$

The Kraus operators of the quantum map associated with the interaction between the system and the measurement device are therefore the projectors on the eigenstates of the system-meter interaction.

Decoherence

The interaction with the measurement device has thus two consequences:

- it gives probabilities of measurement results in agreement with Born's rule for von Neumann measurements, which therefore does not appear in this simplified model as a postulate but as a consequence of the evolution of the whole system treated as a single quantum object;
- according to Equation (5.44) the quantum state of the system is strongly modified by the measurement process in an irreversible way, as expected from a damping process. The evolution leaves the diagonal blocks (in the basis of eigenstates of the observable $\hat{A}$) unchanged and destroys all the off-diagonal blocks of the initial density matrix, i.e. all the coherent connections between the eigenstates. The interaction with a measurement device indeed provokes a strong *decoherence process*.

Such a decoherence is not specific to the simple model used here to make simple calculations. It occurs in any more sophisticated physical model of the measurement, in particular when one tries to precisely model a specific physical device used to measure a specific observable (spin, photon number, energy, etc.). Some realistic models take into account the defects of the detector (not perfect quantum efficiency for a photon detector, for example) and yield probabilities of measurement and a final state of the not post-selected system, which obeys the rules of projective measurements.

5.6.4 The State Collapse Is Not a Physical Evolution

The modeling of the measurement set-up has led us to the expected value for the probability of measurement. However, the quantum state of the system after the interaction with the measurement device that we have been able to derive from this simple model *is not the one given by the state collapse rule*. It rather corresponds to the expression of a not-post-selected state, i.e. the mixed state that one obtains when one takes into account on an equal footing all the recorded results of the measurement (Equation (5.15)). This is expected, as we never used in the derivation the fact that a specific value was obtained, nor the exact physics of the post-selection process.

Looking back at the expression of the conditional state $\rho^{after|a_n}$ obtained after the measurement when this measurement has given the value a_n in the projective and nonprojective cases, we see that *the relations (2.8) and (5.3) giving the expression of the collapsed state are not quantum maps* (Equation (4.17)). Consequently, *they cannot be obtained as the outcome of a physical evolution of the input quantum state* ρ. They are even nonlinear, a rather unexpected property for quantum quantities. In contrast, expression (5.44) for a not post-selected state after the measurement is indeed a quantum map, as expected for the result of the quantum evolution due to the interaction with the measurement device.

In conclusion, we have shown here that expression (5.3) for the conditional state after a measurement has a double origin:

- an irreversible physical evolution induced by the interaction between the system in state ρ and the measurement device, leading to the quantum state $\rho^{after} = \sum_{n=1}^{N_A} \hat{P}_n \rho \hat{P}_n^\dagger$;
- *a change of point of view* concerning the information that the observer wants to extract from the quantum state: Instead of considering the physical system without restriction, and taking into account the data corresponding to all the possible measurements results, *the observer decides to select a smaller quantity of data recorded by the measurement device*, and therefore to restrict their investigations to the subset of data conditioned by the occurrence of a given measurement value a_n. To describe this post-selected subset, the observer only needs the collapsed quantum state $\rho^{after|a_n} = \hat{M}_n \rho \hat{M}_n^\dagger / Tr \rho \hat{\Pi}_n$, where the sum is replaced by the single term associated with a_n. Note that the collapsed state must

be "re-normalized" to 1, a consequence of the change of point of view decided by the observer. This normalization is responsible for the onset of *nonlinearity* in the expression of the collapsed state.

The state collapse is therefore an information-related process. It takes place when the observer decides to make such a post-selection on his data. We will come back to this discussion in Section 5.8.

5.7 Multiple Measurements

We are now interested in performing several measurements on the same quantum state. We expect that, in some instances, *correlations* will manifest themselves between the measurement results. Obviously, by making more and more measurements of different quantities, one will acquire more and more information on the system, so that the Bayesian notion of conditional probabilities as information updating on the considered system is a convenient way to tackle the problem of multiple quantum measurements and of the general characterization of correlations (see, for example, [152]).

5.7.1 The Bayesian Approach of Statistics

Let us consider a physical quantity A and its possible measurement results $\{a_n\}$. The usual way of defining a probability is to consider a great number of copies of the same system on which A is measured: The probability $P(a_n)$ is then the frequency of occurrence of a_n among all the recorded measurement results. This approach does not obviously apply to the prediction of the result of a single measurement unless $P(a_n) = 0$ or 1.

Let us consider the following weather forecast report: "there is a 80% chance that it will rain to-morrow." This is a statistical statement about a single event, and we do not consider it as meaningless because we take our umbrella the next morning. What does this mean? It means the information that meteorologists have gathered from pressure and temperature measurements, satellite photos, etc. and included in their weather forecast code have lead to a prediction of rain, so that it is "very likely" that it rains in these conditions. In this case, probability is related to predictability, or *"likelihood."*

Ultimately, likelihood is a measure of the "degree of belief" in the *information* that we have on the system. Each new piece of information about the state of the system will modify the likelihood of getting a given measurement result. The update of our information takes place when one knows the value of the measurement of another quantity, say B, performed on the same system. The relevant quantity in this point of view is the *conditional probability* of a_n given b_m, $P(a_n|b_m)$, of getting the result a_n for A, given that a previous measurement of another quantity B on the same system has yielded the result b_m.

The approach, consisting of considering probabilities as quantifying information on the system acquired by successive measurements and manipulating conditional probabilities, was introduced by Bayes, and then Laplace, in the eighteenth century and extended further by Fisher. It turns out to be very fruitful in many instances, and especially in quantum mechanics.

5.7.2 Correlations

Correlations between the results of different measurements are at the heart of quantum physics. Some physicists even think that "they constitute the full content of physical reality" [125]. In the next section, we consider different quantities that allow us to precisely characterize correlations.

Correlation Functions, Joint and Conditional Probabilities

Let us assume that we make two successive measurements on a classical system. One first measures the physical quantity B (possible results in the list $\{b_m\}$), which does not affect the system (or affect it in a predictable way), since we are in the classical world. One then measures, in the same system, the physical quantity A (possible results in the list $\{a_n\}$). One then gets a pair of results (a_n, b_m) that gives a point in the (A, B) representation plane. If one repeats the double measurement many times, each time with a new copy of the initial system, one will finally get a 2D histogram giving the number $N(a_n, b_m)$ of occurrences of the couple (a_n, b_m), or, in the case where the a_n and b_m are continuously varying numbers, a "cloud" of points, the density of which gives back the histogram (Figure 5.6). If the points are randomly distributed (Figure 5.6(a)), the knowledge of the specific value b_{m0} for B does not modify the dispersion of results on the quantity A: The two quantities are statistically independent and the function $N(a_n, b_m)$ factorizes into an a_n part and a b_m part. If the points are almost aligned (Figure 5.6(b)), the knowledge of the specific value b_{m0} for B provides an

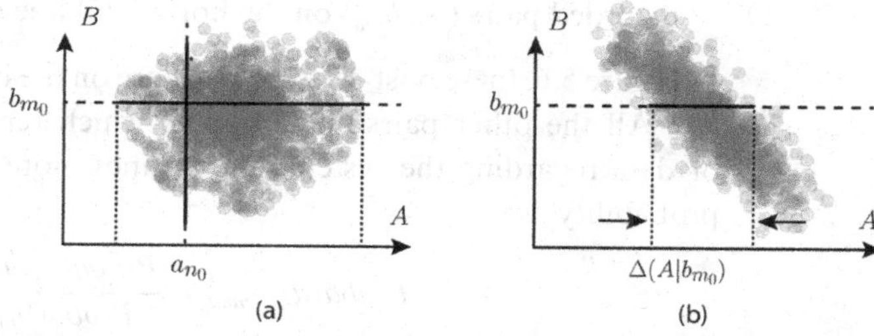

Fig. 5.6 Joint measurement data in the (A, B) plane in classical physics. Each point corresponds to a given measured value of A and a given measured value of B. The conditional variance $\Delta(A|b_{m0})$ gives the width of the statistical distribution of A values corresponding to the value b_{m0} obtained in the measurement of B. If A and B are correlated, this variance is reduced compared to the joint measurement variance.

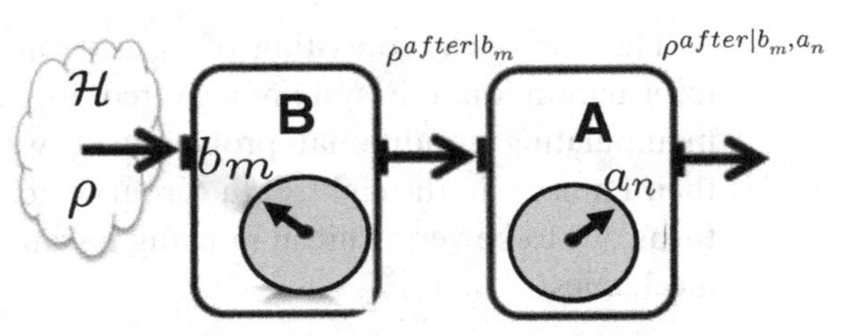

Fig. 5.7 Two successive measurements in quantum physics. The first B measurement collapses the quantum state before the A measurement, giving rise to a loss of information concerning the initial state.

increase in the accuracy with which the value of A can be determined: there is a high degree of correlation between the quantities A and B. Note that in classical physics the first measurement does not modify the system, or it modifies it in a predictable way, so that the temporal order between the A and B measurements does not matter.

The amount of correlation is characterized by the correlation function, given as usual by $\mathbf{C}(A, B) = \overline{AB} - \overline{A}\overline{B}$ (where the bar means the classical average value), or, better, by the *normalized correlation function* $c(A, B)$:

$$c(A, B) = \frac{\mathbf{C}(A, B)}{\sqrt{Var(A)\,Var(B)}}, \tag{5.45}$$

where $c(A, B)$ is a number between -1 and 1, with 1 corresponding to perfect correlation (exactly aligned points with positive slope), and -1 to perfect anticorrelation (exactly aligned points with negative slope). Note that the expectation values or means like $\overline{A}$, and the variance like $Var(A) = \overline{A^2} - (\overline{A})^2$, are calculated over all the N recordings of measurements, i.e. over the total cloud of points in Figure 5.7.

One can also define other quantities characterizing the correlation:

1. the *joint probability* $Proba(a_{n_0}, b_{m_0})$ as the frequency of occurrence of the pair (a_{n_0}, b_{m_0}) within the total set of N recorded pairs (a_n, b_m) in the A-B plane;
2. the *conditional probability* $Proba(a_n|b_{m_0})$ of measuring a_n on A given that the measurement of B on the same system has already given the value b_{m_0}, as the frequency of occurrence of the value a_n within the subset of recorded pairs (a_n, b_{m_0}) on the horizontal line of ordinate b_{m_0}.

In Figure 5.6, these post-selected pairs are on the horizontal line of ordinate b_{m_0}. All the other pairs are dropped, which represents a significant loss of data regarding the system. One defines more precisely the conditional probability as

$$Proba(a_n|b_{m_0}) = \frac{Proba(a_n, b_{m_0})}{Proba(b_{m_0})}, \tag{5.46}$$

where the *marginal probability* $Proba(b_{m_0})$ of measuring b_{m_0} whatever the outcome of the measurement on A, defined as

$$Proba(b_{m_0}) = \sum_n Proba(a_n, b_{m_0}), \tag{5.47}$$

allows us to normalize the conditional probability to 1. Note that this relation is an analog to the operation of partial trace on a quantum system.

The conditional probability distribution $Proba(a_n|b_{m0})$ of A values allows us to define a *conditional mean* $\langle A|b_{m0}\rangle$ and a *conditional variance*, or *scedastic function* $Var(A|b_{m0})$, which gives the mean and the width, respectively, of the horizontal cut in Figure 5.6. If the statistical distribution of A and B values is Gaussian, the conditional variance $Var(A|b_{m0})$ is independent of b_{m0} and given by

$$Var(A|b_{m0}) = Var(A)\left(1 - c(A,B)^2\right). \tag{5.48}$$

It is zero for perfectly correlated or anticorrelated quantities.

5.7.3 Bayes' Rule

It is of course possible to first measure the quantity A, then the quantity B, and use the correlation between these two quantities to determine the conditional probability $Proba(b_m|a_{n0})$ of measuring b_m on B knowing that the previous measurement of A on the same system gave the value a_{n0}, as the probability b_m determined only from the data lying on the vertical line of abscissa a_{n0} of Figure 5.6, given by:

$$Proba(b_m|a_{n0}) = \frac{Proba(a_n, b_{m0})}{Proba(a_{n0})}. \tag{5.49}$$

The two conditional probabilities $Proba(a_n|b_m)$ and $Proba(b_m|a_n)$ are related to the same joint probability $Proba(a_n, b_m)$ by Equations (5.46) and (5.49). The following relation, known as Bayes' rule, relates one conditional probability to the other and to the two marginal probabilities $P(b_m)$ and $P(a_n)$:

$$Proba(a_n|b_m) = Proba(b_m|a_n)\frac{Proba(a_n)}{Proba(b_m)}. \tag{5.50}$$

It can also be written as

$$Proba(a_n|b_m) = Proba(a_n)\frac{Proba(b_m|a_n)}{Proba(b_m)} = Proba(a_n)\ell(a_n|b_m), \tag{5.51}$$

where $\ell(a_n|b_m)$ is called the "Fisher likelihood." It characterizes the change in the probability distribution of A values when one acquires the additional information that the measurement on B has given the value b_m.

The Bayesian approach allows one to relate the conditional probabilities corresponding to the direct and inverse orders. This possibility is often very fruitful for classical probabilities, as can be seen in the exercises at the end of this chapter. It is also a very convenient tool in quantum physics, as it implies a complete inversion of the point of view, which can be illuminating: In many experiments, the quantum state ρ of the system is not known, or is poorly known. What is sure is the result of the measurements that the experimenter has recorded, from which they want to infer the most likely state compatible with these results. This approach is named the maximum likelihood method.

5.7.4 Successive Quantum Measurements

So far, we have considered classical measurements that do not modify the measured system, or that modify it in a predictable way. We now consider the same problem in the quantum world. One possible situation is sketched in Figure 5.7: Given an initial state ρ, one measures successively on it the quantities B and A, assuming no evolution between the two measurements. As the quantum state of the system collapses because of the first measurement, the second measurement is not performed on the same quantum state as the first, though some degree of correlation between the two measurements may remain, as we will see below.

Two von Neumann Measurements

Let us now use the postulates of measurement in the simple case of von Neumann measurements. Associated with the two observables $\hat{A}$, $\hat{B}$, we have projection operators $\hat{P}_n$, $\hat{Q}_m$, with $\hat{A} = \sum_n a_n \hat{P}_n$, $\hat{B} = \sum_m b_m \hat{Q}_m$. After the first measurement giving a result b_m with probability $Tr\rho\hat{Q}_m$, the system enters the second measurement device in the state $\rho^{after|b_m} = \hat{Q}_m \rho \hat{Q}_m / Tr\rho\hat{Q}_m$. One finally gets an expression giving the *the joint probability for successive measurements*, as the product of the probabilities of measuring b_m on ρ by the probability of measuring a_n on $\rho^{after|b_m}$:

$$Proba(a_n, b_m|\rho) = Tr\left(\hat{P}_n \hat{Q}_m \rho\, \hat{Q}_m \hat{P}_n\right) = Tr\left(\rho \hat{Q}_m \hat{P}_n \hat{Q}_m\right). \tag{5.52}$$

This simple formula is known as the Wigner formula.

We observe that $\hat{Q}_m \hat{P}_n \hat{Q}_m$ is not a projector: Two successive von Neumann measurements do not, in general, constitute a von Neumann measurement of the pair (b_m, a_n), but instead a generalized measurement described by the POs $\hat{\Pi}_{m,n} = \hat{Q}_m \hat{P}_n \hat{Q}_m$. In general, $Proba(a_n, b_m) \neq Proba(b_m, a_n)$: the order of the two measurements matters unless the corresponding observables commute.

Applying the definition of the conditional probability to the joint probability (5.49), we obtain, using the closure relation $\sum_n \hat{P}_n = \hat{1}$,

$$Proba(a_n|b_m) = \frac{Tr\left(\rho \hat{Q}_m \hat{P}_n \hat{Q}_m\right)}{\sum_n Tr\left(\rho \hat{Q}_m \hat{P}_n \hat{Q}_m\right)} = \frac{Tr\left(\rho \hat{Q}_m \hat{P}_n \hat{Q}_m\right)}{Tr\left(\rho \hat{Q}_m\right)}. \tag{5.53}$$

Note the similarity with the expression of the joint probability (5.52), which differs only by the normalizing factor. The joint probability is linear in ρ, but not the conditional one, which is normalized only in a subset of results.

The state after the two measurements is given by applying the state collapse postulate to the second measurement:

$$\rho^{after|b_m,a_n} = \frac{\hat{P}_n \hat{Q}_m \rho \hat{Q}_m \hat{P}_n}{Tr\left(\rho \hat{Q}_m \hat{P}_n \hat{Q}_m\right)}. \tag{5.54}$$

We see that the post-measurement operator is, for each pair of results (b_m, a_n),

$$\hat{K}_{m,n} = \hat{P}_n \hat{Q}_m. \tag{5.55}$$

If the two observables $\hat{A}$ and $\hat{B}$ (and hence the projectors $\hat{P}_n$ and $\hat{Q}_m$) commute, the PMO $\hat{K}_{m,n} = \hat{P}_n \hat{Q}_m = \hat{Q}_m \hat{P}_n$ is the projector associated with a von Neumann measurement of the pair (a_n, b_m). It does not depend on the order of the measurements.

Successive von Neumann measurements of commuting PMOs can therefore be considered as a *genuine joint measurement*, from which one can deduce marginals and conditional probabilities like in the classical case. The only major difference is that the state after the joint measurement is modified in an unpredictable way.

Two Generalized Measurements

One can now use the rules of generalized quantum measurements of quantities B then A. We will call respectively $\hat{\Theta}_m$ and $\hat{\Pi}_n$ in the associated POs, $\hat{L}_m$ and $\hat{M}_n$ the associated PMOs.

For the joint probability, one gets

$$Proba(a_n, b_m|\rho) = Tr\rho\hat{\Xi}_{m,n}, \tag{5.56}$$

with a PO equal to $\hat{\Xi}_{m,n} = \hat{L}_m^\dagger \hat{M}_n^\dagger \hat{M}_n \hat{L}_m$. Two successive generalized measurements can thus be described as a joint generalized measurement.

For the conditional probability one obtains

$$Proba(a_n|b_m) = \frac{Tr\rho\hat{\Xi}_{m,n}}{Tr\rho\hat{\Theta}_m}. \tag{5.57}$$

The state of the system after the two measurements is now

$$\rho^{after|a_n,b_m} = \frac{\hat{M}_{m,n}\rho\hat{M}_{m,n}^\dagger}{Tr\rho\hat{\Xi}_{m,n}}, \tag{5.58}$$

with the PMOs $\hat{M}_{m,n} = \hat{M}_n \hat{L}_m$. We obviously have the relation $\sum_{m,n} \hat{M}_{m,n} = 1$.

5.7.5 Repeatability

What happens when one makes the same measurement twice?

Two von Neumann Measurements

From Equation (5.53), for the conditional probability one finds that

$$Proba(a_n|a_m) = \frac{Tr\left(\rho\hat{P}_m\hat{P}_n\right)}{Tr\left(\rho\hat{P}_m\right)} = \delta_{m,n} \tag{5.59}$$

because the projectors $\hat{P}_m$ are orthogonal. These probabilities are actually certainties that the second measurement will always give the result a_n: One is sure that the system, even submitted to several measurements of A, will stay in the collapsed state

$$\rho^{after|a_n} = \frac{\hat{P}_n\rho\hat{P}_n}{Tr\left(\rho\hat{P}_n\right)}. \tag{5.60}$$

As discussed in Appendix A, *quantum mechanics is repeatable with respect to von Neumann measurements*: One is sure to find results confirming that the system under study has not collapsed again to some other state in a second measurement. This very important characteristic of quantum mechanics enforces the statement that a von Neumann measurement is actually a *stochastic state preparation*: One knows for sure which state exits the first measurement device. The price to pay for this certainty is the probabilistic nature of its occurrence. The experimentalist must wait for a random and unpredictable time the occurrence of the measurement result "heralding" the desired state. The success rate of such a preparation is an important parameter to consider when one wants to assess its performance.

Two Generalized Measurements

The conditional probability of measuring a_n after a first measurement giving the result a_m is

$$Proba(a_n|a_m) = \frac{Tr\left(\rho\hat{M}_m^\dagger\hat{M}_n^\dagger\hat{M}_m\hat{M}_n\right)}{Tr\left(\rho\hat{M}_n^\dagger\hat{M}_n\right)}. \tag{5.61}$$

It is no longer δ_{mn}, so that the property of repeatability, and therefore of perfect state preparation, is lost for most generalized measurements.

5.7.6 Simultaneous Quantum Measurements

In successive measurements, the first measurement induces a collapse of the initial state, so that the second measurement is not made on the same state as the first one, and a lot of information is lost because of the state collapse. We would like to find ways to measure the two quantities A and B with a minimum loss of information.

One technique is to measure these quantities *simultaneously*, the list of possible results being the couples a_i, b_j (Figure 5.8). This measurement is now made on the not-collapsed, and therefore not-perturbed, input state, so that it should be less perturbed by the state collapse than by using

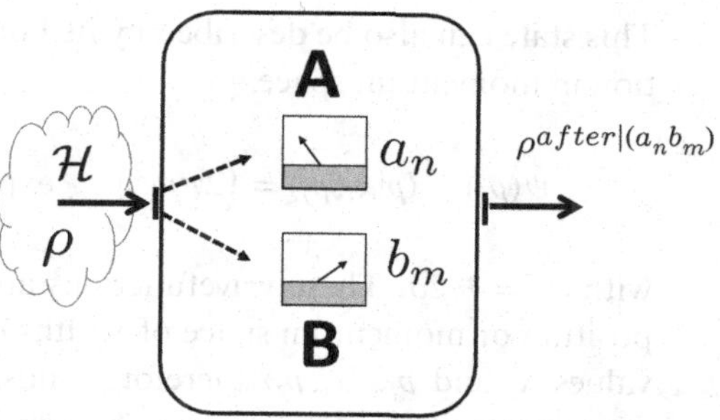

Fig. 5.8 Simultaneous measurement of two quantities A and B, giving as a result a couple of values (a_n, b_m). It requires a kind of cloning, necessarily imperfect, of the input state before it is submitted to the joint measurement. If A and B are noncommuting, quantities such a simultaneous measurement cannot be perfectly precise on the two quantities because of the no-cloning theorem.

successive measurements. We will see in the two following examples that such a measurement cannot be perfect.

5.7.7 Simultaneous Measurement of Conjugate Variables

We know from the first principles of quantum mechanics, namely the no-cloning theorem (Appendix C), that a simultaneous and perfect measurement of noncommuting observables $\hat{A}$ and $\hat{B}$ is impossible. One can instead make a generalized measurement [13] that takes into account an unavoidable addition of excess noise on both quantities, so that this kind of joint measurement is always more noisy and imperfect than the measurements made on $\hat{A}$ or $\hat{B}$ alone. We will come back to the issue of the perturbation of the system by the measurement in Chapter 11.

In this section, we present two ways to analyze such a measurement.

The first is to exhibit a POVM that has all the required mathematical properties for such a simultaneous measurement, without explicitly referring to a specific experimental set-up likely to be described by the POVM. We will apply this method to the simultaneous measurement of position and momentum of a massive 1D particle.

The second approach is to rely on a specific experimental set-up and derive from it the associated POVM. We will apply this method to the simultaneous measurement of the two quadratures of a single mode electromagnetic field.

Let us consider first a generalized measurement (Section 5.1) where the list of possible measurement results is the set of couples (x_i, p_j). We look for a possible POVM associated with such a measurement.

We introduce the state $|x_i, p_j\rangle$, having a wavefunction in position space equal to

$$\psi(x_j) = \langle x|x_i, p_j\rangle = \left(2\pi\sigma^2\right)^{-1/4} \exp\left(-\frac{(x-x_i)^2}{4\sigma^2} + i\frac{p_j x}{\hbar}\right). \tag{5.62}$$

This state can also be described by its Fourier transform, i.e. the wavefunction in momentum space,

$$\tilde{\psi}(p_j) = \langle p|x_i, p_j\rangle = \left(2\pi\sigma'^2\right)^{-1/4} \exp\left(-\frac{(p-p_j)^2}{4\sigma'^2} - i\frac{x_i p}{\hbar}\right), \tag{5.63}$$

with $\sigma' = \hbar/2\sigma$. These wavefunctions are both Gaussian functions in the position or momentum space of widths σ and $\sigma' = \hbar/2\sigma$ around the mean values x_i and p_j. $|x_i, p_j\rangle$ therefore tends, respectively, to the position, or momentum, eigenstate when $\sigma \to 0$, or ∞. These states are also quantum states of minimal uncertainty: they saturate the Heisenberg inequality $Var(X)\,Var(P) = \hbar^2/4$.

One can show that the set of $|x_i, p_j\rangle$ states (which are not orthogonal), form a resolution of the identity

$$\int dx_i dp_j |x_i, p_j\rangle\langle x_i, p_j| = 2\pi\hbar\hat{I}. \tag{5.64}$$

Let us consider the operator $\hat{\Pi}_{x_i,p_j}$ as

$$\hat{\Pi}_{x_i,p_j} = \frac{|x_i, p_j\rangle\langle x_i, p_j|}{2\pi\hbar}. \tag{5.65}$$

It is an operator proportional to a projector, and therefore positive. It satisfies the sum rule (5.2), and as such is a bona fide PO $\hat{\Pi}_{x_i,p_j}$. We will assume here that it is possible to find an experimental protocol that is described by the present POVM.

It is now possible to calculate the separate probability distributions for position and momentum deduced from the measurements described by this POVM. In the case of a measurement performed on the pure state ψ, one finds,

$$Proba(x_i|\psi\rangle) = \int dp_j Proba(x_i, p_j, \psi) = \int dx|\psi(x)|^2 \exp\left(\frac{-(x-x_i)^2}{2\sigma^2}\right), \tag{5.66}$$

$$Proba(p_j|\psi\rangle) = \int dx_i Proba(x_i, p_j, \psi) = \int dp|\tilde{\psi}(p)|^2 \exp\left(\frac{-(p-p_j)^2}{2\sigma'^2}\right). \tag{5.67}$$

When σ, or σ', is small, the probability is close to $|\psi(x_j)|^2$, or $|\tilde{\psi}(p_j)|^2$, so we have indeed exhibited a POVM that gives at the same time probability distributions that are close to the independently measured position and momentum distributions. More precisely, one finds for the variance of these measurements, that

$$Var(X) = Var(X)_\psi + \sigma^2; \quad Var(P) = Var(P)_\psi + \hbar^2/2\sigma^2, \tag{5.68}$$

where $Var(X)_\psi$, or $Var(P)_\psi$, are the variances deduced from separate, regular position, or momentum, single measurements performed on state $|\psi\rangle$.

The simultaneous measurement of position and momentum has indeed given as mean values the exact value of position and momentum mean values without state collapse. However, it has introduced excess noise terms around these means, both on X and P, that cannot be simultaneously zero. They prevent us from exactly reconstructing the state $|\psi\rangle$ from the measurements. According to the needs, one may favour one measurement or the other, but not both. The best trade-off is to take $\sigma^2 = \hbar/2$, the minimum added noise for both measurements.

Let us now consider the second example: the simultaneous measurement of the two quadratures $\hat{X}$ and $\hat{P}$ of a single mode electromagnetic field (Appendix F). It can be implemented by a rather simple set-up, called "double homodyne detection", (sometimes also heterodyne measurement), which is extensively described and discussed in Section 9.4. One shows that such a simultaneous joint measurement gives the correct mean values for the quadratures, but, exactly like in the previous paragraph, at the price of a lower accuracy (meaning larger variance) in the determination of these two quantities.

5.7.8 Measurement of Combinations of Two Operators

If one is interested in the correlations between $\hat{A}$ and $\hat{B}$ measurements, it can be useful to measure *linear combinations* of the noncommuting operators $\hat{A}$ and $\hat{B}$,

$$\hat{D}_\Theta = \hat{A}\cos\theta + \hat{B}\sin\theta, \tag{5.69}$$

for different values of θ. This can be done with field quadratures using homodyne detection (Section 9.4.1) with an adjustable LO phase.

The variance of the measured values is given by

$$Var(D_\Theta) = Var(A)\cos^2\theta + Var(B)\sin^2\theta + C_{AB}\sin 2\theta, \tag{5.70}$$

where the A-B quantum correlation is given by $C_{AB} = \langle\hat{A}\hat{B} + \hat{B}\hat{A}\rangle/2 - \langle\hat{A}\rangle\langle\hat{B}\rangle$. Plotting (in polar coordinates) the angular variation of the variance $Var(D_{\Theta_0})$ around the mean value, one gets the ellipse in Figure 5.9, which can be considered as giving roughly the limit of the uncertainty around the mean value for all values of the angle θ. When the correlation C_{AB} is zero, the axes of the ellipse are the A and B axes. When it is not zero, there is an angle θ_0 such that $\tan(2\theta_0) = 2C_{AB}/(V(A) - V(B))$, different from 0 and $\pi/2$, which gives a variance smaller than $Var(A)$ and $Var(B)$. This minimum variance is equal to

$$Var(D_{\Theta_0}) = \frac{Var(A) + Var(B)}{2} - \sqrt{\left(\frac{Var(A) - Var(B)}{2}\right)^2 + C_{AB}^2}. \tag{5.71}$$

From the values of $Var(A)$, $Var(B)$, and $Var(D_{\Theta_0})$ one can deduce the value of the quantum correlation C_{AB} between A and B before the state collapse. Note that the knowledge of the minimum variance requires the

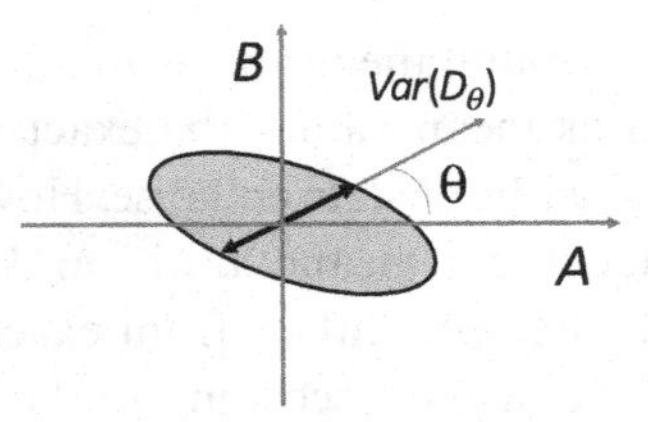

Fig. 5.9 Variation with angle Θ of the variance $Var(D_\Theta)$ of the measurement of the linear combination $\hat{D}_\Theta = \hat{A}\cos\Theta + \hat{B}\sin\Theta$. The minimum of this variance is a measure of the amount of squeezing (see Section 8.2). The eccentricity of the ellipse is a measure of the correlation between these quantities.

measurement of the variances $Var(D_\Theta)$ for all values of θ. It is necessary to make many measurements on identically prepared initial states and different θ values. This is an example of *quantum tomography*. In the case of electric field quadratures, such an analysis can be extended to the full multimode case [66].

5.8 The State Collapse: An Update about the Piece of Information One Wants to Get about the System

5.8.1 Is State Collapse Indispensable?

Let us go back to the case of von Neumann successive measurements of B then A. Joint probabilities $Proba(a_n, b_m|\rho)$ for all values of a_n and all values of b_m are experimentally obtained by recording the results of many successive measurements of the quantities B and A, storing them in a computer memory, and deriving from them the joint probabilities. We saw in Section 5.7 that these joint probabilities are given by the Wigner formula (5.52):

$$Proba(a_n, b_m|\rho) = Tr\left(\hat{P}_n\hat{Q}_m\rho\,\hat{Q}_m\hat{P}_n\right). \tag{5.72}$$

This expression is directly derived from the state collapse postulate after the first measurement. It allows us to predict the statistics of all possible successive measurements.

From the Wigner formula one deduces the conditional probability (5.53):

$$Proba(a_n|b_m) = \frac{Tr\left(\rho\hat{Q}_m\hat{P}_n\hat{Q}_m\right)}{Tr\left(\rho\hat{Q}_m\right)}, \tag{5.73}$$

which can also be written as

$$Proba(a_n|b_m) = Tr\left(\rho_m\hat{P}_n\right), \tag{5.74}$$

with

$$\rho_m = \frac{\hat{Q}_m \rho \hat{Q}_m}{Tr\left(\rho \hat{Q}_m\right)}. \tag{5.75}$$

Let us assume for a moment that we do not want to use the state collapse postulate. Instead, we use the Wigner formula (5.72) to calculate correlations between successive measurements, applied to the initial quantum state ρ, as a postulate.

Assume we are interested only in the cases when the first measurement has given the specific value b_{m_0}, discarding all cases that have given a result different from b_{m_0} (this process is called "post-selection"). To calculate conditional probabilities, we can use formula (5.74), which is directly derived from the Wigner formula. Note that this expression looks like Born's rule (5.1), but applied to a different quantum state, "a conditional state" ρ_m, instead of the initial state ρ.

Let us go back to the Bayes' conception that the relevant probabilities are the conditional probabilities, because they reflect our degree of information about a system, and apply it to quantum probabilities: Each time one measures something on a quantum system, one acquires a piece of information, and as a result the conditional probabilities change. To account for this change, the tool used to calculate the probabilities, i.e. the density matrix, must be updated accordingly: it changes from ρ to ρ_m. This result has been deduced from Wigner formula, and does not require us to refer to a somewhat mysterious state collapse. In particular, the problem of the exact time of the state collapse is eliminated.

We have therefore shown, using the Wigner formula, and not the state collapse as a postulate, that one can calculate conditional probabilities by using the regular Born's rule for "normal" probabilities, but applied on a "conditional state," which is precisely the one given by the state collapse postulate.

The state collapse postulate appears in this point of view[1] as a way to describe the effect of information updating resulting from the first measurement. It is then quite natural that the updated state cannot be written as a the result of an evolution described by a quantum map because it is not a physical evolution, but the result of a choice by the observer of the relevant data. In this point of view, *there is no physical collapse of the quantum state, but rather a change of point of view concerning the statistical ensemble* on which one calculates probabilities, variances, correlations: the whole set of data, or only the ones containing the precise value b_m as a first result of measurement.

[1] A point of view that is not shared by all quantum physicists (for a review of the different point of views, see [107]).

5.8.2 An Example of Information Update: The No-Jump Conditional Evolution in the Shelving Technique

Because of state collapse, any measurement strongly modifies the quantum state of a system. At the same time, it causes an information update that affects, in a conditional way, its further evolution. So, in many instances, it is interesting to consider the evolution of a system as a succession of measurements and conditional post-measurement evolutions. We illustrate this approach on the example of the shelving technique, presented in Sections 1.3 and 1.6. It will allow us to understand jumps under a new light.

Consider the three-level system configuration of Figure 1.4. The excited state $|B\rangle$ has an extremely "fast" photon emission rate γ_B, while the other one $|D\rangle$ has an extremely "slow" photon emission rate γ_D. Let us consider the evolution during a small time interval Δt. The Kraus operators describing the relaxation of the excited states are $\hat{K}_D = \sqrt{\gamma_D \Delta t}|G\rangle\langle D|$ and $\hat{K}_B = \sqrt{\gamma_B \Delta t}|G\rangle\langle B|$, whose action is to bring the system to the ground state every time a photon is emitted. One can now describe the evolution in two ways.

The first method is to use the POVM formalism directly in terms of a quantum map between times t and $t + \Delta t$ involving Kraus PMOs $\hat{K}_j$ (Equation (4.47)). This requires knowledge of the Kraus operator $\hat{K}_0$ ("no-jump operator"), which one can obtain by using the sum rule (Equation (4.55)),

$$\hat{K}_0^\dagger \hat{K}_0 + \hat{K}_D^\dagger \hat{K}_D + \hat{K}_B^\dagger \hat{K}_B = 1, \tag{5.76}$$

so that

$$\hat{K}_0 = 1 - \frac{\gamma_B}{2}\Delta t|B\rangle\langle B| - \frac{\gamma_D}{2}\Delta t|D\rangle\langle D|. \tag{5.77}$$

The second method is to use the Lindblad differential equation, analog to relation (4.76),

$$\begin{aligned}\frac{d\hat{\rho}}{dt} = & \frac{1}{i\hbar}\left[\hat{H}_{at}, \hat{\rho}\right] \\ & - \frac{\gamma_B}{2}(\rho|B\rangle\langle B| + |B\rangle\langle B|\rho) + \gamma_B|G\rangle\langle B|\rho|B\rangle\langle G| \\ & - \frac{\gamma_D}{2}(\rho|D\rangle\langle D| + |D\rangle\langle D|\rho) + \gamma_D|G\rangle\langle D|\rho|D\rangle\langle G|.\end{aligned} \tag{5.78}$$

We now distinguish two different situations:

1. The first situation involves the emission of photons. The further evolution of the system is described by the stochastic application of the jump operators $\hat{K}_D$ and $\hat{K}_B$ that are represented in the Lindblad equation by the jump terms $\gamma_B|G\rangle\langle B|\rho|B\rangle\langle G| + \gamma_D|G\rangle\langle D|\rho|D\rangle\langle G|$, which project the wave function to the ground state.
2. The second situation is when no photons are emitted. The evolution is then governed by $\hat{K}_0$. This conditional evolution is described by removing the jump terms from the Lindblad equation, which then reads

$$\left(\frac{d\hat{\rho}}{dt}\right)_{cond} = \frac{1}{i\hbar}\left[\hat{H}, \hat{\rho}\right] + \frac{\gamma_B}{2}\{|B\rangle\langle B|, \hat{\rho}\} + \frac{\gamma_D}{2}\{|D\rangle\langle D|, \hat{\rho}\}. \quad (5.79)$$

This equation can be written as the Schrödinger equation,

$$i\hbar\frac{\partial}{\partial t}|\psi\rangle_{cond} = \hat{H}_{eff}|\psi\rangle_{cond}, \quad (5.80)$$

with the non-Hermitian effective Hamiltonian

$$\hat{H}_{eff} = \hat{H} - i\hbar\frac{\gamma_B}{2}|B\rangle\langle B| - i\hbar\frac{\gamma_D}{2}|D\rangle\langle D|. \quad (5.81)$$

In the shelving experiment, two coherent external pump fields are resonant on the transitions from the ground state to the excited states, one "strong," one "weak." They are described by adding to the Hamiltonian $\hat{H}_{eff}$ the new term:

$$\hat{H}_{Rabi} = i\hbar\frac{\Omega_{BG}}{2}|B\rangle\langle G| + i\hbar\frac{\Omega_{DG}}{2}|D\rangle\langle G| + \text{Hermitian conjugate}, \quad (5.82)$$

where Ω_{BG} and Ω_{DG} are the corresponding Rabi frequencies.

The evolution, *conditioned by the recording of no clicks on the photodetector*, for the state

$$|\psi\rangle_{cond} = c_G|G\rangle + c_D|D\rangle + c_B|B\rangle, \quad (5.83)$$

obeys the following equations, derived from Equation (5.80) with the Hamiltonian $\hat{H}_{eff} + \hat{H}_{Rabi}$:

$$\dot{c}_G = -\frac{\Omega_{BG}}{2}c_B - \frac{\Omega_{DG}}{2}c_D; \quad \dot{c}_B = \frac{\Omega_{BG}}{2}c_G - \frac{\gamma_B}{2}c_B; \quad \dot{c}_D = \frac{\Omega_{DG}}{2}c_G - \frac{\gamma_D}{2}c_D. \quad (5.84)$$

One finds, after a straightforward derivation, that in the conditions of the shelving technique $c_B = \Omega_{BG}c_G/\gamma_B \simeq 0$, the set of dynamical equations is reduced to the $|D\rangle - |G\rangle$ pair. One also finds that the state is not normalized, its time-dependent norm $||\psi\rangle_{cond}|^2$ giving the probability of recording no click during the time interval $[0, t]$.

One obtains the updated conditional evolution of the system by renormalizing its norm to 1. Being a two-level system $|D\rangle, |G\rangle$, it is better described by its Bloch vector $\vec{U}_{DG}$. Its three coordinates (X_{DG}, Y_{DG}, Z_{DG}) are found to be, with the initial condition $c_G(t = 0) = 1$,

$$X_{DG}(t) = \text{sech}\left(\frac{\Omega^2_{BG}}{2\gamma_B}t - constant\right), \quad Y_{DG}(t) = 0,$$

$$Z_{DG}(t) = \tanh\left(\frac{\Omega^2_{BG}}{2\gamma_B}t - constant\right). \quad (5.85)$$

These hyperbolic functions describe the conditional evolution of the state vector during time intervals when no jump from fluorescence-on to fluorescence-off is observed after the last click (Figure 1.5). The observation of such a conditional evolution requires that all photons emitted by the system give rise to a recorded click. The experiment has been done in

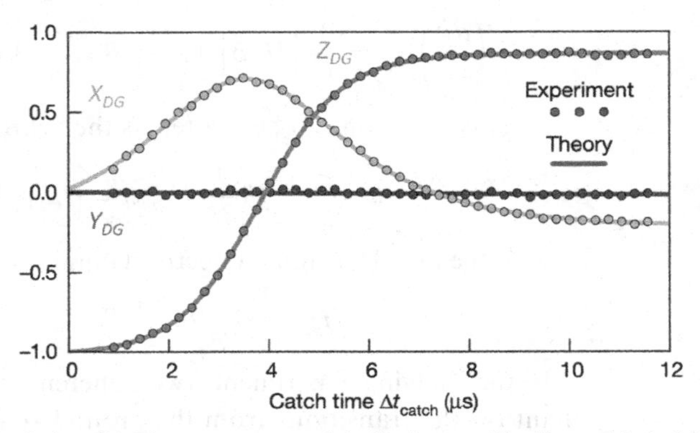

Fig. 5.10 Conditional evolution of the three components of the Bloch vector (X_{DG}, Y_{DG}, and Z_{DG}) of the two-level system $|G\rangle$, $|D\rangle$. The time Δt_{catch} is the time since the last click. From Z. Minev et al. To catch and reverse a quantum jump mid-flight, *Nature*, 2019 [127].

superconducting circuits [127] (see Figure 5.10), where the fluorescence can be totally channeled in a wave guide connected to the detector. To focus on the rare no-jump evolution, the authors defined the variable $t = \Delta t_{\text{catch}}$ to measure the time after the last click.

In absence of clicks, the continuous projection is "smooth" ($\hat{K}_0|\psi\rangle = |\psi\rangle + \mathcal{O}(dt)$) and does not produce a strong collapse. It drives the state from $|G\rangle$ to $|D\rangle$. This can be understood with the following reasoning: If the system was in $|G\rangle$ the strong Rabi drive Ω_{BG} would bring it into $|B\rangle$ and cause a subsequent photon emission (and detection), thus – in absence of clicks – one grows confidence (smoothly) in that the state is "likely" be in $|D\rangle$. In addition, the conditional evolution occurs along a meridian of the Bloch sphere, which is in the X-Z plane when one chooses Ω_{DG} real.

In Figure 5.10, the three components of the Bloch vector are measured independently as a function of Δt_{catch} showing that the jumps from bright to dark are in fact very rapid ($\Omega_{DG} \sim 2\pi/50$ μs in this experiment), which is orders of magnitude too weak to account for a deterministic transition from $|G\rangle$ to $|D\rangle$ in a few microseconds. Note also that the measured hyperbolic tangent evolution, which is very close to the theoretical predictions, is qualitatively very different from the usual Rabi oscillation.

This experiment is a good example of quantum process: The initial state is precisely prepared, the evolution that is monitored is a mixing of Hamiltonian evolution, spontaneous emission and measurement-induced Bayesian update, and the QND measurement is performed with high fidelity.

5.9 Retrodictive Measurements

In many domains of science, one tries to infer the future from the present. But in some of them, like history, paleontology, or archaeology, the issue is to infer the past from the present. In physics, and especially in quantum physics, dealing with processes consisting of state preparation, evolution, and measurements, these two complementary approaches are useful. Quantum mechanics usually uses a predictive approach: Knowing the initial quantum state, what can we expect from measurements on the system after its evolution? But the so-called *retrodictive approach* is also relevant and interesting: Knowing the measurement result, what is the state, prior to evolution, that is the most compatible with the measurement? This point of view is especially useful when the measurement is destructive, in which case there is "no future" for the system.

Let us consider the following situation: A measurement of property B on a quantum system giving results b_m leads to the generation of a series of collapsed states $\rho^{after|b_m}$ with probabilities of occurrence $Proba(b_m|\rho)$ (Equations (5.1) and (5.3)).

The system is then subject to a measurement of quantity A. In order to keep the discussion simple, we assume no evolution between the measurements, and a von Neumann measurement of A. The predictive probability of getting the result a_n on state ρ_m is given by the usual Born's rule,

$$Proba_{pred}(a_n|b_m) = Tr\left(\rho_m \hat{P}_n\right). \tag{5.86}$$

Now consider the following situation [14]. We know that the measurement of A has given the result a_n. What is the retrodictive probability $Proba_{retro}(b_m|a_n)$ that allows us to determine the probability of being before the measurement in one of the states ρ_m? To answer this question, one relies on Bayes' rule, which precisely establishes a connection between these two probabilities,

$$Proba_{retro}(b_m|a_n) = Proba_{pred}(a_n|b_m)\frac{Proba(b_m)}{\sum_p Proba_{pred}(a_n|b_p)Proba(b_p)}, \tag{5.87}$$

which gives, using Equations (5.86) and (5.47), the final expression for the retrodictive probability,

$$Proba_{retro}(b_m|a_n) = \frac{p_m Tr\left(\rho_m \hat{P}_n\right)}{Tr\rho\hat{P}_n}, \tag{5.88}$$

where ρ is the not-post selected state (Section 5.4.3),

$$\rho = \sum_m p_m \rho_m. \tag{5.89}$$

This expression simplifies when the not post-selected state is unbiased, i.e. contains no information, so that $\rho = \hat{1}/d$, where d is the dimension of the Hilbert space. In such a case, the retrodicted probability is

$$Proba_{retro}(b_m|a_n) = Tr\left(d\, p_m \rho_m \frac{\hat{P}_n}{Tr\hat{P}_n}\right) = Tr\left(\rho_m^{retro}\hat{\pi}_n\right), \tag{5.90}$$

with $\rho_m^{retro} = d\, p_m \rho_m$, $\hat{\pi}_n = \hat{P}_n/Tr\hat{P}_n$. The retrodicted probability exhibits the structure required by Gleason's theorem for a probability (Equation (5.6)).

When the initial state is not unbiased, this structure is also present, but with a more complicated expression [14].

5.10 Quantum Process Tomography

We saw in Chapters 2, 4, and 5 that the generalized rules of quantum mechanics (recalled in Appendix B), involve a small number of linear operators acting on the Hilbert state $\mathcal{H}$: the density matrix ρ characterizing the initial state, the Kraus operators $\hat{K}_\ell$ characterizing the quantum map, and the post-measurement operators $\hat{M}_n$ characterizing the detection-induced conditional evolution. These operators exhaustively describe the different parts of the quantum process. They can be accessed experimentally using techniques called *quantum process tomography*, which involves a series of different measurements performed on identical copies of the input quantum state that test the system under study [11].

5.11 Exercises for Chapter 5

Exercise 5.1 A bell in your neighborhood tolls every hour on the hour with probability 1/2 on regular days. You also know that the mechanism is initialized at midnight and that the process fails about once per year. If it does, the bell does not work at all that day.

What would you think if by 8:00 pm you have not heard it toll even once? Formalise this using the Bayesian view of probability theory.

Exercise 5.2 You need to decide if your one-euro coin is biased or not. You have no idea if it is, so your a priori is 50:50. The first time you flip it you get H.

(a) Does your degree of belief change?
(b) You flip the coin a second time and get H again. Does your degree of belief change?
(c) Generalize to n flips resulting in H.

Exercise 5.3 Show that unitary evolution can be stopped by repeated measurements. This is known as the quantum Zeno effect.

Exercise 5.4

(a) Does a quantum system decay if it is isolated?
(b) Consider a cavity losing energy at rate γ in a zero temperature environment. What is the state at time $t \gg 1/\gamma$?
(c) Describe the state's trajectory if the evolution happens when monitoring the emitted photon number in the environment.

(d) Describe the state's trajectory if the evolution happens when monitoring the intracavity photon number.
(e) Describe the state's trajectory if the evolution happens when monitoring the value of one quadrature using heterodyne detection.

The moral of this exercise is that the trajectories an open quantum system takes depend on how it is interacting with the surrounding world.

Exercise 5.5 Convince yourself that conditional probability reflects *logical implication* and not *physical causation* [100, 101].

(a) Do this by computing the probability of first drawing a white ball W_1 and second a red ball R_2 from an urn initially containing five white balls and six red balls and verifying that $Proba(W_1, R_2) = Proba(W_1)Proba(R_2|W_1) = Proba(R_2)Proa(W_1|R_2)$. That is, as far as probability is concerned, the first draw has as much influence on the second draw as the second one has on the first. Can the order subindices can be dropped in this example?
(b) Consider now the scenario in which as soon as a red ball is drawn, all balls left in the urn become red. Convince yourself that now $Proba(W_1, R_2) \neq Proba(R_1, W_2)$ and that the order of subindices cannot be dropped.
(c) Convince yourself that *successive* measurement of noncommuting observables q and s cannot be described by a *simultaneous* probability $Proba(q, s)$ since the order of the measurements matters.
(d) How strongly measuring a property q implies a property s? From the logical tautology $(q \Rightarrow s) \Leftrightarrow -(-q$ and $p)$ and that $Proba(q \text{ and } p) = Proba(q|p)Proba(p)$, show that $Proba(q \Rightarrow s) = 1 - (1 - Proba(s|q))Proba(q)$. That is, probability theory can be used to measure the validity of a logical statement (an implication in this case).
(e) Consider q ="it is raining" and s ="the floor is wet." The implication $q \Rightarrow s$ manifests causality, but it is nonetheless formally equivalent to $-s \Rightarrow -q$. Reflect upon this.
(f) Consider a pair of spin-1/2 particles in the singlet state and the observables $\hat{Q} = \hat{\sigma}_z \otimes 1$ and $\hat{S} = 1 \otimes (-\hat{\sigma}_x - \hat{\sigma}_z)/\sqrt{2}$. Is it fair to talk about the simultaneous probability $Proba(q, s)$? Does this have to do them being "space-like" separated or not? Compute the correlator $K(q, s) = \langle \hat{Q}\hat{S} \rangle = \langle q \times s \rangle$? What is the relationship with $Proba(q = s) = Proba(q \Leftrightarrow s)$?

Exercise 5.6 About the impossibility of simultaneous *measurement* of non-commuting quantities [150]: What does it mean?

(a) Independently of measurement, do quantum states with a well-defined position and a momentum exist?

(b) What is the uncertainty bound between two successive measurements of position and momentum over a particle in state $\psi(x)$? Are these simultaneous measurements? Use as a measurement of position and momentum transmission filters with transmitivity $f(x)$ and $f(p)$. That is, if the particle is detected, $f(x)$ weights its position distribution. For example, if one considers a Dirac delta $f(x) = \delta(x - x_0)$ and the particle is detected, one has measured the particle in the eigenstate of $\hat{X}$ labeled $|x_0\rangle$. How would you implement the momentum filter?
(c) What is the probability of finding a particle initially in state $\psi(x)$ in a minimal uncertainty coherent state. Take $\alpha = (x + ip)/\sqrt{2}$. What is the bound on the variance for x and p?
(d) Consider beam-splitting the state ψ into two "copies" and using one state to measure $\hat{x}$ and the other one to measure $\hat{p}$. What is the uncertainty for the joint measurement here?

Exercise 5.7 Take a qubit in state $|\psi_0\rangle = \alpha_0|0\rangle + \beta_0|1\rangle$ and make it "collide" successively with a stream of $N \gg 1$ "environmental" ancilla qubits (indexed with n) all in initial state $|+x\rangle = \frac{1}{\sqrt{2}}(|0\rangle + |1\rangle)$. The collision is such that $\hat{U} = e^{-i\epsilon\hat{H}/\hbar}$, where $\hat{H}/\hbar = \hat{\sigma}_z \otimes \hat{\sigma}_z$ and $\epsilon \ll 1$.

(a) Find a set of Kraus operators describing the map of the collisions.
(b) Find a recurrent expression for the purity of the qubit after n collisions. Is there entanglement?
(c) Take $\epsilon = \omega\delta t$ and derive the Lindblad equation. Discuss.
(d) Take $\epsilon = \sqrt{\gamma\delta t}$ and derive the Lindblad equation. Discuss.
(e) Suppose that after a collision you measure the corresponding environmental qubit in basis $|\pm y\rangle = \frac{1}{\sqrt{2}}(|0\rangle \pm i|1\rangle)$. Describe the state of the qubit at stage n conditional to the series of observed results. Discuss.
(f) Show that this nonunitary evolution describes a measurement in the $\{|0\rangle, |1\rangle\}$ basis. Show that for an initial state $\alpha|0\rangle + \beta|1\rangle$ the final state tends with probability $|\alpha|^2$ to $|0\rangle$ and with probability $|\beta|^2$ to $|1\rangle$.
(g) Convince yourself that the shrinking of the norm in this nonunitary equation gives the probability of the conditional evolution.
(h) Find a linear differential equation (keeping the normalisation factor to order 0 in δt) for the state vector under this continuous measurement condition over the ancillas (they are measured instantaneously after the collision and the result of the measurement is made available). Discuss the trajectories. Are they continuous? Are they differentiable? Identify the quantum jumps in the equation.
(i) Find a nonlinear differential equation expanding the normalisation factor up to order δt for the state vector and show that it preserves the normalisation. Is it unitary?

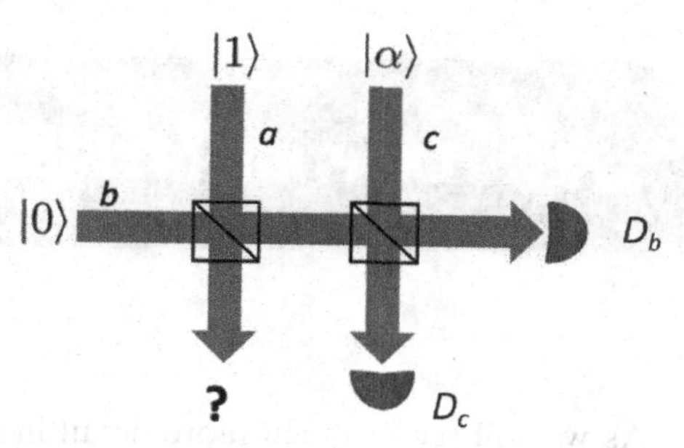

Fig. 5.11 A quantum scissor [13].

(j) Discuss the fact that a differential jump is achieved over the qubit at the price of a discrete collapse on the ancillas ($\sim\langle \pm y|x\rangle$). Does this model allow us to bypass the discontinuous evolution during measurement?

Exercise 5.8 Show that the evolution of the subsystem is, in the most general case, independent of the coherence between the system and the environment and that it is irrelevant whether the environment is measured or not.

Exercise 5.9 Consider a medical test that detects COVID infection. It gives two answers, either "yes" or "no". But it is not perfect: Statistical studies performed on a large set of contaminated people, labeled (C), show that it gives the answer "yes" for 90% of them. Studies performed on noncontaminated persons, labeled (NC), show that it gives the answer "no" for 99% of them. What is the likelyhood that a patient is contaminated given that it has "yes" result? This example is obviously a single-event situation that may turn to drama in case of the wrong decision.

Exercise 5.10 Conditional preparation of a single photon qubit state, or "quantum scissors" [141]:

A single photon state is often approximated by a coherent state $|\alpha\rangle$ of weak α value. This state has components mainly on $|n = 0\rangle$ and $|n = 1\rangle$ Fock states, but it has also weak but nonzero components on the Fock states of higher n values, which can be detrimental in some applications. "Quantum scissors" are needed to conditionally get rid of these spurious projections.

Consider the set-up sketched in Figure 5.11, which has three input modes, labeled a, b, c, which are filled by states $|1\rangle$, $|0\rangle$, and $|\alpha\rangle$, respectively. Determine the conditional state $|\Psi^{out}_{acond}\rangle$ of the system in mode a after an ideal photon number measurement by photon counters D_c and D_d has been performed on output modes b and c and has given the result: one photon in b and no photon in c.

What is the probability of such an event? For which value of α do we obtain the conditional qubit state $(|0\rangle + e^{i\theta}|1\rangle)/\sqrt{2}$ in output mode a? What is the expression of the probability operator and of the post-measurement operator?

6 Experiment: Bipartite Systems

As we will see in much more detail in Chapter 7, an entangled state is a quantum state describing a bipartite system that cannot be written as a tensor product $|\psi_A\rangle \otimes |\psi_B\rangle$ of individual quantum states for the two parties (usually named A and B, or Alice and Bob). In order to generate such states and characterize them, two well-separated parties on which one can make independent measurements are required, as well as a prior coupling between these two parties that has ceased when they are subject to measurements, so that the state collapses caused by the measurements taking place in A and B can be assumed not to be related by a physical process. This situation is encountered in many different physical systems. In this experimental chapter, we briefly describe a few of them, keeping in mind that we could have chosen many more examples.

6.1 Light

6.1.1 Parametric Down Conversion

Spontaneous parametric down-conversion, or parametric fluorescence, described in Section I.2, provides the bipartite system that is the most often chosen by experimentalists. It takes place in a nonlinear crystal pumped by laser light of frequency ω_p and wavevector $\mathbf{k}_p$ that generates pairs of signal and idler photons of frequencies ω_s and ω_i, and wave vectors $\mathbf{k}_s$ and $\mathbf{k}_i$. The energy conservation ensures that $\omega_p = \omega_s + \omega_i$, whereas the momentum conservation, or phase matching relation, implies $\mathbf{k}_s + \mathbf{k}_i \simeq \mathbf{k}_p$. This latter relation is rigorously valid for infinitely long interaction lengths, and only approximate for finite length crystals. The signal and idler beams form a cone of fluorescent light around the direction of the pump beam (Figure 6.1).

At low intensities of the pump beam the electromagnetic field quantum state $|\psi_{out}\rangle$ produced by such a process can be written, within a normalization factor, as the vacuum state plus a sum of pairs of "twin" photons:

$$|\psi_{out}\rangle = |0\rangle + \int d\omega_s d\mathbf{k}_s f(\omega_s, \mathbf{k}_s)|1 : \omega_s, \mathbf{k}_s\rangle \otimes |1 : \omega_i, \mathbf{k}_i\rangle, \qquad (6.1)$$

where $|0\rangle$ is the vacuum state and f is a coupling factor depending on the pumping light and on the characteristics of the nonlinear material.

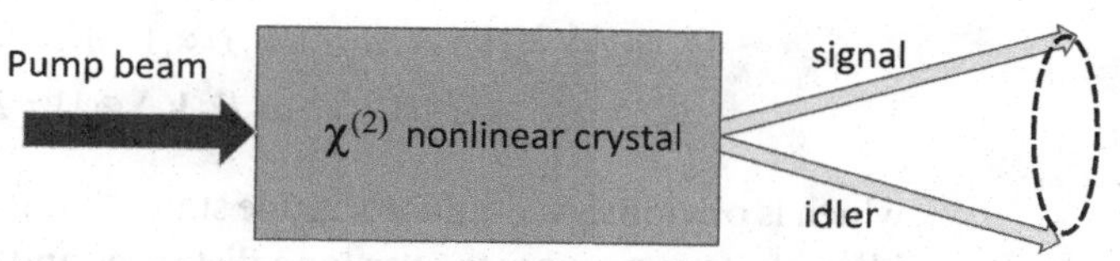

Fig. 6.1 Couples of down-converted light beams are spontaneously emitted by a nonlinear crystal submitted to an intense pump laser beam. The signal and idler beams produced by this process are emitted on two symmetrical directions of a cone. The sum of the frequencies of the signal and idler beams is equal to the pump frequency.

It is usually very small, so that the brightness of the two-photon source is low. The signal and idler photons are emitted along wavevectors that are symmetrical with respect to the propagation, forming a cone of down-converted light (Figure 6.1).

In order to select a single bipartite system, experimentalists use irises or single mode fibers to select a single transverse mode and frequency filters and Fabry–Perot etalons to select a single frequency mode. The bipartite state is made of a single Alice photon around $\mathbf{k}_s$, ω_s, and of a single Bob photon around the associated symmetrical zone of the idler light around $\mathbf{k}_i$, ω_i. The twin photons then propagate in different directions and quickly become far away from each other. When they have the same polarization, vertical (V) for example, the quantum state describing this situation is (see also Equation (I.9))

$$|\psi_{out}\rangle = |0\rangle + f|1 : V, \omega_s, \mathbf{k}_s\rangle \otimes |1 : V, \omega_i, \mathbf{k}_i\rangle. \tag{6.2}$$

Some crystals generate signal and idler light in orthogonal polarizations H and V, generating the state

$$|\psi_{out}\rangle = |0\rangle + f|1 : H, \omega_s, \mathbf{k}_s, \rangle \otimes |1 : V, \omega_i, \mathbf{k}_i\rangle. \tag{6.3}$$

In this case Alice and Bob can easily separate their parties using a polarizing beamsplitter.

When the detection is conditioned by a click on a photon counter, the vacuum part of the states (6.1), (6.2) and (6.3) is absent in the post-selected state, as it gives no signal: the two-photon state is factorized, but not entangled. It is characterized by the existence of perfect correlations between Alice and Bob photons, in emission time, frequency, and directions. Such pairs of single photons are perfectly suited to conditionally generate single signal photons heralded by the detection of one idler photon (Section 5.5).

In order to generate entangled states, one must use more elaborate schemes. One solution is to use, instead of a single crystal, two joined thin crystals rotated by $\pi/2$ along the propagation direction, and an optical configuration for which the detector cannot distinguish pairs of photons produced by one crystal or the other. The experimental parameters are chosen so that the signal and idler photons have the same frequency ($\omega_A = \omega_B = \omega_p/2$). The quantum state describing the system, discarding the vacuum part of the state, is now the two-photon state:

$$|\psi_p\rangle = |1 : H, \omega_p/2, \mathbf{k}_s\rangle \otimes |1 : V, \omega_p/2, \mathbf{k}_i\rangle + e^{i\phi}|1 : V, \omega_p/2, \mathbf{k}_s\rangle \otimes |1 : H, \omega_p/2, \mathbf{k}_i\rangle, \quad (6.4)$$

which is obviously a nonfactorizable state. Though the generated signal and idler photons propagate over long distances and can be very far apart, one cannot speak of specific quantum states for Alice and Bob. In order to fully account for their properties, one needs "nonlocal" states that describe very extended physical systems as a single global quantum object.

When one increases the pump light intensity, more than one photon can be found in the signal and idler modes. To increase the brightness of the source one can also insert the nonlinear crystal in an optical cavity that is resonant for the signal and idler modes: this device, called an optical parametric oscillator (OPO) is somehow similar to a laser, the parametric gain (Equation (I.6)) replacing the stimulated emission gain. This configuration is better described in terms of fluctuations of continuously varying field quadratures than in terms of photons (see Chapter 9). Below a threshold for the pump intensity, the OPO emits a field of zero mean with modified fluctuations. One can show that these fluctuations are correlated on the $\hat{X}$ quadrature and anticorrelated on the $\hat{P}$ quadrature [153], witnessing the existence of Einstein–Podolsky–Rosen (EPR) entanglement [62]. Above the threshold, the OPO emits intense and coherent signal and idler beams. The EPR quadrature entanglement existing below the threshold translates into an amplitude-phase EPR entanglement above the threshold, a configuration studied in Chapter 9. Note that entanglement is by no means restricted to the single or few photon regime.

6.1.2 Cascaded Spontaneous Emission

A second example of bipartite states is provided by the cascaded spontaneous emission of a three-level atom, with levels $|f\rangle$, $|e\rangle$, and $|g\rangle$, which is at time $t = 0$ in its upper excited state $|f\rangle$. It decays to its ground state $|g\rangle$ in a two-step process, $|f\rangle \to |e\rangle$, $|e\rangle \to |g\rangle$ by emitting two photons corresponding to wavevectors $\mathbf{k}$ and $\mathbf{k}'$. It can be shown [161] that the photon state $|\psi_c\rangle$ produced by such a process can be written as

$$|\psi_c\rangle = |0\rangle + \int d\mathbf{k}d\mathbf{k}' \frac{g(\omega, \mathbf{k}, \omega', \mathbf{k}')}{(\omega + \omega' - \omega_\alpha - \omega_\beta + i\gamma_\alpha)(\omega - \omega_\beta + i\gamma_\beta)} \times |1 : \omega, \mathbf{k}\rangle \otimes |1 : \omega', \mathbf{k}'\rangle, \quad (6.5)$$

where ω_α and ω_α are, respectively, the Bohr frequencies of transitions $|f\rangle \to |e\rangle$ and $|e\rangle \to |g\rangle$, and g is a function depending on the properties of the atom. γ_α and γ_β are the line widths of the two transitions. This state is similar to the one described by Equation (6.1). It encompasses strong temporal correlations between photon counts, but no spatial correlations between the spontaneously emitted photons, as $\mathbf{k}$ and $\mathbf{k}'$ can have arbitrary directions.

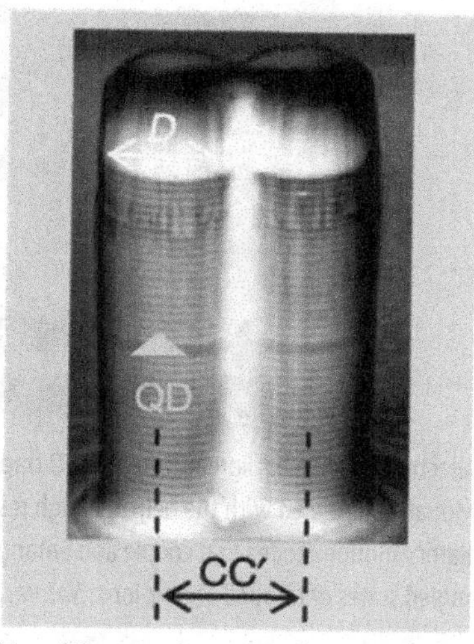

Fig. 6.2 Two identical pillar microcavities of diameter D are grown very close to each other ($CC' \simeq D \simeq 3\ \mu m$). Only one contains a quantum dot (QD), which can be excited to the bi-exciton state, which then decays by two indistinguishable cascades. From A. Dousse et al. Ultra-bright source of entangled photon pairs, *Nature*, 2010 [59].

The state (6.5), discarding the vacuum part, is not entangled. To generate entanglement one needs to create a quantum state with two different and indistinguishable paths leading to the ground state. This has been implemented in a calcium atom, for example [9], which features the cascade $|f, J = 0, m = 0\rangle \rightarrow |e, J = 1, m = \pm 1\rangle \rightarrow \ |g, J = 0, m = 0\rangle$. In such a configuration one can show that the generated state is entangled and that the linear polarizations of the cascading photons are perfectly correlated.

One can also use quantum dots instead of atoms or ions to implement intense sources of entangled photon pairs using a two-photon cascade of a bi-exciton. In [59] (see Figure 6.2) for example, two identical pillar microcavities made of two Bragg mirrors are fabricated very close to each other, so that they are coupled by mode overlap. Only one cavity contains a quantum dot, the other being used to modify the frequency of the mode. The system can be specifically designed and fabricated to generate the required cascading three-level excitonic system with the possibility of two indistinguishable paths for the decay.

6.2 Matter

6.2.1 Ions

In Chapter 3 and Appendix J, we described methods to trap ions and cool them down to their vibrational and electronic ground state. It is also possible, using a linear trap (Figure 6.3), to confine in a row several ions that can

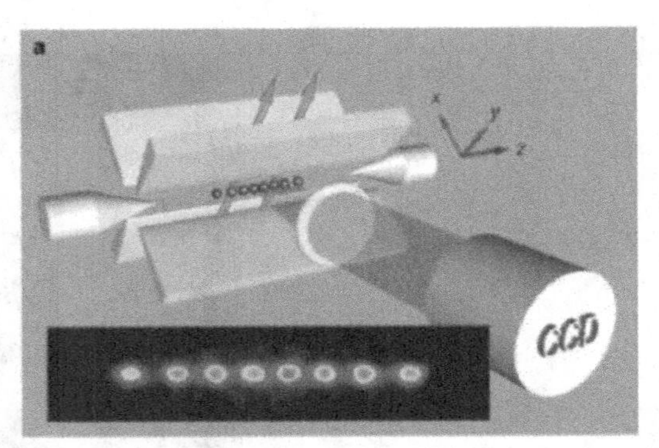

Fig. 6.3 The electrodes in the background create a 1D trap that is filled by ground state ions that can be individually addressed with the help of a high resolution CCD camera. The ions have collective oscillatory motion modes that couple and entangle all of them. From R. Blatt and D. Wineland. Entangled states of trapped atomic ions, *Nature*, 2008 [20].

be individually submitted to laser irradiation and imaged on a CCD camera, this forming a kind of quantum register. Because of the small distance between the trapped ions (around 8 μm), their Coulomb coupling is strong, resulting in a collective motion that is better described as a quantized normal mode to which all ions participate. It is possible not only to steer at will the Bloch vector of a chosen ion by sequences of resonant pulses (see Section G.2) but also, using a 2π pulse on an auxiliary transition, to change the sign of the component of the wavefunction on the state $|-, n_v = 1\rangle$, where $|n_v = 1\rangle$ is the single phonon collective motional state in the normal mode. We will not go into the details of the experimental technique to generate 2-qubit entangled states, and to implement a CNOT quantum gate, one of the building blocks of universal quantum computing [20, 130]. Entangled ionic qubits can also be prepared by the simultaneous irradiation of the two qubits.

6.2.2 Macroscopic Samples

Macroscopic atomic samples, even containing 10^9 atoms or so, may also exhibit full quantum properties provided that they are coherently interacting with light. Whereas a single two-level atom behaves as a spin 1/2 (Appendix D), a set of N two-level atoms interacting with coherent light can be shown to behave like a single quantum system, more precisely like a spin $N/2$, during the coherence time of the system. In addition, the sample can be dense enough that its optical depth is smaller than its length, so that all the light is absorbed in the medium.

Figure 6.4 shows the three-level configuration that is often used. With only the signal field switched on, the quantum state of polarization encoded single photon qubit can be transferred into the qubit state of the two hyperfine levels $|g\rangle$ and $|s\rangle$ (Figure 6.4). In a reverse way, using a Raman configuration where the difference between the frequencies of the applied

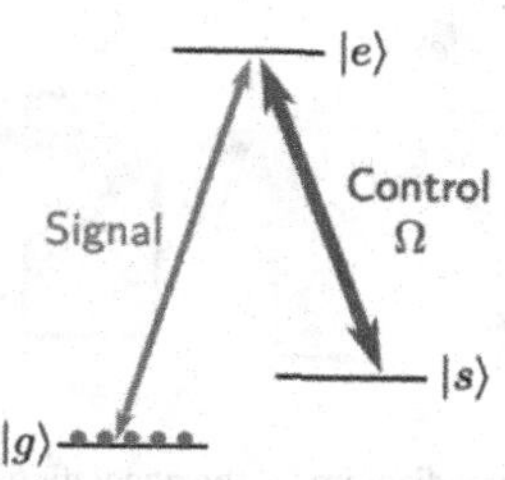

Fig. 6.4 Energy level diagram for an atomic memory for light. The recorded qubits are linear superposition of the two ground-state levels. They are read by injecting in the sample signal light, which induces by stimulated Raman scattering the emission of quantum light on the control beam that reflects the qubit state.

Fig. 6.5 Entanglement of two macroscopic atomic samples, situated inside the two cylindrical magnetic shields, through which a common beam (dashed beam) is sent. From K. Hammerer et al. Quantum interface between light and atomic ensembles, *Reviews of Modern Physics*, 2010 [85].

signal and control fields is equal to the energy difference (within the factor $\hbar$) between the two lower states, the atomic sample can transfer back its qubit state to the light that is emitted by a stimulated Raman effect: this is the basic idea of an atomic quantum memory for the light quantum state. We will not go into the details of such light–matter interaction, which can be found in [85].

It is, in particular, possible to entangle two macroscopic samples of atoms contained in glass cells by sending an appropriate laser beam through the two samples successively (Figure 6.5). One produces in this way entanglement between macroscopic systems separated by macroscopic distances (of the order of a meter). The cylindrical metallic boxes are magnetic shields needed in order to increase the coherence time.

6.3 Conditional Entanglement

It is also possible to create entanglement using post-selection, an experimental technique called heralding. It consists in using, in addition to the Hamiltonian evolution of the two interacting systems, a projective

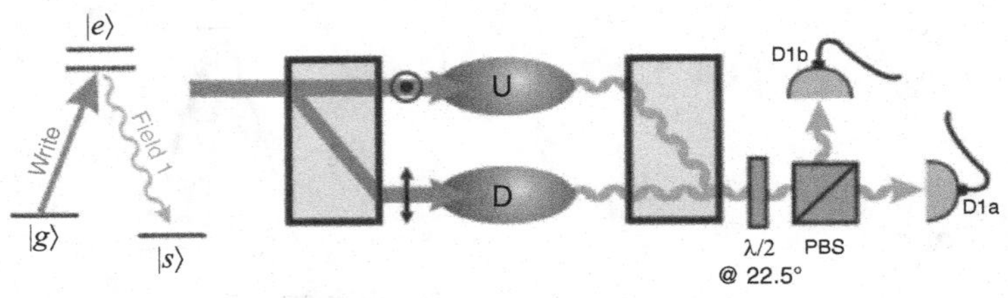

Fig. 6.6 Experimental diagram for the generation of heralded entanglement. One does not know from which sample the photon detected at the output of the set-up is coming from. This possibility of two paths gives rise to entanglement between them. From J. Laurat et al. Heralded entanglement between atomic ensembles: preparation, decoherence and scaling, *Physical Review Letters*, 2007 [111].

measurement that collapses the factorized Alice–Bob state onto an entangled state. This kind of conditional generation is a random process that has often a small probability, which reduces the efficiency of the method.

Let us give an example, taken among many other implementations, of remote conditional entanglement of two ensembles of caesium atoms, U and D, that are illuminated with independent light beams of orthogonal polarizations (Figure 6.6). The two resulting Raman fields are collected by single mode fibers. These two beams are then mixed on a polarizing beamsplitter rotating the polarization by $\pi/4$. When a photon is detected by the photodetectors on one of the two output ports, it is impossible to know which atomic sample has generated it, so that the post-selected state of the whole system is an entangled state that is a combination of the two possibilities, namely,

$$|\psi\rangle = \frac{1}{\sqrt{2}}(|U:0\rangle \otimes |D:1\rangle \pm |U:1\rangle \otimes |D:0\rangle), \tag{6.6}$$

where 0 means all atoms in ground state and 1 means one excitation shared by all the atoms of the sample.

6.4 Cavity-Mediated Interactions in Circuit QED

In the superconducting qubit platform, coupling mediated by a cavity is used to entangle two circuits that are five millimeters apart from each other [119]. Note that ideas originating from simple atomic systems were successfully extended to macroscopic circuits. This is again an outstanding illustration of the universality of quantum mechanics.

In this experiment, the cavity mediating the interaction is a $\lambda/2$ ($\sim$12.3 mm) strip-line resonator and the solid state qubits are placed close to the antinodes. We reproduce the set-up in Figure 6.7(a). The upper panel shows a diagram of the circuit and the half-wavelength mode coupling the qubits. The resonator is made of aluminum printed on a chip and

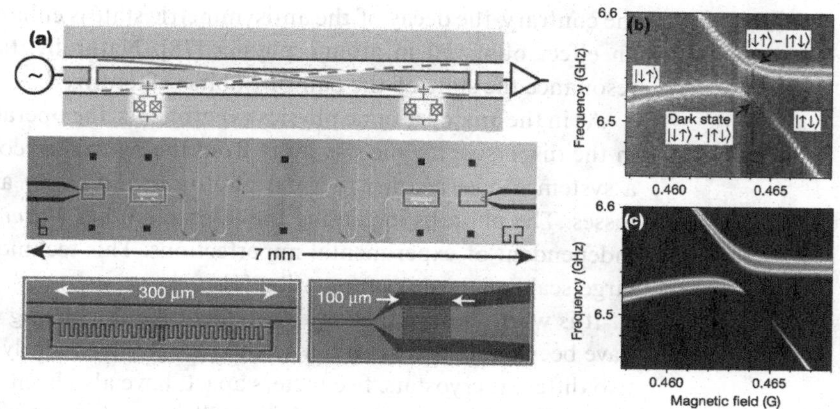

Fig. 6.7 (a) The upper panel is a sketch of the circuit used to entangle two SQUID-transmons using a virtual cavity excitation in [119]. The lower panels present details of the circuit architecture. The meandering aluminum line (middle panel) is the $\lambda/2$ resonator mediating the interaction. The qubits are in the rectangular boxes and enlarged in the lower panel. (b) Microwave spectroscopy of the system showing the vacuum Rabi splitting on resonance (Figure G.1). The data are displayed as a function of the magnetic field controlling the qubit frequencies. (c) Theoretical prediction. From J. Mayer et al. Coupling superconducting qubits via a cavity bus, *Nature*, 2007 [119].

the qubits are SQUID-transmons (SQUID stands for "superconducting quantum interference device"; see Section 10.5). The resonator is coupled to an on-chip transmission line.

The frequency of the SQUID-transmons $\omega_{1,2}$ can be continuously tuned in situ by changing a magnetic field applied to the sample. One has

$$\omega_{1,2} = \omega_{1,2}^{\max} \sqrt{|\cos\left(\pi\Phi_{1,2}/\Phi_0\right)|}, \tag{6.7}$$

where $\Phi_{1,2}$ is the magnetic flux through the SQUID loops and Φ_0 is the flux quantum.

By tuning their frequency while applying a microwave probe field to the transmission line and collecting the transmitted signal, one can perform spectroscopy of the system. Figure 6.7(b) displays the spectroscopic data as a function of the magnetic field over the sample, and the theory predictions are plotted in Figure 6.7(c). We recognize the typical "vacuum Rabi splitting" (Figure G.1) between the two nonlinear SQUID modes showing the long-range interaction at resonance and demonstrating that the system has reached the so-called strong coupling limit (Section G.4.1). An oscillation in the time domain data back has been also observed [119].

According to theory, the symmetric state is a "dark state": It cannot be excited by the antisymmetric mode mediating the interaction. This is direct evidence of the coherence of the superposition and thus of entanglement. The observation has far reaching consequences: The dark state is protected against decay, an effect known as the inverse-Purcell effect, first predicted [104] and then observed [98] in the context of cavity QED with Rydberg atoms. This entangled state thus has an enhanced lifetime. Very much on the contrary, the decay of the antisymmetric state is enhanced by the cavity, an effect observed in atomic physics [78]. Naturally, far away from the resonance the state of the pair of qubits is factorized.

As in the analog atomic physics experiments, the operation of this set-up in the dispersive regime (far away from the resonance condition) provides a system robust against thermal photon in the cavity and against cavity losses. The photons mediating the interaction are *virtual* and thus greatly independent of experimental imperfections. This technique is the basis of large-scale quantum computation [6].

It is worth noting that superconducting qubits living on different chips have been entangled via other techniques [134]. Recently, qubits located in two different cryostats, five meters apart, have also been entangled [118].

The list of experimental schemes allowing experimentalists to entangle various quantum systems is very long, and here we have just given a rough idea of this flourishing domain. We could also have mentioned NV centers in diamond and thin oscillating membranes. There is also the possibility of creating *hybrid entanglement* between different quantum systems, such as light and atoms. Finally, entanglement is not restricted in bipartite systems, and exists also in *multipartite systems* with more than two parties.

6.5 Exercises for Chapter 6

Exercise 6.1 What is a parametric process?

Exercise 6.2 Consider a cascaded ladder system of three atomic levels $f \to e \to g$. Compute the variation in time of the $e \to g$ fluorescence signal as a function of the transition lifetimes τ_{fe} and τ_{eg} taking the initial state to be f. Discuss the condition $\tau_{fe} = \tau_{eg}$.

Exercise 6.3 Consider a bipartite system generated by the explosion of a rock into two pieces. Each of the pieces, due the conservation of angular momentum, has an angular momentum $\vec{J}_i$ that is equal in magnitude and opposite in direction but otherwise random. This is, they have a strong correlation but not "independent properties" (i.e. each rock piece is in a completely random $\vec{J}_i$). Two observers receiving the pieces perform binary observations given by

$$r_{\vec{n}_i} = \text{sign}(\vec{n}_i \cdot \vec{J}_i).$$

Here, $\vec{n}_i$ are unit vectors determining the direction of the measurement. The angle in between the observation directions is given by $\vec{n}_1 \cdot \vec{n}_2 = -\cos\theta$. Show that the mean values of $r_{\vec{n}_i}$ are zero but the correlation $\langle r_{\vec{n}_1} r_{\vec{n}_2} \rangle$ is not. Show that if $\theta = 0$ the correlation is saturated. Compute its analytical expression as a function of θ. Repeat the previous calculation for the polarization of a pair of photons generated in an SPDC crystal in a Bell state. The observables are $\hat{r} = \vec{n}_i \cdot \vec{s}$, where $\vec{s} = (\hat{\sigma}_x, \hat{\sigma}_y, \hat{\sigma}_z)$. Do the computed correlations as a function of θ agree? See [142]. At what angle are the predictions equal and maximally different?

Exercise 6.4 Consider a nonlinear crystal of length L generating SPDC entangled pair of photons of polarization H and V with index of refraction n_H and n_V. What is the condition on the coherence of the pump laser to generate a pair of photons entangled in polarization? What would happen if the coherence time of the laser is shorter than $\frac{L}{c}(n_H - n_V)$? Is the answer dependent on the integration time of the detector?

7 Entanglement

In Chapter 6 we gave several examples of bipartite systems in which the two parties (for example two light beams, two ions, two superconducting circuits) have interacted in the past and are at the moment t of consideration separated by a distance that forbids any interaction between them. Even better, one can organize the experimental set-up so that the two parties are separated by a space-like interval so that they cannot be connected by any signal propagating at the velocity of light.

In the traditional way we will call "Alice" and "Bob" the two parties, and $\mathcal{H}_A$ and $\mathcal{H}_B$ the two corresponding Hilbert spaces (Figure 7.1).

In this chapter, we will characterize entangled states, introduce entanglement witnesses, discuss the "Einstein–Podolsky–Rosen paradox" and the Bell inequalities, and present some applications of entangled states in quantum information processing and quantum teleportation.

7.1 Entangled Pure States

7.1.1 Definition

A quantum state $|\Psi_{AB}\rangle$ belonging to a Hilbert space $\mathcal{H}_{AB}$, which is the tensor product of two Hilbert spaces $\mathcal{H}_{AB} = \mathcal{H}_A \otimes \mathcal{H}_B$, is an entangled state if there is no basis $\{|u_i\rangle\}$ of space $\mathcal{H}_A$ and no basis $\{|v_j\rangle\}$ of space $\mathcal{H}_B$ on which it can be written as a tensor product $|\Psi_A\rangle \otimes |\Psi_B\rangle$. This concept was introduced by Schrödinger in 1935 [158].

Note that in this definition one considers only "local" bases, which do not mix the two parties $\mathcal{H}_A$ and $\mathcal{H}_B$. In general, it is not easy to use this definition to show that a bipartite state is an entangled state, as one must consider all possible local basis choices. Here, we will give more practical entanglement criteria.

The simplest bipartite system is the two qubit case. One finds that the so-called four *Bell states*,

$$|\Psi_\pm\rangle = \frac{1}{\sqrt{2}}(|+-\rangle \pm |-+\rangle), \quad |\Phi_\pm\rangle = \frac{1}{\sqrt{2}}(|++\rangle \pm |--\rangle), \tag{7.1}$$

are all entangled.

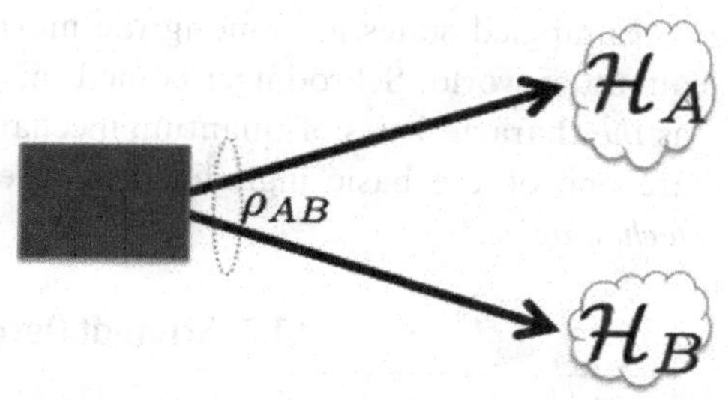

Fig. 7.1 Generation of an entangled state describing a bipartite system globally, with no possibility of separately describing a quantum state for party *A* and another for party *B*, even though *A* and *B* are remote and cannot interact at the time of the measurement.

States that are not factorized on a given basis may be factorized when one uses another basis of $\mathcal{H}_A$ and another basis of $\mathcal{H}_B$. For example, the state

$$|\psi_1\rangle = (|++\rangle - |--\rangle - |+-\rangle + |-+\rangle)/2 \tag{7.2}$$

is factorizable in the basis $(|+\rangle \pm |-\rangle)/\sqrt{2}$ of both spaces, whereas

$$|\psi_2\rangle = (|++\rangle - |--\rangle + |+-\rangle + |-+\rangle)/2, \tag{7.3}$$

differing only by one change of sign, cannot be not factorized in any couple of local bases.

We noted in Section 2.2 that, even though the overall state $|\Psi_{AB}\rangle$ is entangled, local properties, measured on Alice's side, for example, using a "local" measurement set-up, can be derived from the knowledge of the local reduced density matrix $\rho_A = Tr_B\rho_{AB}$, where $\rho_{AB} = |\Psi_{AB}\rangle|\langle\Psi_{AB}|$. The same is true for Bob on his side using $\rho_B = Tr_A\rho_{AB}$.

Let us take the example of the Bell state $|\Psi_-\rangle$:

The reduced density matrix on Alice's side ρ_A is equal to $(|+\rangle\langle+| + |-\rangle\langle-|)/2$. It is proportional to the identity. It therefore has the same expression in any basis. It describes a mixed state with a minimal information content. For example, it gives equal probabilities 1/2 for spin measurements in any direction.

The joint probabilities of couples $(+,+)$, $(+,-)$, $(-,+)$, $(-,-)$ are, respectively, 0, 0.5, 0.5, 0. The conditional probabilities $P(+|+)$, $P(+|-)$, $P(-|+)$, $P(-|-)$ calculated by formula (5.53) are, respectively, 0, 1, 1, 0 probabilities, which actually correspond to certainties.

So, Alice has no information about her own qubit, but knows Bob's qubit value with certainty once she has measured her qubit. In the entangled state $|\Psi_-\rangle$, *correlations* between Alice and Bob's measurements are maximal, though Alice and Bob may be far away from each other and not able to communicate. These correlations have no classical analog. Pure entangled states featuring strong Alice–Bob correlations that reduce to local states obtained by partial trace that have no information content are called *maximally entangled states*.

Entangled states are among the most striking specific features of the quantum world. Schrödinger coined this property "not as *one*, but rather as *the* characteristics of quantum mechanics." Nowadays, entangled states are one of the basic ingredients of the promising domain of *quantum technologies*.

7.1.2 Schmidt Decomposition

E. Schmidt showed that any bipartite state can be written as

$$|\Psi_{AB}\rangle = \sum_{n=1}^{S} \sqrt{p_n} |\phi_{nA}\rangle \otimes |\phi_{nB}\rangle, \tag{7.4}$$

where p_n is real, with $0 \leq p_n \leq 1$, $\sum p_n = 1$, and $\{\phi_{nA}\}$ and $\{\phi_{nB}\}$ are two orthonormal sets of wavevectors, respectively, belonging to spaces $\mathcal{H}_A$ and $\mathcal{H}_B$. S is called the "Schmidt number."

From Equation (7.4), one gets

$$\rho_{AB} = \sum_{n,n'} \sqrt{p_n p_{n'}} |\phi_{nA}\rangle |\phi_{nB}\rangle \langle \phi_{n'A}| \langle \phi_{n'B}| \tag{7.5}$$

and the reduced density matrix in space $\mathcal{H}_A$ is

$$\rho_A = Tr_B \rho_{AB} = \sum_n p_n |\phi_{nA}\rangle\langle\phi_{nA}|. \tag{7.6}$$

Therefore, p_n is an eigenvalue, and $|\phi_{nA}\rangle$ an eigenstate, of the reduced density matrix ρ_A, whereas p_n is also an eigenvalue, and $|\phi_{nB}\rangle$ an eigenstate, of the reduced density matrix ρ_B. Being positive operators, they have real positive eigenvalues with $\sum p_n = 1$.

The Schmidt number S is smaller than or equal to the smaller dimension of the two Hilbert spaces. If $S = 1$ the state is factorized. If $S > 1$ the state is entangled. The condition $S > 1$ is an example of what is often called an *entanglement witness*: It is a simple test that signals the property "entanglement" of a given pure state. Moreover, S itself gives a kind of "measure" of the "strength" of entanglement in the case of a pure two-partite state. Note that the Bell states (7.1) are directly written as a Schmidt sum with $S = 2$ and $p_n = 1/2$.

Another entanglement witness is given by the purity of the reduced state $P_A = Tr\rho_A^2$: If the two-partite pure state is factorized, then the local density matrix $|\phi_{1A}\rangle\langle\phi_{1A}|$ is also pure and $P_A = 1$. Reciprocally, if P_A is one, then $\sum p_n = 1$ and $\sum p_n^2 = 1$, which happens only when only one p_n is nonzero so the state is factorized.

7.1.3 Alice–Bob Correlations

Alice and Bob can predict the probabilities of local measurement results on their own without caring about the other's result using their reduced density matrices. What happens if they now compare their results by sending them to a common place? They will be now able to discover possible correlations between what they have independently measured. To predict such correlations, they now need to use the full quantum state $|\Psi_{AB}\rangle$.

Let us consider local physical parameters, A on Alice's side and B on Bob's side, associated with observables $\hat{A}$ acting on $\mathcal{H}_A$ and $\hat{B}$ acting on $\mathcal{H}_B$, with respective eigenvalues and projectors $\{a_\ell, \hat{P}_\ell\}$ and $\{b_{\ell'}, \hat{Q}_{\ell'}\}$. One obviously has $[\hat{P}_\ell, \hat{Q}_{\ell'}] = 0$, and therefore $[\hat{A}, \hat{B}] = 0$. For the sake of simplicity, we assume that both are von Neumann measurements.

We use the formalism around joint measurements introduced in Section 5.7. Then the joint probability of getting the couple of results $(a_\ell, b_{\ell'})$ is

$$Proba(a_\ell, b_{\ell'}) = \langle \hat{P}_\ell \hat{Q}_{\ell'} \hat{P}_\ell \rangle = \langle \hat{P}_\ell^2 \hat{Q}_{\ell'} \rangle = \langle \hat{P}_\ell \hat{Q}_{\ell'} \rangle \tag{7.7}$$

because the projection operators commute. Let us consider the two possible situations:

- If the state is factorized, then

$$Proba(a_\ell, b_{\ell'}) = \langle \phi_{1A}|\hat{P}_\ell|\phi_{1A}\rangle\langle \phi_{1B}|\hat{Q}_{\ell'}|\phi_{1B}\rangle = Proba(a_\ell)Proba(b_{\ell'}), \tag{7.8}$$

  and therefore the two measurements are uncorrelated. One can also deduce it from the wave collapse postulate: When Alice obtains the result a_ℓ of her prior measurement, the state collapse associated with the measurement modifies $|\phi_{1A}\rangle$ and leaves Bob's part unchanged. So the measurement performed by Bob will not be affected by Alice's measurement. The reasoning is much simpler than the previous one, but seems to depend on the order of measurements. It is not the case in the reasoning involving joint probabilities.
- If the state is entangled, then the previous argument is no longer valid: The state collapse due to Alice's measurement will modify Bob's side, inducing a correlation between the measurement results.

  More precisely, let us consider specific observables $\hat{F}_A$ and $\hat{F}_B$ *that are diagonal in the Schmidt bases*, with eigenvalues α_ℓ and $\beta_{\ell'}$. It is straightforward to show that the joint probabilty of getting the couple of results $(\alpha_\ell, \beta_{\ell'})$ is

$$Proba(\alpha_\ell, \beta_{\ell'}) = p_\ell \delta_{\ell,\ell'}, \tag{7.9}$$

  so that the conditional probability of measuring $\hat{F}_B$ knowing the result on $\hat{F}_A$ is

$$Proba(\alpha_\ell|\beta_{\ell'}) = \frac{Proba(\alpha_\ell, \beta_{\ell'})}{Proba(\beta_{\ell'})} = \delta_{\ell,\ell'}. \tag{7.10}$$

  Therefore, the two measurements results for this particular choice of observables are *perfectly correlated*: After her measurement, Alice knows with certainty the result that Bob will obtain, even though they are separated by millions of kilometers!

It must be stressed here that *the existence of such a perfect correlation between remote measurement results is not a specific quantum effect* but also exists classically. Let us consider, for example, a classical random process that draws integers ranging from 1 to S. For a given drawing lot yielding the integer ℓ, an operator sends an envelope to Alice with value α_ℓ written inside, and at the same time an envelope to Bob with value β_ℓ written inside. When Alice, who knows the rules of the game but not the result of the draw, opens her envelope and "measures" the value α_ℓ, she knows for sure what Bob will measure when he opens his envelope, whatever the distance between them, the temporal order of their openings, and in the absence of any physical link between Alice and Bob.

Such a correlation between remote locations does not violate the relativistic causality: The information that Alice has on Bob's results is "local" to Alice's space. It cannot be transferred to Bob for comparison at a velocity larger than c. In addition, it cannot be used to transfer information from Alice to Bob, because Alice does not master the outcome of her measurement.

7.2 Entangled Mixed States

7.2.1 Entanglement and Separability

The situation here is more complicated than for a pure state as there are now three possibilities:

- A *factorized state* is described by a density matrix of the form $\rho_{AB} = \rho_A \rho_B$, where ρ_A and ρ_B are density matrices in the two spaces $\mathcal{H}_A$ and $\mathcal{H}_B$. The same reasoning as above shows that the measurements performed by Alice and Bob are not correlated.
- A *separable state* is described by a density matrix that is a statistical superposition of factorized states. It is therefore of the form

$$\rho_{AB} = \sum_n p_n \, \rho_{nA} \otimes \rho_{nB}, \tag{7.11}$$

where ρ_{na} and ρ_{nb} are density matrices in the two spaces $\mathcal{H}_1$ and $\mathcal{H}_R$. To simplify notations we will omit the tensor product symbol $\otimes$ and write $\rho_{na}\rho_{nb}$ the tensor product $\rho_{na} \otimes \rho_{nb}$. We will do the same for factorized quantum states. ρ_{AB} can be produced by making a classical draw with probability p_n of uncorrelated factorized states, like in the classical example of the envelopes given in the previous section. It is easy to show that one has

$$\begin{aligned} Proba(a_\ell, b_{\ell'}|\rho_{AB}) &= \sum p_n Proba(a_\ell|\rho_{nA}) Proba(b_{\ell'}|\rho_{nB}) \\ &\neq Proba(a_\ell) Proba(b_{\ell'}), \end{aligned} \tag{7.12}$$

and therefore that the Alice and Bob measurements have some degree of correlation, the classical or quantum origin of which is still an active debate between researchers.

- Finally, an *entangled state* is described by a density matrix that cannot be written in the form of Equation (7.11).

7.2.2 An Entanglement Witness: The Positive Partial Transpose Criterion

We note that the definition of an entangled mixed state is more subtle that the definition of an entangled pure state. In addition, there is no unique Schmidt decomposition for a mixed bipartite state. So it is not a simple task to decide whether a given mixed state is entangled or not. So far, only sufficient conditions can be found in the literature, except for specific classes of systems. Here, will describe one of the most widely used entanglement witnesses [97], called the "positive partial transpose" test (PPT).

Let us introduce the basis $|u_i, v_j\rangle$ of the bipartite Hilbert space $\mathcal{H}_A \otimes \mathcal{H}_B$. It is easy to show that the transpose ρ^T of a positive operator ρ, defined as usual by $\langle u_i, v_j|\rho^T|u_{i'} v_{j'}\rangle = \langle u_{i'}, v_{j'}|\rho|u_i v_j\rangle$, is also a positive operator: The transpose operation is a particular case of a quantum map.

Let us now define the *partial transposed (PT) density matrix* ρ^{PT} by

$$\langle u_i, v_j|\rho^{PT}|u_{i'} v_{j'}\rangle = \langle u_{i'}, v_j|\rho|u_i v_{j'}\rangle, \tag{7.13}$$

in which only the space $\mathcal{H}_A$ is affected by the transposition. If the state is separable, one can write it as Equation (7.11), and therefore,

$$\rho_{AB}^{PT} = \sum_n p_n \, \rho_{nA}^T \rho_{nB}, \tag{7.14}$$

where ρ_{AB}^{PT} describes another separable state. It is a bona fide density matrix, and is therefore a positive operator.

Reciprocally, we have shown that, *if ρ_{AB}^{PT} is not a positive operator, then ρ_{AB} cannot be written like Equation* (7.11) *and therefore describes an entangled state.*

The PPT criterion is only a sufficient condition: Some entangled states have a positive partial transpose operator, so that one cannot say anything about the entanglement of a quantum state described by a density matrix ρ_{AB} with a positive ρ_{AB}^{PT} (except for Hilbert spaces of dimensions 2×2 and 2×3, for which the condition has been shown to be necessary and sufficient). This situation is called "bound entanglement."

Let us give a simple example. The Bell state $|\Psi_-\rangle$, which is described by the following density matrix in the basis $(|++\rangle, |+-\rangle, |-+\rangle, |--\rangle)$,

$$\begin{bmatrix} 0 & 0 & 0 & 0 \\ 0 & 1 & -1 & 0 \\ 0 & -1 & 1 & 0 \\ 0 & 0 & 0 & 0 \end{bmatrix}, \tag{7.15}$$

which has positive eigenvalues (1,0,0,0), like any pure state. Its partial transpose reads:

$$\begin{bmatrix} 0 & 0 & 0 & -1 \\ 0 & 1 & 0 & 0 \\ 0 & 0 & 1 & 0 \\ -1 & 0 & 0 & 0 \end{bmatrix} \tag{7.16}$$

It has eigenvalues (−1,1,1,1) and is therefore not positive: $|\Psi_-\rangle$ is an entangled state.

7.3 Conditional Nondestructive Preparation of a Quantum State

Let us consider a generic bipartite system, like those described in Chapter 6, described by the density matrix ρ_{AB}, which may be entangled, separable, or factorized. Alice then makes a measurement of her side that gives results in a probabilistic way. We want to determine the state of Bob's system when Alice has performed a generalized measurement that has given the specific result a_n (Figure 7.2).

Let $\hat{K}_n$ be the corresponding PMO, which acts only on $\mathcal{H}_A$, and $\hat{\Pi}_n = \hat{K}_n^\dagger \hat{K}_n$ the associated PO. The bipartite state after the result a_n has been obtained is, according to Equation (5.3), $\hat{K}_n \rho_{AB} \hat{K}_n^\dagger / Tr(\hat{K}_n \rho_{AB} \hat{K}_n^\dagger)$. The state of subsystem B alone is obtained by a partial trace on space $\mathcal{H}_A$:

$$\rho_B^{after|a_n} = \frac{Tr_A\left(\hat{K}_n \rho_{AB} \hat{K}_n^\dagger\right)}{Tr\left(\hat{K}_n \rho_{AB} \hat{K}_n^\dagger\right)}. \tag{7.17}$$

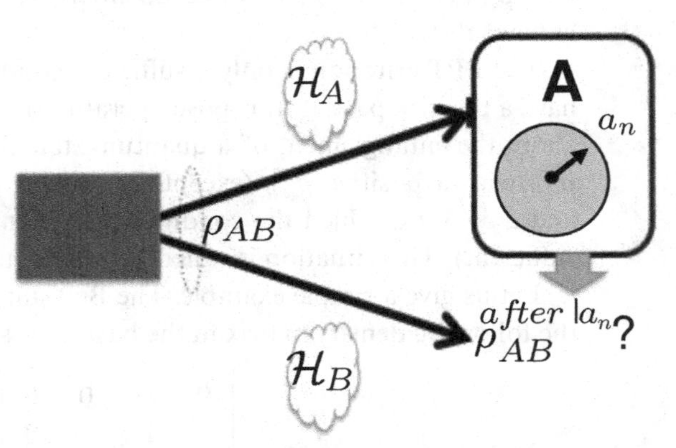

Fig. 7.2 Conditional nondestructive generation of a specific "heralded" state on Bob's side from an entangled AB state, by post-selection of a given measurement output a_n on Alice's side.

Using the properties of the partial trace with respect to commutation of operators, one finally finds the state of Bob's party when Alice has measured on her side the specific value a_n as

$$\rho_B^{after|a_n} = \frac{Tr_A\left(\rho_{AB}\hat{\Pi}_n\right)}{Tr\left(\rho_{AB}\hat{\Pi}_n\right)}. \tag{7.18}$$

Let us consider the two cases:

- If the state is factorized, so $\rho_{AB} = \rho_A\rho_B$, then

$$\rho_B^{after|a_n} = \frac{Tr_A\left(\rho_A\hat{\Pi}_n\right)\rho_B}{Tr_A\left(\rho_A\hat{\Pi}_n\right)Tr_B(\rho_B)} = \rho_B. \tag{7.19}$$

  As intuitively expected, in the factorized and therefore uncorrelated case a measurement on A has no influence on B.
- If the state is separable, so $\rho_{AB} = \sum_i p_i\, \rho_{iA}\rho_{iB}$, then

$$\rho_B^{after|a_n} = \frac{\sum_i p_i Tr_A\left(\rho_{iA}\hat{\Pi}_n\right)\rho_{iB}}{\sum_i p_i Tr_A\left(\rho_{iA}\hat{\Pi}_n\right)Tr_B(\rho_{iB})} = \frac{1}{\sum_i p_i q_{ni}}\sum_i p_i q_{ni}\,\rho_{iB}, \tag{7.20}$$

  where $q_{ni} = Tr_A(\rho_{iA}\hat{\Pi}_n)$ is the probability of getting the result a_n when party A is in state ρ_{iA}. The state of party B, and therefore our knowledge of its properties, is changed in a conditional way depending on the specific outcome of the measurement on A. This change takes place without any physical action on B, hence the name "nondestructive" generation. In addition, if the measurement is chosen in such a way that there is a single value i_n for which q_{ni} is different from zero (and therefore equal to 1), then the conditionally generated state is simply

$$\rho_B^{after|a_n} = \rho_{i_nB}. \tag{7.21}$$

  This state is said to be "heralded" by the result of the measurement on the other party.

The state (7.21) is generated when the initial state ρ_{AB} is separable, which includes entangled states as a particular case. Full entanglement is therefore not necessary in this situation, only correlations. The heralded state is pure if ρ_{i_nB} is pure.

Finally, note that the heralded state is not generated in a deterministic way, but in a not-controlled, probabilistic way. In this context, a very important parameter is the "success rate," i.e. the probability of obtaining the desired state, which is equal to the probability p_{i_n} of measuring a_n.

Let us now be more specific and consider the case of twin photon generation by spontaneous parametric down conversion, as described in Section 6.1 (see Figure 7.3), in which a pure, entangled state is generated,

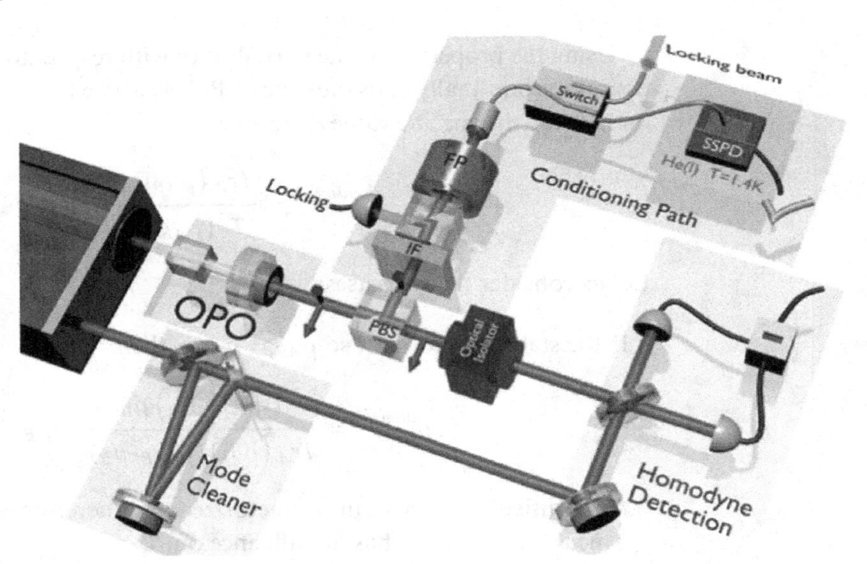

Fig. 7.3 Characterization of a heralded single photon state produced from parametric down conversion: An optical parametric oscillator (OPO) produces pairs of temporally correlated signal and idler photons of orthogonal polarizations. The idler photon is separated from the signal photon by a polarizing beamsplitter. It is detected by an SSPD (superconducting single photon detector) after proper frequency and spatial filtering in order to detect photons in a single light mode. The signal photon is sent to an homodyne detection device (see Section 9.4). The homodyne measurement result is recorded only when there is a "click" on the SSPD. This set-up allows for a heralded generation of a pure single photon state in a mastered electromagnetic field mode. Reprinted with permission from O. Morin et al. High fidelity single photon source based on a type II optical parametric oscillator, *Optics Letters*, 2012 [131]. Copyright The Optical Society.

described by the Schmidt sum $|\psi_{AB}\rangle = \sum_i \sqrt{p_i}|1:i,A\rangle|1:i,B\rangle$. The corresponding density matrix is

$$\rho_{AB} = |0\rangle\langle 0| + \sum_{i,i'} \sqrt{p_i p_{i'}}|1:i,A\rangle\langle 1:i',A| \otimes |1:i,B\rangle\langle 1:i',B|, \qquad (7.22)$$

where $|1:i,A\rangle$, for example, describes a single photon in the optical mode labeled by i.

We first suppose that the detector is not selective and counts all photons of beam A with the same efficiency η, and has dark counts with probability d. The PO corresponding to a click, according to Chapter 5, is $\hat{\Pi}_{click} = \eta \sum_\ell |1:\ell,A\rangle\langle 1:\ell,A| + d|0\rangle\langle 0|$. The heralded state is, in this case,

$$\rho_B^{after|click} = \frac{1}{\eta + d}\left(\eta \sum_\ell p_\ell |1:\ell,B\rangle\langle 1:\ell,B| + d|0\rangle\langle 0|\right). \qquad (7.23)$$

It is a mixed state, with a possibility of no photons because of dark counts on A, and of a statistical mixture of single photon states. When there are no dark counts, the heralded state does not depend on the quantum efficiency of the detection, whereas the success rate to produce, equal to η, depends on it.

Finally, let us note that if the single photon detector is mode selective, $\hat{\Pi}_{click} = \eta|1 : \ell_0, A\rangle\langle 1 : \ell_0, A|$, the heralded state, equal to $|1 : \ell_0, B\rangle\langle 1 : \ell_0, B|$, is the pure single mode single photon state $|1 : \ell_0, B\rangle$.

7.4 EPR Quantum Correlations in Discrete Variable Systems

In 1935, Einstein, Podolsky, and Rosen (EPR) published their famous "EPR" paper [62], in which they pointed out the puzzling character of the quantum correlations between two remote systems that have interacted in the past and are described by an entangled state. They used this situation to question the "completeness" of quantum mechanics.

They took as an example a continuous variable entangled state featuring position and momentum correlations, a case that we will consider in Section 9.5. A few years later, Bohm [21] extended the EPR argument to systems characterized by discrete variables, such as spins or polarization. We will consider the simplest of such systems, namely the two qubit system, in the following sections of this chapter.

7.4.1 Entangled States of Two Qubits

Let us assume that two spin 1/2 particles have been created in the Bell state $|\Psi_-\rangle$. From an angular momentum point of view, this state is actually the singlet state $|S = 0, M_S = 0\rangle$ of the total spin $\hat{\mathbf{S}} = \hat{\mathbf{S}}_A + \hat{\mathbf{S}}_B$. Being a state of angular momentum 0, it is invariant with respect to rotations, so that

$$|\Psi_-\rangle = \frac{1}{\sqrt{2}}(|+_x -_x\rangle - |-_x +_x\rangle) = \frac{1}{\sqrt{2}}(|+_y -_y\rangle - |-_y +_y\rangle), \tag{7.24}$$

where $|\pm_x\rangle$, for example, is the eigenstate of $\hat{S}_x$ with eigenvalue $\pm 1/2$.

On Alice's side, the reduced quantum state is $\rho_A = (|+\rangle\langle+| + |-\rangle\langle-|)/2$, which contains the minimal information about the state: A measurement of the angular momentum $\hat{\mathbf{S}}_A$ projected over axes Ox, Oy, or Oz will give random results so that $\langle\hat{S}_{xA}\rangle = \langle\hat{S}_{yA}\rangle = \langle\hat{S}_{zA}\rangle = 0$, and the same is true on Bob's side.

As shown in Section 7.1, the pure entangled state $|\Psi_-\rangle$ is characterized by very strong correlations between the measurements performed by Alice and Bob: An argument based on the state collapse postulate shows that when Bob measures $\hat{S}_{xA}$ and finds 1/2, he knows for sure that Alice will find the value $-1/2$ for $\hat{S}_{xB}$, and when Bob finds $-1/2$, he knows for sure that Alice will find the value $+1/2$. The same is true for their measurements of the other components of the spin components of the two particles like and $\hat{S}_{yA}$ and $\hat{S}_{yB}$). This implies that $\langle\hat{S}_{xA}\hat{S}_{xB}\rangle = \langle\hat{S}_{yA}\hat{S}_{yB}\rangle = \langle\hat{S}_{zA}\hat{S}_{zB}\rangle = -1/4$, and for the normalized correlation functions (Equation (5.45)),

$$c(S_{xA}, S_{xB}) = c(S_{yA}, S_{yB}) = c(S_{zA}, S_{zB}) = -1. \tag{7.25}$$

There is a perfect anticorrelation between the measurements. More generally, it is a simple exercise to show that

$$c(\mathbf{S}_A.\mathbf{e}_A, \mathbf{S}_B \cdot \mathbf{e}_B) = -\mathbf{e}_A.\mathbf{e}_B, \tag{7.26}$$

where $\mathbf{e}_A$ and $\mathbf{e}_B$ are two unit vectors of arbitrary direction.

7.4.2 Bell Inequality

As already noted in Section 5.8, classical correlations between remote objects also exist. Let us adapt the example given above for the two qubit case. Our operator is actually a shoemaker fabricating left and right shoes, which are either red or blue. Side and color are two two-valued properties of the shoes that we call, respectively, S and S', with the code $S = 1$ for red, $S = -1$ for blue, $S' = 1$ for left, $S' = -1$ for right. The shoemaker has a supply containing the same number of blue and red pairs of shoes, from which he chooses randomly a pair. He sends then one shoe of the pair to Alice, and the other to Bob. When Alice opens her parcel and finds for example a right red shoe, she knows for sure that Bob will receive a parcel containing a left red shoe, whatever the distance between them and the order of their openings.

Let us note by an upper bar the average over many draws. One has $\bar{S}_A = \bar{S}_B = \bar{S'}_A = \bar{S'}_B = 0$ and $\overline{S_A S_B} = 1$, $\overline{S'_A S'_B} = -1$ and therefore a perfect correlation (or anticorrelation) between the measurements: the physical situation seems to be very similar with the spin case or EPR case.

Bell [15] made the striking discovery that the correlations between classical systems, like the ones existing in the example of the shoemaker given previously, are actually subject to a limit. He considered the following combination of measurement results:

$$B = S_A S_B + S_A S'_B + S'_A S_B - S'_A S'_B, \tag{7.27}$$

Being a combination of four ± 1, the mean value $\overline{B}$ is a priori comprised between -4 and 4. But the bound is actually tighter. Let us write B as

$$B = S_A(S_B + S'_B) + S'_A(S_B - S'_B). \tag{7.28}$$

If Bob finds identical values for S_B and S'_B in his draw, then $B = \pm 2S_A = \pm 2$. If he finds opposite values then $B = \pm 2S'_A = \pm 2$. Then $\overline{B}$ is an average of values ± 2 and therefore comprised between -2 and 2:

$$-2 \leq \overline{B} = c(S_A, S_B) + c(S_A, S'_B) + c(S'_A, S_B) - c(S'_A, S'_B) \leq 2. \tag{7.29}$$

This is the famous *Bell inequality* [15, 41] (or better, Bell–Clauser–Horne–Shimony–Holt inequality) that constrains all classical correlations. Note that $\overline{B}$ is equal to 2 in the example with the shoes, because $\overline{S_A S'_B} = \overline{S'_A S_B} = 0$.

Note that in the present case of classical correlations, the properties to be measured are fixed for each draw, without possibility of change between the generation and the detection, and are randomly varying from one draw to the other. These properties are indeed locally carried by some kind of label (left-right, red-blue) that is randomly changing and not known by the experimentalists, hence the name "hidden variable" given to this label.

7.4.3 Quantum Correlations Violate Bell Inequality

Let us go back to quantum physics, and more precisely to the two spin 1/2 case, and consider that the two measurements made by Alice are spin component measurements $\mathbf{S}_A.\mathbf{e}_A$ and $\mathbf{S}_A.\mathbf{e}'_A$ with unit vectors $\mathbf{e}_A$ and $\mathbf{e}'_A$ taken along the Ox and Oy directions, and the two measurements made by Bob are spin component measurements $\mathbf{S}_B.\mathbf{e}_B$ with unit vectors $\mathbf{e}_B$ and $\mathbf{e}'_B$ taken along the diagonal and antidiagonal directions of the xOy plane. Using relation (7.26), we get

$$c(\mathbf{S}_A.\mathbf{e}_A, \mathbf{S}_B \cdot \mathbf{e}_B) + c(\mathbf{S}_A.\mathbf{e}'_A, \mathbf{S}_B \cdot \mathbf{e}_B) + c(\mathbf{S}_A.\mathbf{e}_A, \mathbf{S}_B \cdot \mathbf{e}'_B) - c(\mathbf{S}_A.\mathbf{e}'_A, \mathbf{S}_B \cdot \mathbf{e}'_B)$$
$$= \frac{1}{\sqrt{2}} + \frac{1}{\sqrt{2}} + \frac{1}{\sqrt{2}} - \left(-\frac{1}{\sqrt{2}}\right) = 2\sqrt{2}. \tag{7.30}$$

This quantity, which is the quantum analogue of the classical $\overline{B}$ defined in Equation (7.28), is larger than 2 for this specific choice of geometry: *Probabilities derived from quantum mechanics lead to a violation of the Bell inequality*. Bell has thus shown that there are physical situations where quantum mechanics predicts correlations between Alice and Bob measurements that cannot be attributed to correlations existing between classical remote objects having interacted in the past.

Bell's discovery triggered a series of experiments that extended over several decades ([82], complement 5.C) that can be all sketched out in Figure 7.4: A two-photon source S produces pairs of polarization entangled photons. The polarization states are measured by Alice and Bob using polarizers a and b that can be rotated around the propagation axis. Alice and Bob measure the polarization of the photons using the two different directions of polarization. They gather their results in a common place CM, calculate the correlations between their measurements, and derive the experimental value of quantity B from Equation (7.26).

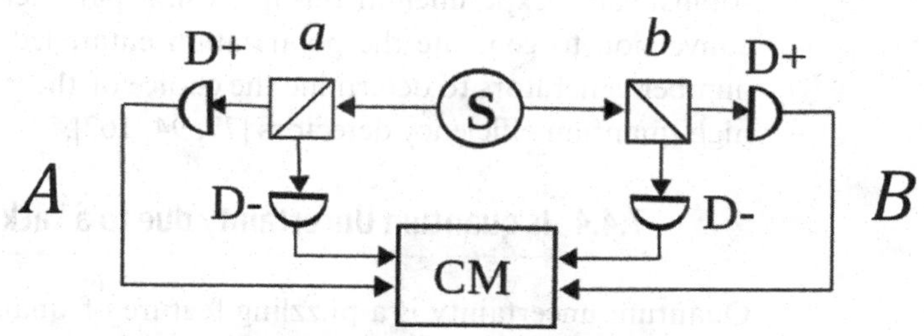

Fig. 7.4 Diagram of an experiment testing Bell inequality N entangled polarization photon states: S: entangled state source; a, b: adjustable polarizers; D: single photon detectors; CM: correlation measurement.

The first experiment was performed by Clauser and Freedman [70] using cascaded spontaneous emission in Calcium atoms excited by a discharge (see Section 6.1.2), with an acquisition time of 200 h. It showed a clear violation of the Bell inequality. Subsequent experiments, using cascaded spontaneous emission excited by tunable lasers, then parametric down-conversion in a nonlinear crystal, reduced the acquisition time more and more drastically and increased the signal-to-noise ratio [9, 40, 123]. They were all unambiguously in favor of quantum mechanics.

A natural question that remained was whether the verification of quantum mechanics was an artifact created by possible "loopholes" in the experiments. Bell himself considered one, called the locality loophole, as particularly important: It is obvious that if there is some sort of communication among the parties, whatever happens on one side can be communicated to the other side. In this case one finds that $\bar{B} \leq 4$ and a local hidden variable model can easily account for the result predicted by quantum theory ($\bar{B} = 2\sqrt{2}$). In order to test quantum theory more stringently, one needs to close this loophole and the way to do it is to rely on *relativity* and take distant enough Alice and Bob locations to guarantee that there will be no causal influence from one particle to the other. This landmark experiment was first done by A. Aspect in 1982 [9]. In this experiment, the direction of polarizers was changed "on the flight" rapidly enough, and Alice and Bob detectors distant enough, that Alice and Bob measurements could not be causally connected. Aspect's experiments showed "beyond reasonable doubt" that local hidden variable theory cannot explain the observations but that quantum mechanics could. Many other loopholes remained open: In Aspect's experiments the choice of the polarizer directions was effected by a deterministic process, and thus the information about the measurement choice pre-existed to the measurement. This could be exploited by a local hidden variable model to explain the results. These kinds of "conspiracy models" seem very unlikely from a physical point of view, and thus defeat the purpose of finding a reasonable classical explanation. Regardless, due to their importance they received well-deserved attention from the community and this "freedom of choice" loophole, as well as several others, were progressively closed in the subsequent decades by A. Zeilinger and others, with the help of increasingly sophisticated experimental set-ups, using parametric spontaneous down conversion to generate the polarization entangled photons, true random number generators to determine the choice of the polarizer directions, and high quantum efficiency detectors [77, 94, 162].

7.4.4 Is quantum Uncertainty due to a Lack of Knowledge?

Quantum uncertainty is a puzzling feature of quantum mechanics, a limitation to our knowledge of nature. A important question then arises: Is this uncertainty intrinsic, or is it provisional and due to the fact that quantum mechanics may not completely describe microscopic objects? We have seen that the introduction of local physical supplementary parameters leads to consequences that are in contradiction with quantum mechanics predictions. In addition, experiments have validated quantum mechanics predictions and not the models based on local hidden parameters. The explanation of quantum randomness by our ignorance of these parameters is rejected. We must admit that quantum uncertainty is intrinsic, and that we cannot predict, for example, the exact time when a photon will be spontaneously emitted by an excited atom and, more generally, the exact time of a quantum jump.

7.5 An Example of Quantum Information Processing with Entangled States: Database Search

Entangled states are characterized by correlations that are different from the classical ones. They can be used to perform new tasks, especially in the domain of quantum information and quantum computing. More detailed accounts of these applications can be found in [13, 112, 136].

In this section we give a striking example of improvement by quantum means of a widely used protocol: the search of a given entry in a list, for example, in a telephone book. This is the simplest example showing that the rules of quantum mechanics can lead to the resolution of a problem in a smaller number of operations than by using the classical rules. It is known as the *Grover algorithm* [136].

Consider a given database, for example a list of N names X_i in alphabetical order, and corresponding telephone numbers $N(X)$. The problem to solve is the "inverse problem," namely to find the name X_i corresponding to a given telephone number N_i, so that $N_i = N(X_i)$. This is not straightforward, especially for big databases, as the telephone numbers appear in a disordered way. To simplify the following explanation, we suppose that N is a power of 2: $N = 2^p$. We will then write names as binary numbers. For example, for $p = 3$ the telephone book looks like:

$$\begin{aligned}
X_0 &= 000 \to N_0 \\
X_1 &= 001 \to N_1 \\
X_2 &= 010 \to N_2 \\
X_3 &= 011 \to N_3 \\
X_4 &= 100 \to N_4 \\
X_5 &= 101 \to N_5 \\
X_6 &= 110 \to N_6 \\
X_7 &= 111 \to N_7.
\end{aligned} \tag{7.31}$$

The list of names is ordered and easily manageable. The list of numbers looks random. The classical way consists of browsing through the name list in a random or deterministic way. Since one can find N_i by chance at any place in the list with equal probability, the average number of trials leading to the searched X_i is $N/2 = 2^{p-1}$.

The quantum algorithm permitting us to solve the search problem consists of:

- A Hilbert space of "quantum registers" of dimension $p + 1$ spanned by a tensor product of $p + 1$ qubits, each qubit being either $|0\rangle$ or $|1\rangle$, allowing us to use binary numbering. The first p qubits are used to write the binary number x, which is associated with the name X in the list (7.31) as the qubit quantum state $|x\rangle$. For example, $|x\rangle = |100\rangle$ is associated with the name X_4 in our telephone book. We will call $|x_i\rangle$ the quantum state associated with the searched name X_i. The last qubit, called the "ancillary qubit" $|\phi\rangle$ will be used in the computation.
- A unitary quantum evolution operator, $\hat{U}_{N_i}$, acting on this Hilbert space, that can be physically implemented on the qubits and depends on the given telephone number N_i. Like any quantum map, it is a *linear operator obeying the superposition principle*, a property that is not necessary for the different gates used in classical computing. This operator will be used repeatedly in the calculation. The number of iterations of $\hat{U}_{N_i}$ corresponds to the number of trials in the classical search.
- An initial state $|\psi_i\rangle$ such that $(\hat{U}_{N_i})^m|\psi_i\rangle \simeq |x_i\rangle \otimes |\phi\rangle$ gives the solution x_i of the problem after m iterations.

More precisely, $\hat{U}_{N_i}$ is the product of two unitary evolutions $\hat{U}_{N_i} = \hat{U}_2\,\hat{U}_{1,N_i}$:

- $\hat{U}_{1,N_i}$ acting on a state $|x\rangle \otimes |0\,or\,1\rangle$ gives the same state if $x \neq x_i$, and gives $|x\rangle \otimes |1\,or\,0\rangle$ if $x = x_i$. This transformation, which depends on the input "question", is straightforward to implement as it uses the telephone book in the "right way," looking for the xth name in the list and checking whether it is the right one or not. One can find in Ref. [136] the details of the series of quantum gates needed to implement such a unitary transformation.
- The second evolution operator $\hat{U}_2$ is equal to $(2|s\rangle\langle s| - \hat{1}) \otimes \hat{1}$, where $|s\rangle$ is a symmetrical linear combination of the N different quantum states $|x\rangle$:

$$|s\rangle = \frac{1}{\sqrt{N}} \sum_{x=1}^{N} |x\rangle. \tag{7.32}$$

It is simple to show that $\hat{U}_2$ is unitary. Acting on a given quantum state $|x'\rangle \otimes |\phi\rangle$, it gives

$$\hat{U}_2|x'\rangle \otimes |\phi\rangle = \left(\frac{2}{N}\sum_{x=1}^{N} |x\rangle - |x'\rangle\right) \otimes |\phi\rangle, \tag{7.33}$$

so that, apart from being the identity for the last qubit $|\phi\rangle$, its action on a general state $|\psi\rangle = \sum c_{x'}|x'\rangle$ is such that

$$\langle x'|\hat{U}_2|\psi\rangle = \frac{2}{N}\sum_{x=1}^{N} c_x - c_{x'}, \tag{7.34}$$

or, written in a different way,

$$\frac{1}{2}\left(\langle x'|\hat{U}_2|\psi\rangle + \langle x'|\psi\rangle\right) = \bar{c}_x, \tag{7.35}$$

where $\bar{c}_x = \frac{1}{N}\sum_{x=1}^{N}$ is the mean value of the quantum states coefficients. The transformation $\hat{U}_2$ therefore consists of replacing each coefficient $c_{x'}$ of a quantum state $|\psi\rangle$ by its symmetrical value with respect to the mean $\bar{c}_x$.

Now we use a specific input state as Ψ_i. The input state is chosen as $|\psi_i\rangle = |s\rangle \otimes |\phi'\rangle$, where $|s\rangle$ is the symmetrized state already mentioned and $|\phi'\rangle = \frac{1}{\sqrt{2}}(|0\rangle - |1\rangle)$. The state $|s\rangle$ is obviously entangled with respect to the different qubits composing the quantum register. One notes that $\hat{U}_{1,N_0}$ acting on $|x\rangle \otimes |\phi'\rangle$ gives the same state if $x \neq x_0$ and minus this state if $x = x_0$.

The calculation consists of applying the product of the two previous transformations m times to get the final state $|\psi_f\rangle = (\hat{U}_2\hat{U}_1)^m|\psi_i\rangle$.

Let us consider the effect of these successive transformations when $N = 8$ and $p = 3$ in the expression (7.36). Each line contains the coefficients of the decomposition of the state over the 3-qubit-register basis starting from the initial state and after each step. We assume, for example, that the searched name is the fourth in the list:

$$\left|\begin{array}{cccccccc} 1 & 1 & 1 & 1 & 1 & 1 & 1 & 1 \\ 1 & 1 & 1 & -1 & 1 & 1 & 1 & 1 \\ 4/8 & 4/8 & 4/8 & 20/8 & 4/8 & 4/8 & 4/8 & 4/8 \\ 4/8 & 4/8 & 4/8 & -20/8 & 4/8 & 4/8 & 4/8 & 4/8 \\ -2/8 & -2/8 & -2/8 & 22/8 & -2/8 & -2/8 & -2/8 & 2/8 \\ -2/8 & -2/8 & -2/8 & -22/8 & -2/8 & -2/8 & -2/8 & 2/8 \end{array}\right| \times 1/\sqrt{8} \tag{7.36}$$

One observes that the coefficient of the fourth column (the right "answer") grows step after step. After three steps the right answer has a probability weight of 95%. One can show that the coefficient "points" to the right answer after an average number of steps equal to $\sqrt{N} = 2^{p/2}$ instead of $N/2$, which is quite a speed-up when N is very large. For example, a search of one million items needs one thousand quantum operations instead of 500,000 classical browsing steps.

Roughly speaking, the quantum advantage is due to the fact that one presents as an input the state $|s\rangle$, which is the linear superposition of all possible solutions. The algorithm processes all the possibilities simultaneously, in a kind of parallel way. What is exploited here are the two quantum resources of entanglement and superposition.

There are other algorithms for which the quantum advantage is even larger. This is, for example, the case of the *Shor algorithm* [136], a quantum algorithm able to factorize numbers in polynomial time instead of subexponential time for the corresponding classical algorithm.

7.6 Quantum Teleportation

Quantum teleportation, which nowadays is a real physical task implemented in the laboratory and routinely used as a tool for quantum computation and quantum communications, should more properly be named "quantum scanner printer." A classical scanner reads a page of unknown text at Alice's position and encodes it in a communication channel that transmits the information by internet to another scanner printer located at Bob's position. Bob's scanner decodes the scan and prints the recovered text on a blank page. In classical physics, there is no fundamental limitation to the accuracy with which the reading is made, and Bob's copy can be perfectly identical to the input text. This is not the case in quantum physics because of the no-cloning theorem (see Appendix C), which forbids the existence of two strictly identical copies of an unknown input state. Quantum teleportation is a technique for producing in Bob's place a perfect copy of Alice's quantum state. It is not quite what we would expect from science fiction – it is a slower-than-light transfer of information, not of matter.

Consider more precisely the following situation: Alice receives a qubit state unknown to her, $|\psi_{Ain}\rangle = \alpha|-\rangle + \beta|+\rangle$, that she wants to send to Bob. Alice can telephone to him but there is no mail service between them. In other words, they share a classical communication channel but no way to transfer the particle itself. According to classical physics, Alice could observe her state in detail and describe it to Bob over the phone so he could reconstruct it on his side. In this way both would have a perfect copy of it. But we know that quantum mechanics forbids cloning (see Appendix C)! A necessary condition to bypass the no-cloning theorem is that Alice sacrifices her state in the process in order for Bob alone to have it. It turns out that in order to accomplish the task, Alice and Bob will need to *share an extra pair of entangled qubits*. As shown in Figure 7.5, in this sense, entanglement will be *the* resource consumed in this *quantum communication channel.*

More precisely, we assume that Alice and Bob have previously shared the entangled Bell state $|\Psi^-_{A,B}\rangle = \frac{1}{\sqrt{2}}(|+_A -_B\rangle - |-_A +_B\rangle)$. The whole system is then described by a three-qubit state (two on Alice's side, one on Bob's) $|\psi_{Ain,A,B}\rangle = |\psi_{Ain}\rangle \otimes |\Psi^-_{-AB}\rangle$. Let us introduce the four Bell states of the two qubits on Alice's side:

$$|\Psi^-\rangle_{A_{in},A} = \frac{1}{\sqrt{2}}(|+_{Ain}\rangle|-_A\rangle - |-_{Ain}\rangle|+_A\rangle), \tag{7.37}$$

$$|\Psi^+\rangle_{A_{in},A} = \frac{1}{\sqrt{2}}(|+_{Ain}\rangle|-_A\rangle + |-_{Ain}\rangle|+_A\rangle), \tag{7.38}$$

$$|\Phi^-\rangle_{A_{in},A} = \frac{1}{\sqrt{2}}(|+_{Ain}\rangle|+_A\rangle - |-_{Ain}\rangle|-_A\rangle), \tag{7.39}$$

$$|\Phi^-\rangle_{A_{in},A} = \frac{1}{\sqrt{2}}(|+_{Ain}\rangle|+_A\rangle + |-_{Ain}\rangle|-_A\rangle). \tag{7.40}$$

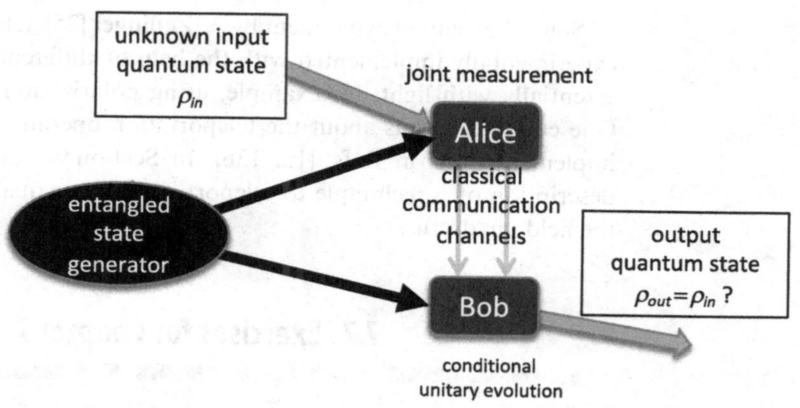

Fig. 7.5 Quantum teleportation. Alice makes measurements on a mixture between the state to teleport and the first part of an entangled state and sends the results to Bob through a classical communication channel. The teleported state results from the processing of the second part of the entangled state that is conditioned by Alice's measurement results.

The state $|\psi_{Ain,A,B}\rangle$ can also be written on the Bell state basis. A simple calculation gives the decomposition

$$|\psi_{Ain,A,B}\rangle = \frac{1}{2}\big(|\Psi^-_{Ain,A}\rangle|\psi^1\rangle_B + |\Psi^+_{Ain,A}\rangle|\psi^2\rangle_B + |\Phi^-_{Ain,A}\rangle|\psi^3\rangle_B + |\Phi^+_{Ain,A}\rangle|\psi^4\rangle_B\big), \tag{7.41}$$

with

$$|\psi^1\rangle_B = -\alpha|-_B\rangle - \beta|+_B\rangle, \tag{7.42}$$

$$|\psi^2\rangle_B = -\alpha|-_B\rangle + \beta|+_B\rangle, \tag{7.43}$$

$$|\psi^3\rangle_B = \alpha|+_B\rangle + \beta|-_B\rangle, \tag{7.44}$$

$$|\psi^4\rangle_B = \alpha|+_B\rangle - \beta|-_B\rangle. \tag{7.45}$$

Alice makes a Bell state measurement on her two-qubit Hilbert space. The POVM associated with such a measurement consists of the projectors on the four Bell states:

$$\hat{P}_1 = |\Psi^-_{Ain,A}\rangle\langle\Psi^-_{Ain,A}| \quad \hat{P}_2 = |\Psi^+_{Ain,A}\rangle\langle\Psi^+_{Ain,A}|, \tag{7.46}$$

$$\hat{P}_3 = |\Phi^-_{Ain,A}\rangle\langle\Phi^-_{Ain,A}| \quad \hat{P}_4 = |\Phi^+_{Ain,A}\rangle\langle\Phi^+_{Ain,A}|. \tag{7.47}$$

For each of the four answers $i = 1, 2, 3, 4$ obtained on Alice's side, the state collapse postulate implies that Bob gets a well-defined state $|\psi^i\rangle_B$. These four states are connected to the state to teleport by a well-defined unitary operator $\hat{U}^i_B$ acting only on Bob's Hilbert space. Alice then sends to Bob by a classical channel the result of her measurement $i = 1, 2, 3, 4$. Finally, Bob applies to his part of the entangled state the unitary transformation $\hat{U}^i_B$ and gets a state that is exactly the input state $|\psi_{Ain}\rangle$. Note that Alice's measurement destroys the input unknown qubit state, so that the no-cloning theorem is not violated.

Since the seminal experiment by A. Zeilinger [25], teleportation has been experimentally implemented with the help of different quantum systems, essentially with light, for example, using polarization states of photons. One can find details about the teleportation operation and experimental implementations in Refs. [13, 136]. In Section 9.7, one can also find a description of a technique to teleport continuous quantum variables like the field quadratures.

7.7 Exercises for Chapter 7

Exercise 7.1 Consider four binary variables such that Q, R, S, $T = \pm 1$. Convince yourself that the Bell inequality ($\langle QS + RS + RT - QT\rangle < 2$) violation demands a high positive logical correlation between Q and S, between R and S, and between R and T, while demanding a highly negative correlation between Q and T. Show that there is a formal contradiction by computing its truth table.

Exercise 7.2 Consider the following practical gambling scenario.

Alice and Bob enter a casino. The casino will flip a fair coin for Alice and another fair coin for Bob simultaneously. Alice and Bob can only see their own coin and not the other one. Each of them has a red and a green button. They are asked to press one button after each flip. If they choose different colors, they play. If they play, they earn one euro if they both got heads (HH) and they loose one euro in any other of the three remaining situations (TT, HT, TH). If they choose the same colors then they don't play, no money is lost or won, and the game restarts. A new pair of coins is flipped.

(a) Convince yourself that if they cheat by smuggling cellphones into the casino they can win by communicating.
(b) Show that there is no classical strategy to beat the casino if no communication is allowed.
(c) Devise a quantum strategy to beat the casino. Can the quantum strategy be used when communication is forbidden by relativity?

Exercise 7.3 Consider a single beam of photons traversing two successive polarizers named Alice and Bob. Each polarizer can have its transmission axis set to one of two angles $\theta_A = \{0°, 45°\}$ and $\theta_B = \{-22.5°, 22.5°\}$. The observables are designated Q (0°) and R (45°) for Alice, S (−22.5°) and T (22.5°) for Bob, and the observation result is labeled +1 if the particle is transmitted and −1 if it is not.

(a) Compute $\langle QS + RS + RT - QT\rangle$. Assume the most general probability distribution $Proba(q, r, s, t)$ (which stands for the probability of having a realization of Q, R, S, T random variables with values q, r, s, and t) and verify the Bell inequality for this chronologically ordered pair of measurements over the same particle. Is there any entanglement? Discuss.

(b) Show that the quantum prediction for this quantity if the initial beam is circularly polarized is $\langle QS+RS+RT-QT\rangle = 2\sqrt{2}$. What if it is completely unpolarized? Can you measure this quantity with a single photodiode at the output of the last polarizer?

(c) Show that because the measurements are causally connected one can build a trivial hidden-variable-theory to get $\langle QS + RS + RT - QT\rangle = 4$. Why is the inequality then not violated as strongly as it in principle could be?

(d) Discuss the importance of the non-locality condition in the experiments violating Bell inequalities.

Exercise 7.4 Consider the prototypical Bell inequality violation. Write a system of equations constraining the most general probability distribution $Proba(q, r, s, t)$ to fulfill the quantum prediction and show that the system of linear equations is incompatible.

Exercise 7.5 Show from classical probability theory that if one has two bits a and b with bidirectional correlation $Proba(a|b) = Proba(b|a)$, the individual systems are completely random and that $Proba(a \Rightarrow b) = Proba(b \Rightarrow a)$. How does this observation apply to entangled spins and our understanding of causality (see exercises in Chapter 5)?

Exercise 7.6 Gull's theorem, an alternative to Bell's theorem, states that "the EPR-B correlations cannot be programmed in two independently running computers" (the -B stands for David Bohm who extended the continuous variable argument of Einstein, Podolsky, and Rosen to spin 1/2). Prove it as follows:

(a) Convince yourself that in the presence of a pair of particles in a singlet state the probability of measuring a pair of +1 when Alice measures the spin direction along θ_A and Bob measures along θ_B is $Proba(+1, +1|\theta_A, \theta_B) = \frac{1}{4}(1 - \cos(\theta_A/2 - \theta_B/2))$.

(b) Convince yourself that the probability of measuring +1 on either side and regardless of the angle is 1/2.

(c) Consider the computer functions $f_A^{(m)}(\theta_A)$ and $f_B^{(m)}(\theta_B)$ describing the mth result (± 1) for Alice and Bob's measurements. Convince yourself that, if a total of M measurements is to be made, the value of the functions for any given angle needs to be +1 on about $M/2$ of the trials.

(d) Observe that for $\theta_A = \theta_B$, the probability of measuring a pair of +1 is zero and thus the function must be deterministic and respect $f_A^{(m)}(\theta) = -f_B^{(m)}(\theta)$.

(e) Convince yourself that for any three angles, one has, for at least two among them – say θ' and θ'' – about $M/8$ trials for which $f_A^{(m)}(\theta') = f_A^{(m)}(\theta'') = +1$.

(f) Show that during these $M/8$ trials the computer functions must be constrained to simultaneously satisfy

$$Proba(+1,+1|\theta,\theta') = \frac{1}{4}(1-\cos(\theta/2-\theta'/2))$$

and

$$Proba(+1,+1|\theta,\theta'') = \frac{1}{4}(1-\cos(\theta/2-\theta''/2)).$$

Show that this is not possible for arbitrary θ.

8 Experiment: Continuous Quantum Fluctuations

8.1 From Discrete Random Jumps to Continuous Quantum Noise

Let us consider a semiconductor photon detector, such as those described in Section 1.1, that is illuminated by a weak light source. It delivers a photocurrent $i(t)$ made of "clicks" occurring at random times (Figure 8.1) that witness the presence of individual quantum jumps for the electrons in the detector from the valence band to the conduction band. These clicks have a constant height i_0 and a duration τ set by the finite response time of the detector. If one increases the light intensity, the density of clicks increases, but not their height. As the random clicks get closer to each other, they may overlap: The photocurrent takes values 0, i_0, and sometimes $2i_0$. Further increasing the light intensity, the clicks always overlap and the photocurrent takes values that are multiples of i_0. The randomness of the occurrence of quantum jumps translates into the randomness of the photocurrent values. At high light intensities (1 mW of visible light corresponds to roughly 10^{15} photons per second), it is not possible to see the discrete steps due to light energy quantization. The photocurrent intensity $i(t)$ undergoes continuously varying temporal *quantum fluctuations*, also called *quantum noise*, around a nonzero mean value $\langle i(t) \rangle$.

Continuously varying quantum noise, in which the exact value of the photocurrent intensity is unpredictable, is another avatar of the uncertainty inherent to the quantum world. Note that quantum noise affects quantities other than the light intensity such as, for example, the phase of a monochromatic electromagnetic field, the position and momentum of mechanical harmonic oscillators, or the flux and charge in a superconducting circuit.

If the quantum efficiency of the detector is close to unity, every incoming photon gives birth to a single electron in the photodiode electrical circuit: The photocurrent intensity $i(t)$ is proportional to the number of incident photons on the detector during the measurement time ΔT, though the photons are not isolated one by one.

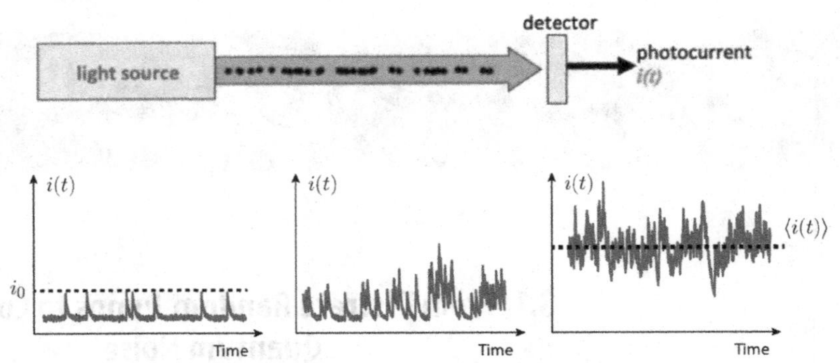

Fig. 8.1 Photocurrent $i(t)$ for different levels of the photon flux. (Left) At low flux, one observes at random times "clicks" of constant heights, which (Center) begin to overlap for intermediate fluxes; (Right) At high flux, all clicks overlap, so that one observes random temporal variations of the photocurrent intensity (the vertical scale of the last recording on the right is strongly reduced with respect to the two left ones).

8.2 Experimental Manipulation of Quantum Noise

8.2.1 Quantum Noise Reduction

Let us briefly describe one of the first experiments that generated light with tailored quantum fluctuations: A frequency doubled YAG laser produces a coherent beam at 0.53 μm. It pumps an optical parametric oscillator (OPO) containing a $LiNbO_3$ nonlinear crystal, which is chosen in such a way that the signal and idler photons generated by parametric down conversion have the same wavelength, 1.06 μm, and the same polarization. One is therefore in the configuration of degenerate down conversion, which leads to squeezed vacuum state generation (Section I.2). The output beam from the OPO is analyzed by a balanced homodyne detection set-up (Section 9.4) using the fundamental wave at 1.06 μm from the YAG laser as the local oscillator. The difference between the two photocurrents gives a signal proportional to the instantaneous value of a phase quadrature component $\hat{X}_\theta$. This signal is sent to a spectrum analyzer, which Fourier analyzes it and directly displays its variance $V(\theta)$ at a chosen Fourier frequency Ω (usually a few MHz). The phase θ of the local oscillator can be scanned by a piezoelectric device. Figure 8.2 shows that for some values of θ the variance $V(\theta)$ is less that the vacuum state, or shot noise, variance (dotted line obtained by blocking the light coming from the OPO), whereas it is above this line for intermediate values: This is the signature of a squeezed vacuum state. Its purity, equal to the product of the minimum and maximum values of the quantum noise, turns out to be close to 1 before it is degraded by losses and the finite quantum efficiency of the detectors.

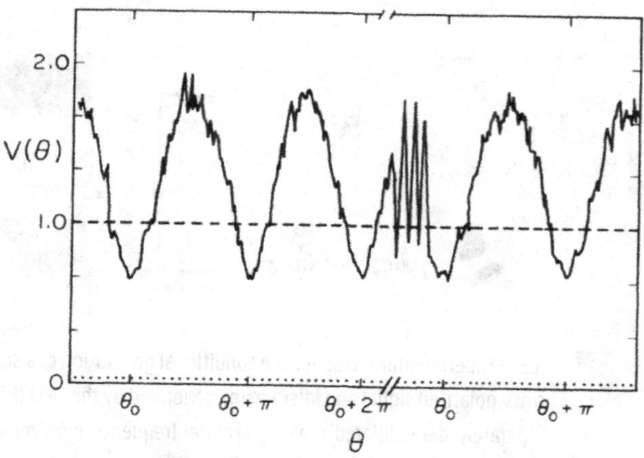

Fig. 8.2 Variance $V(\theta)$ of the quantum noise measured on the field generated by a sub-threshold, frequency degenerate, parametric oscillator using homodyne detection, as a function of the phase θ of the local oscillator. Squeezing is present when the experimental trace goes below the dotted line giving the vacuum noise level. From Ling-An Wu et al. Generation of squeezed states by parametric down conversion, *Phys. Rev. Letters*, 1986 [180].

In such a pioneer experiment, the quantum noise reduction is rather weak (by a factor 2 on the variance). A much stronger noise reduction, by factor 30 in [169], was obtained in the 2010s. This kind of reduced quantum noise source has even been successfully implemented on the gravitational wave detectors (see Section 10.2 and reference [2]) to improve their sensitivity.

8.2.2 Conditional Generation of Sub-Poissonian Light from Twin Beams

Conditional measurements can be also performed in the continuous variable regime. Here, we give a brief account of one of the first experiments showing this possibility by generating a conditional state with reduced intensity fluctuations from intensity correlated beams in a way analogous to the conditional generation of a single photon state from the correlated two-photon state (Section 5.5).

Figure 8.3 shows the set-up and some results. A doubled YAG laser pumps an optical parametric oscillator, inside which temporally correlated, cross-polarized signal and idler "twin" photons are generated. Above the threshold, the system generates two intense (mW) "twin" beams that are separated by a polarizing beamsplitter and have almost identical intensity fluctuations: Measuring the instantaneous intensity fluctuation of the idler beam gives information about the instantaneous intensity of the idler beam, with an uncertainty that is smaller than the shot noise. One keeps the recorded values of the signal intensity conditioned by the fact that the idler intensity falls inside a band much narrower that its standard deviation. This simple technique allows us to reduce the intensity noise of the signal beam

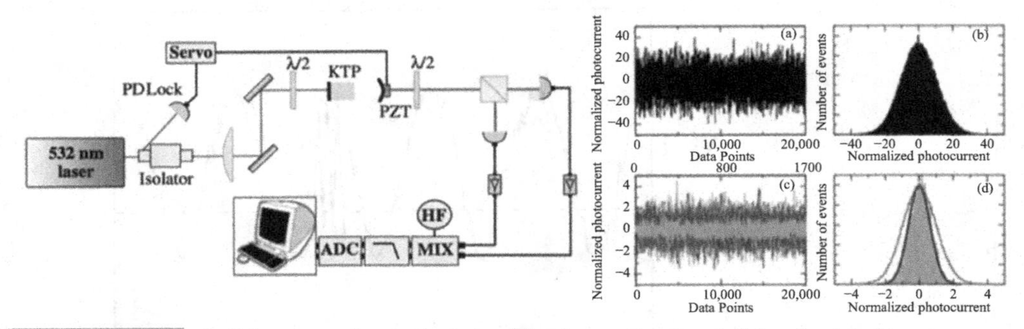

Fig. 8.3 (Left) Experimental set-up for the conditional generation of a sub-Poissonian state. The cross-polarized signal and idler beams generated by the OPO (KTP) above threshold are detected separately, demodulated at a given Fourier frequency, digitized and simultaneously recorded on a computer. (Top right) Signal beam instantaneous fluctuations, with the corresponding histogram. (Bottom right) Unselected (broad fluctuations) and conditionally selected idler beam data (reduced fluctuations), with the corresponding histogram, displaying a reduced variance of the selected signal below the standard quantum noise level. From J. Laurat et al. Conditional preparation of a quantum state in the continuous variable regime: Generation of a sub-Poissonian state from twin beams, *Phys. Rev. Letters*, 2003 [110].

64% below shot noise from twin beams exhibiting 82% of noise reduction in the intensity difference.

8.3 Squeezing of Motion in a Micromechanical Resonator

The position of a massive, harmonically bound object is a continuous variable characterized by two quadrature operators $\hat{X}_1$ and $\hat{X}_2$. As explained in Appendix F, these operators do not commute, and their variances obey a Heisenberg inequality:

$$\mathrm{Var}(X_1)\mathrm{Var}(X_2) \geq \left(\frac{\hbar}{2m\omega_m}\right)^2 \tag{8.1}$$

In the ground state, the spatial extension of the zero point motion can be written as $\Delta x = \sqrt{\hbar/2m\omega_m}$. In a motional squeezed state, the uncertainty is reduced below this value for one quadrature at the expenses of the other quadrature, which is affected by larger fluctuations.

Figure 8.4 shows an example of motional squeezing of a micromechanical oscillator of a nearly macroscopic body (a few *micrometers*) coupled to a microwave (6.9 GHz) superconducting cavity located in a dilution refrigerator. A regular radiative sideband cooling sequence cools the oscillator close to its ground state. Here, we will not detail the experimental technique, called dissipative squeezing, that leads to a phase-sensitive cooling and, ultimately, to a squeezed state, with a squeezing measured by homodyne detection of 1.1 dB below the noise level of a coherent state (standard quantum noise).

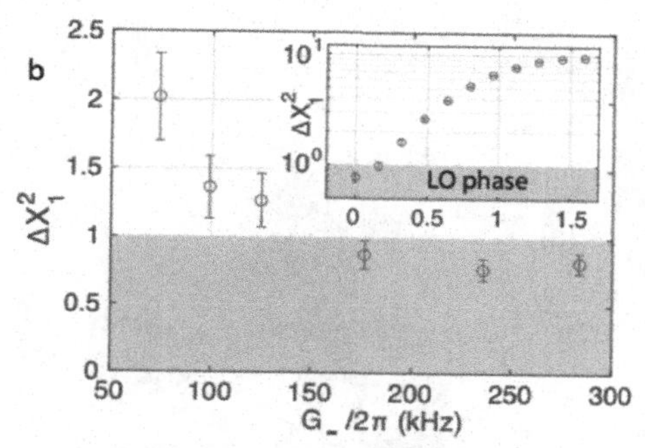

Fig. 8.4 (Inset) Quadrature variance as a function of LO phase. (Large box) Best variance as a function of pump power. The gray region below 1 means squeezing below the quantum noise limit. From J.-M. Pirkkalainen et al. Squeezing of quantum noise of motion in a micromechanical resonator, *Phys. Rev. Letters*, 2015 [145].

8.4 Nonlinear cQED Oscillators

As described in Appendix K, the strong anharmonicity of the Josephson potential allows us to study quantum optics effects in the highly nonlinear regime [176]. As we saw in Chapter 3, the strong Kerr effect accessible at the single photon level in superconducting circuits can be used to dynamically generate Schrödinger cats [182]. We now show that by adding $\hat{a}^{\dagger 2}$ and $\hat{a}^2$ terms to the Kerr Hamiltonian, the Schrödinger cats can be directly generated as steady energy eigenstates [69].

When one adds a pump field in the parametric configuration (pump frequency equal to twice the system frequency), the system is described by the nonlinear "Kerr-cat" Hamiltonian [80]:

$$\begin{aligned}\hat{H}_{KC}/\hbar &= -K\hat{a}^{\dagger 2}\hat{a}^2 + \epsilon_2\left(\hat{a}^{\dagger 2} + \hat{a}^2\right)\\ &= -K\left(\hat{a}^{\dagger 2} - \frac{\epsilon_2}{K}\right)\left(\hat{a}^2 - \frac{\epsilon_2}{K}\right) - \frac{\epsilon_2^2}{K},\end{aligned} \tag{8.2}$$

where ϵ_2 is proportional to the pump field amplitude. Figure 8.5(a) gives the phase space representation of this Hamiltonian, obtained by taking $\hat{a} \to (x + ip)/\sqrt{2}$. It appears as a double well,

$$H_{KC}(x,p)/\hbar = -\frac{K}{4}\left(x^2 + p^2\right)^2 + \epsilon_2\left(x^2 - p^2\right),$$

with a saddle point of energy $U_0 = \hbar\epsilon_2^2/K$. Equal energy contours appear in Figure 8.5(b). The factorized form of the Hamiltonian makes it evident that the coherent states $|\alpha_c\rangle$ of amplitude $\alpha_c = \pm\sqrt{\epsilon_2/K}$ are degenerate eigenstates [148] of minimal energy. These states are harmonic oscillator ground states in each of the wells. The linear combination $|C_\pm\rangle \propto |\alpha_c\rangle \pm e^{i\phi}|-\alpha_c\rangle$ are steady Schrödinger cat states with a mean number of photons $\bar{n} \sim \epsilon_2/K$.

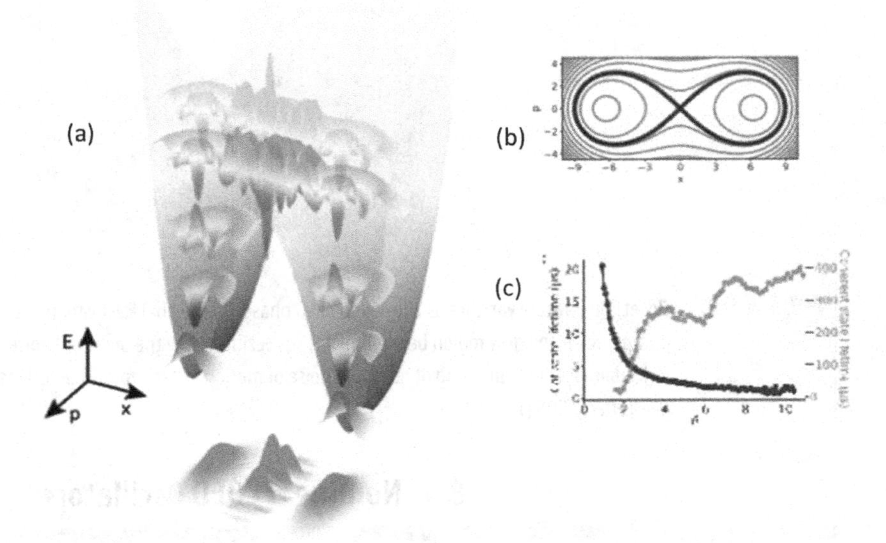

Fig. 8.5 (a) Phase-space representation of the Hamiltonian H_{KC}, including the Wigner functions of the lower lying eigenstates. The doubly degenerate ground state is formed from two coherent states localized in one or other of the wells. The steady-state Kerr-cat states are linear combinations of these coherent states, the Wigner function of which is displayed below the double well. One sees the two symmetrical bumps corresponding to the classical mixture, and the oscillating coherent interference part in the middle. The figure shows also some excited states. The lower lying ones are concentrated in one of the wells. When the energy is above the one of the saddle point separatrix (Figure 8.5(b)), the quantum states extend over the whole available phase space. (b) Equal energy contours. The bold contour is the separatrix. (c) Measurement of the (Right) coherent state lifetime and (Left) cat state lifetime as a function of $\bar{n}$. The structure of the Hamiltonian reveals itself directly in the staggered increase of coherent state lifetime while the cat state lifetime decreases smoothly with the cat size $\propto 1/\bar{n}$. From N. Frattini. *Three-Wave Mixing in Superconducting Circuits: Stabilizing Cats with SNAILs*, Yale University, PhD thesis, 2021 [69].

Energy eigenstates with an energy smaller than U_0 must be approximately degenerate since the tunnel splitting is exponentially suppressed by the barrier. To estimate the number of bound orbits inside the wells one can use Bohr's action quantization. The separatrix between the in-well states and the out-of-well states is given by a "8 shaped" curve $\mathscr{C}$ (Figure 8.4(b)) known as a Bernoulli's lemniscate. Its area is computed analytically to be

$$\frac{1}{2\pi} \oint_{\mathscr{C}} P dX = \frac{\bar{n}}{\pi} \hbar,$$

where X has the dimension of position and P the dimension of momentum. Then the semiclassical prediction is that energy pairs will effectively become degenerate every time the quantity $N = \bar{n}/\pi$ achieves an integer value. It is in agreement with the numerical diagonalization of $\hat{H}_{KC}$.

The pairwise degeneracies correspond to the decoupling of the Kerr oscillator into two independent harmonic wells and this has important consequences over the decoherence properties of the system. Decoherence processes (like the absorption of thermal photons, for example) inducing transition to excited states will not induce a well-flip if the final states are confined in a given well. The prediction is that the coherent state lifetime will be increased in *steps* as a function of $\bar{n}$, taking place every $\bar{n}/\pi$, every time a new pair of excited states "falls" into the wells. The experimental observations validate this semiclassical picture for small quantum numbers. The additional protection gained every $\bar{n}/\pi$ allows us to encode quantum information in the wells of this system, giving rise to the so-called Kerr-cat qubit. Finally, note that the lifetime measurement of the superposition of these coherent states, the cat states, shows a coherence time that decreases with the cat state mean photon number $\bar{n}$, in exactly the same way as any other Schrödinger cat state (see Chapter 9).

8.5 Exercises for Chapter 8

Exercise 8.1 Show that the Bernoulli distribution of probability for having k positive events in N binary trials $Proba(k) = \binom{N}{k} p^k (1-p)^{N-k}$ tends to the Poissonian distribution of mean value μ $\left(Proba(k) = \mu^k e^{-\mu}/k!\right)$ in the limit $N \to \infty$, $p \to 0$, and $pN \to \mu$. Here, p is the success probability for a single binary trial. What is the expected distribution for photon-shot noise?

Exercise 8.2 Show that driving a quantum harmonic oscillator by an external field, or drive, using Hamiltonian $\hat{H}_{drive} = \epsilon_1(\hat{a}^\dagger + \hat{a})$ and under single photon dissipation, stabilizes a single coherent state. Discuss the parallel with the classical scenario. How is the result modified in the case of a detuned drive?

Exercise 8.3 Compute the mean photon number and the normalization constant for cat states $|C^{\varphi}_{\pm\alpha}\rangle = (|\alpha\rangle \pm e^{i\varphi}|-\alpha\rangle)/N^{\varphi}_{\pm}$.

Exercise 8.4 Show by Taylor expanding around the minima (maxima?) of the Hamiltonian $\hat{H}_{KC}/\hbar = -K\hat{a}^{\dagger 2}\hat{a}^2 + \epsilon_2(\hat{a}^{\dagger 2} + \hat{a}^2)$ that the anharmonicity of each well is independent of ϵ_2 but that the small oscillation frequency grows linearly. Show that the energy spacing inbetween the energy levels grows linearly with ϵ_2, up to a nonlinear correction proportional to K.

(a) What is the splitting inbetween the ground state and the first excited state? How does this quantity constrain the adibaticity condition to operate the two-degenerate ground state as a qubit [80]?

(b) Compute the area inside the Bernoulli's lemniscate defined by the curve $K\alpha^2(x^2 + p^2)^2 = 2\epsilon_2(x^2 - p^2)$. Discuss the relationship between the bound state in a potentail well, its depth U_0, and its phase-space area. Gain intuition by computing the number of bound levels of a particle of mass m in a 1D square well of depth V_0 and length L. Find an example of a potential well where the amount of bound states is completely independent of U_0.

9 Continuous Variable Systems

In this chapter, we introduce tools that are specific to quantum systems depending on physical quantities that can take any value in a given interval, and thus are called continuous variable (CV) systems, in contrast to qubits and spin systems that depend on physical quantities varying by finite steps, called digital variable (DV) systems. These tools rely on the description of the CV systems in the (X, P) phase space, in particular by the Wigner function. Different systems can be treated with these tools:

1. The motion of massive particles, characterized by their position and momentum x and p. The corresponding quantum observables, $\hat{X}$ and $\hat{P}$, satisfy

$$\left[\hat{X}, \hat{P}\right] = i\hbar. \tag{9.1}$$

2. Electromagnetic fields, characterized by quadrature operators X_ℓ and P_ℓ defined in each mode ℓ. The corresponding observables, $\hat{X}_\ell$ and $\hat{P}_\ell$, satisfy

$$\left[\hat{X}_\ell, \hat{P}_{\ell'}\right] = i\delta_{\ell\ell'}. \tag{9.2}$$

3. The voltages and currents generating these electric fields also enjoy the same status. In superconducting circuits, the phase variable $\hat{\varphi}(t) = \int^t \hat{V}(t')dt'$ and the charge variable $\hat{Q}(t) = \int^t \hat{I}(t')dt'$ satisfy

$$\left[\hat{\varphi}, \hat{Q}\right] = i\hbar. \tag{9.3}$$

4. The polarization state of light is classically characterized by the three Stokes parameters S_i $(i = 1, 2, 3)$ that can be measured using simple combinations of intensity measurements. It is described at the quantum level by operators $\hat{S}_i$ [26] obeying the angular momentum commutation relations

$$\left[\hat{S}_1, \hat{S}_2\right] = 2i\hat{S}_3, \quad \left[\hat{S}_2, \hat{S}_3\right] = 2i\hat{S}_1, \quad \left[\hat{S}_3, \hat{S}_1\right] = 2i\hat{S}_2. \tag{9.4}$$

   This quantum formalism is useful at the single photon level and also for macroscopic light, which contains a great number of photons. As for the intensity of a macroscopic light beam, the granular structure of the light can be neglected in this case. The length of the vector on the Poincaé sphere is very large, so that one can consider it to vary in a continuous way. For the sake of brevity, we will not treat this case in this textbook.

The three first systems are introduced, respectively, in Appendices E, F, and K. In this chapter, we will write expressions in terms of dimensionless operators $\hat{x}$ and $\hat{p}$ (Equation (F.10)) that obey the dimensionless commutator $[\hat{x},\hat{p}] = i$ (Equation (F.13)). They can be easily translated into expressions valid for the position and momentum of a particle, for field quadratures $\hat{X}_\ell$ and $\hat{P}_\ell$, and for phase and charge variables $\hat{\varphi}$ and $\hat{Q}$.

9.1 Wigner Function

A CV system, as with any other quantum system, can be described by a density matrix ρ. One can write it on the Fock state basis, which emphasizes its corpuscular features but hides its wave features. A classical wave is characterized by a point in phase space, having phase and amplitude as polar coordinates, or quadratures X and P as Cartesian coordinates. To describe CV quantum systems in a way analogous to classical waves, Wigner introduced a real function of phase space variables x and p that contains all the quantum properties of the state, but somehow looks like a joint probability density for the two quadratures. It is the appropriate tool to bridge the gap, as much as is possible, between the classical and quantum descriptions. For this reason, the Wigner function is more "readable" than the density matrix. Other phase-space, classical looking functions exist, namely the Glauber–Sudarshan P-function, and the Husimi Q-function, that we briefly introduce in Section 9.1. For the sake of simplicity, we have chosen to concentrate most of this presentation on the Wigner function, which is the most practical one to describe CV quadrature measurements, whereas the P-function, for example, is more convenient for DV photon number or coincidence measurements. Some basic properties are found in this section. The reader interested in the detailed properties of the W-, P-, and Q- functions can refer to more specialized books and review papers [3, 11, 66, 113, 117, 139, 157, 175, 177].

9.1.1 Definition

The Wigner function of a CV quantum state that is described by its density matrix ρ is the function $W_\rho(x,p)$ of two real phase-space variables x and p defined as

$$W_\rho(x,p) = \frac{1}{2\pi}\int dp \left\langle x - \frac{p}{2}\middle|\rho\middle|x + \frac{p}{2}\right\rangle e^{ipx}, \tag{9.5}$$

where $|x\rangle$ is the eigenstate of operator $\hat{X}$ with eigenvalue x defined by Equation (F.34) for the field, and Equations (E.1) and (E.2) for the particle. It is often convenient to consider the Wigner function as a function $W(\alpha)$ of the complex phase-space variable $\alpha = x + ip$.

It can easily be shown from Equation (9.5) that W is real. It is also linear in ρ, and therefore the Wigner function has the same convexity property as the density matrix (Equation (2.46)). In particular, because any mixed state is a convex linear superposition of pure states, any mixed state Wigner function is a convex superposition of pure state Wigner functions.

The Wigner function can be also written in an analogous way in the p-representation instead of the x-representation:

$$W_\rho(x,p) = \frac{1}{2\pi}\int dx \left\langle p - \frac{x}{2}\right|\rho\left|p + \frac{x}{2}\right\rangle e^{ipx} \tag{9.6}$$

where $|p\rangle$ is the eigenstate of operator $\hat{P}$ with eigenvalue p.

Relation (9.5) can be inverted: The density matrix, in the X-representation for example, can be calculated from the Wigner function by the relation

$$\langle x|\rho|x'\rangle = \int dp\, e^{-ip(x-x')}\, W_\rho\left(\frac{x+x'}{2}, p\right), \tag{9.7}$$

whereas in a pure case, the wavefunction $\psi(x)$ can be recovered from the Wigner function by the simple formula [34]

$$\psi(x) = \frac{1}{\psi^*(0)}\int dp\, e^{-ipx}\, W\left(\frac{x}{2}, p\right), \tag{9.8}$$

where $\psi^*(0)$ is determined by the normalization condition.

The two descriptions of a CV quantum state in terms of a density matrix or of a Wigner function are therefore equivalent.

Note that relation (9.5) can also be used to define a Wigner function $W_A(x,p)$ associated with any observable $\hat{A}$ function of $\hat{X}$ and $\hat{P}$ by replacing ρ in relation (9.5) by the Hermitian operator $\hat{A}$:

$$W_A(x,p) = \mathfrak{W}(\hat{A}) = \frac{1}{2\pi}\int dx \left\langle p - \frac{x}{2}\right|\hat{A}\left|p + \frac{x}{2}\right\rangle e^{ipx}. \tag{9.9}$$

The transformation $\mathfrak{W}$, which maps an operator into a phase-space function, is known as the *Wigner transformation*, whereas the inverse transformation $\mathfrak{W}^{-1}$, relating a phase-space function to the corresponding operator is called the *Weyl transformation*. For example,

$$\mathfrak{W}\left(\hat{1}\right) = W_1(x,p) = \frac{1}{2\pi}, \quad \mathfrak{W}\left(\hat{X}\right) = W_X(x,p) = \frac{x}{2\pi},$$
$$\mathfrak{W}\left(\hat{P}\right) = W_P(x,p) = \frac{p}{2\pi}. \tag{9.10}$$

We also have

$$\langle x|\hat{A}|x'\rangle = \int dp e^{-ip(x-x')}\, W_A\left(\frac{x+x'}{2}, p\right). \tag{9.11}$$

There are alternatives to Equation (9.5) for defining Wigner functions. One is to relate it to the Fourier transform of the displacement operator $\hat{D}(x,p)$ (Equation (F.16)),

$$W_\rho(x,p) = \frac{1}{2\pi}\iint dx'dp'\,Tr\left(\rho\hat{D}(x',p')\right)e^{-i(xx'+pp')}, \tag{9.12}$$

where the double integral extends over the whole phase space (x,p).

There is another simple expression, which can be experimentally implemented, in terms of the photon number parity operator $\hat{\mathcal{P}} = \exp i\pi\hat{a}^\dagger\hat{a}$, such that,

$$\hat{\mathcal{P}}|n\rangle = (-1)^n|n\rangle, \tag{9.13}$$

or, equivalently, by

$$\hat{\mathcal{P}}|x\rangle = |-x\rangle, \quad \hat{\mathcal{P}}|p\rangle = |p\rangle. \tag{9.14}$$

One can show that [116]

$$W_\rho(x,p) = \frac{2}{\pi}Tr\left(\rho\hat{D}((x+ip)/\sqrt{(2)})\mathcal{P}\hat{D}^{-1}((x+ip)/\sqrt{(2)})\right), \tag{9.15}$$

where $\hat{D}$ is the displacement operator. It implies, in particular, that $W(0,0)$ is, within the pre-factor $2/\pi$, the mean value of the photon parity operator. $W(0,0)$ will therefore be negative in all number states $|n\rangle$ with n odd.

At last, it can be shown that a Wigner function cannot take any real value, more precisely,

$$|W(x,p)| \leq \frac{1}{\pi}. \tag{9.16}$$

Definition (9.5) for the Wigner function can be extended in a compact form to the case of N modes u_ℓ (Equation (F.1)) by introducing the quadrature column vectors of dimension N, $\mathbf{X} = (x_1, ..., x_N)^T$ and $\mathbf{P} = (p_1, ..., p_N)^T$. W is now the real function of $2N$ variables given by the N multiple integral

$$W_\rho(\mathbf{X},\mathbf{P}) = \frac{1}{(2\pi)^N}\int ... \int d\mathbf{P}\left\langle \mathbf{X} - \frac{\mathbf{P}}{2}\right|\rho\left|\mathbf{X} + \frac{\mathbf{P}}{2}\right\rangle e^{i\mathbf{X}^T\mathbf{P}}, \tag{9.17}$$

where the vector $|\mathbf{X}\rangle$ is the tensor product of X-quadrature eigenstates $|x_1\rangle \otimes \cdots \otimes |x_N\rangle$ in all modes.

Partial tracing is straightforward in this context: The Wigner function of the quantum system traced over the Hilbert space associated to the first mode u_1, for example, is obtained by integrating the Wigner function over the phase space variables (x_1, p_1).

9.1.2 Mathematical Properties and Consequences

Trace Product Formula

An important theorem allows us to calculate a trace product $Tr\left(\hat{A}\hat{B}\right)$, an expression which is often encountered in quantum physics under different forms, for example in mean values or Born's rule, as the overlap of the two associated Wigner functions. More precisely, one can prove that for any operators $\hat{A}$ and $\hat{B}$,

$$Tr\left(\hat{A}\hat{B}\right) = \iint dpdx\, W_A(x,p)\, W_B(x,p). \tag{9.18}$$

- If one takes $\hat{A} = \rho$, $\hat{B} = \hat{1}$, one finds that W is normalized to 1,

$$\iint dpdx\, W(x,p) = 1. \tag{9.19}$$

- If one takes $\hat{A} = \rho_1$, $\hat{B} = \rho_2$, one finds,

$$Tr(\rho_1\rho_2) = \iint dpdx\, W_{\rho_1}(x,p)\, W_{\rho_2}(x,p). \tag{9.20}$$

  The trace product $Tr(\rho_1\rho_2)$ is zero for two orthogonal pure states. In this case, the overlap between the two associated Wigner functions is zero as well. This is possible only if at least one of the two Wigner functions has a negative part: *Wigner functions can take positive and negative values as well.*
- If one takes $\hat{A} = \rho$, $\hat{B} = \rho$, one finds a compact expression for the purity of the state:

$$P_u = Tr\rho^2 = \iint dpdx\, (W(x,p))^2. \tag{9.21}$$

Link with Probabilities

- If one takes $\hat{A} = \hat{\Pi}_n$ as a PO and $\hat{B} = \rho$, one finds another expression of the generalized Born's rule for the probability of getting a result a_n,

$$Proba(a_n|\rho) = \iint dpdx\, W(x,p)\, W_{\Pi_n}(x,p). \tag{9.22}$$

- If one takes $\hat{A} = \rho$, $\hat{B} = \hat{X}$ or $\hat{B} = \hat{P}$, one finds that the mean value of the quantum operators $\hat{X}$ or $\hat{P}$ appears as the integral of the corresponding classical quantity x or p weighted by the Wigner function,

$$\langle\hat{X}\rangle = \iint dpdx\, x\, W(x,p), \quad \langle\hat{P}\rangle = \iint dxdp\, p\, W(x,p). \tag{9.23}$$

  These equations can be written as

$$\langle\hat{X}\rangle = \int dx\, x\, Proba(x) \quad \text{with} \quad Proba(x) = \int W(x,p)dp, \tag{9.24}$$

$$\langle\hat{P}\rangle = \int dp\, p\, Proba(p) \quad \text{with} \quad Proba(p) = \int W(x,p)dx. \tag{9.25}$$

$Proba(x)$ and $Proba(p)$ can be shown to be quantities that are positive and normalized to 1. In the pure state case they are, respectively, equal to $|\psi(x)^2|$ and $|\tilde{\phi}(p)|^2$, where $\psi(x)$ and $\phi(p)$ are, respectively, the wavefunctions in position or momentum space. They are therefore the *regular probability densities* of measuring single quadratures, X (or P), in the interval $[x, x+dx]$

(or $[p, p+dp]$). They appear in the present formalism as *marginal probabilities* deduced from the Wigner function by integrating it over the conjugate variable.

The phase-rotated quadratures x_θ, p_θ defined by Equation (F.15) constitute another basis for the phase-space x, p. They are associated with phase-rotated operators $\hat{X}_\theta$ and $\hat{P}_\theta$ that can also be measured by homodyne detection with appropriate choices of the LO phase. One easily shows that

$$\langle \hat{X}_\theta \rangle = \int dx_\theta x_\theta \, Proba(x_\theta), \tag{9.26}$$

with

$$Proba(x_\theta) = \int W(x_\theta = x\cos\theta + p\sin\theta, p_\theta = p\cos\theta - x\sin\theta) dp_\theta. \tag{9.27}$$

$Proba(x_\theta)$ is, like $Proba(x)$ and $Proba(p)$, a bona fide probability density for the rotated quadrature, obtained as a marginal by integration of W over the orthogonal quadrature in phase space.

More generally, one can write for any function $f(x)$ of the single quadrature x,

$$\langle f(\hat{X}) \rangle = \iint dxdp f(x) \, Proba(x). \tag{9.28}$$

This means that $Proba(x)$ *as introduced in Equation (9.24) is useful not only to determine mean values, but also to calculate the higher moments*. It permits us to determine the whole probability distribution of the x-quadrature. It is, of course, the same for the statistical distribution of p-quadrature fluctuations, and of any single rotated quadrature x_θ or p_θ.

It is important to note that, among the three phase-space functions P (Glauber-Sudarshan), W (Wigner), and Q (Husimi), we will introduce at the end of this section, *the Wigner function is the only one to have this property*, while the P- and Q-functions can be used to calculate mean values, but not the higher moments directly. The Wigner function is therefore perfectly suited to determining the probability distribution of the quadratures that are measured by homodyne detection.

From Phase Space to Hilbert Space

It is possible to write a general relation between a classical phase-space function $f(x,p)$ and its Weyl transform $\hat{F}(\hat{x},\hat{p}) = \mathfrak{W}^{-1}(f(x,p))$, operating in the Hilbert space of quantum operators, by using the McCoy procedure: One first calculates a classical function F given by the relation

$$F(x,p) = e^{-i\partial_x\partial_p/2} f(x,p) \tag{9.29}$$

by expanding the exponential in series. One then moves, in all the terms of $F(x,p)$, the x factors to the left-hand side of these terms, and all the p

factors to their right-hand side ("x normal ordering"). The quantum operator $\hat{F} = \mathfrak{W}^{-1}(f)$ is then obtained by replacing x and p by the corresponding operators $\hat{X}$ and $\hat{P}$. For example, if $f = px$, then one finds by developing the exponential in Equation (9.29) that $F = px - 1/2$, so that $\mathfrak{W}^{-1}(f) = \hat{F} = \hat{x}\hat{p} - 1/2$.

The inverse transformation is simply given by $\mathfrak{W}(f(x,p)) = e^{i\partial_x\partial_p/2}F(x,p)$.

It is also possible to write Equation (9.29) in terms of the complex variable $\alpha = x + ip$. One first calculates the function F,

$$F(\alpha, \alpha^*) = e^{-i\partial_\alpha\partial_{\alpha*}/2}f(\alpha, \alpha*). \tag{9.30}$$

In the expression for $F(\alpha, \alpha^*)$, one then moves the α^* factors to the left-hand side of these terms, and all the α factors to the right-hand side ("$\hat{a}^\dagger$ normal ordering"). Finally, the quantum operator is obtained by replacing α^* and α by the corresponding operators $\hat{a}^\dagger$ and $\hat{a}$.

There is another method for implementing Weyl transforms. When the phase-space function is of the form $g = x^{n_1}p^{n_2}$, its Weyl transform is equal to $\mathfrak{W}(g) = S(\hat{X}^{n_1}\hat{P}^{n_2})$, where S means the symmetrization operation, obtained by adding all the possible terms equal to the product of n_1 operators $\hat{X}$ and n_2 operators $\hat{P}$ and dividing the result by the number of terms. For example, $\mathfrak{W}(xp) = (\hat{X}\hat{P} + \hat{P}\hat{X})/2$.

9.1.3 Examples

Example of Weyl Transform

We now consider the problem of finding the quantum Hamiltonian operator describing a system whose energy is known. It amounts to determining the Weyl transform of the expression of the energy. We take, as an example, the Kerr anharmonic oscillator, the energy of which reads $H = \delta a^* a + Ka^{*2}a^2$, where δ is the photon frequency.

We use the McCoy procedure on expression $f = a^{*2}a^2$, which yields $F = a^{*2}a^2 + 2a^*a + \frac{1}{2}$, and finally,

$$\hat{F} = \hat{a}^{\dagger 2}\hat{a}^2 + 2\hat{a}^\dagger\hat{a} + \frac{1}{2}. \tag{9.31}$$

The Hamiltonian is therefore

$$\hat{H} = (\delta + 2K)\hat{a}^\dagger\hat{a} + K\hat{a}^{\dagger 2}\hat{a}^2 + \frac{K}{2}. \tag{9.32}$$

We observe that the oscillator frequency is renormalized by $2K$ as a consequence of the ordering choice. This Lamb-like shift is a consequence of vacuum fluctuations, revealed here as due to the noncommutativity of the operators.

One can also determine the quantum Hamiltonian operator by using the fully symmetric expression in creation and annihilation operators. One obtains

$$\hat{H} = \frac{\delta}{2}\left(\hat{a}^{\dagger}\hat{a} + \hat{a}\hat{a}^{\dagger}\right)$$
$$+ \frac{K}{8}\left(\hat{a}^{\dagger 2}\hat{a}^{2} + \hat{a}^{2}\hat{a}^{\dagger 2} + 2\hat{a}\hat{a}^{\dagger}\hat{a}\hat{a}^{\dagger} + 2\hat{a}^{\dagger}\hat{a}\hat{a}^{\dagger}\hat{a} + \hat{a}\hat{a}^{\dagger 2}\hat{a} + \hat{a}^{\dagger}\hat{a}^{2}\hat{a}^{\dagger}\right), \quad (9.33)$$

which is the symmetrized expression of the same quantity.

Examples of Wigner Functions

We will give some examples of Wigner functions, which are written for the field quadratures of quantum optics, because it is the domain where they have been mostly measured and characterized.

Let us begin by considering the Wigner functions of vacuum and of a coherent state $|\alpha\rangle$:

$$W_0(x,p) = \frac{1}{\pi}e^{-(p^2+x^2)}, \quad W_\alpha(x,p) = \frac{1}{\pi}e^{-(p-p_0)^2-(x-x_0)^2}, \quad (9.34)$$

with $\alpha = (x_0 + ip_0)/\sqrt{2}$. All these Wigner functions are 2D gaussian functions, everywhere positive (Figure 9.1). In particular, the vacuum Wigner function is everywhere positive, implying that the Wigner functions of all number states orthogonal to the vacuum have negative parts.

Moreover, the Wigner functions of number states $|n\rangle$ are

$$W_n(x,p) = \frac{(-1)^n}{\pi}e^{-(x^2+p^2)}L_n\left(2\left(x^2+p^2\right)\right), \quad (9.35)$$

where L_n is the Laguerre polynomial of order n. In particular,

$$W_1(x,p) = \frac{-1}{\pi}e^{-(x^2+p^2)}\left(1 - 2\left(x^2+p^2\right)\right). \quad (9.36)$$

One notes that $W_1(0,0) = -\frac{1}{\pi}$. The Wigner function of a single photon state, as expected, is negative at the origin (Figure 9.2).

The Wigner function of a squeezed vacuum state, introduced in Appendix F, Equation (F.23), is

$$W_s(x,p) = \frac{1}{\pi}e^{-(x^2e^{-2S}+p^2e^{2S})}. \quad (9.37)$$

It is the product of two Gaussians of different widths along x and p. Like the vacuum and coherent states, it is everywhere positive.

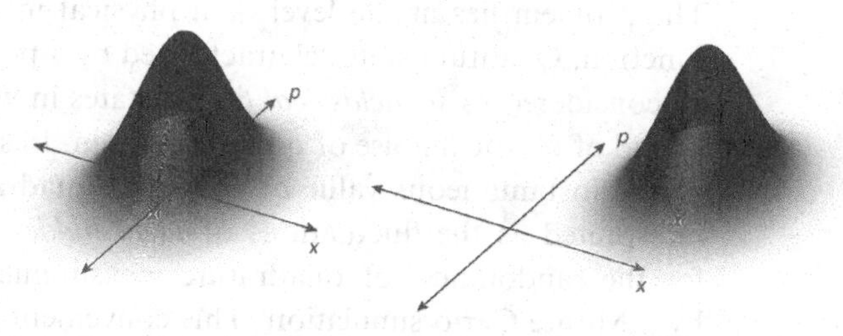

Fig. 9.1 (Left) Wigner function of the vacuum state. (Right) Wigner function of a coherent state, which is a displaced vacuum state.

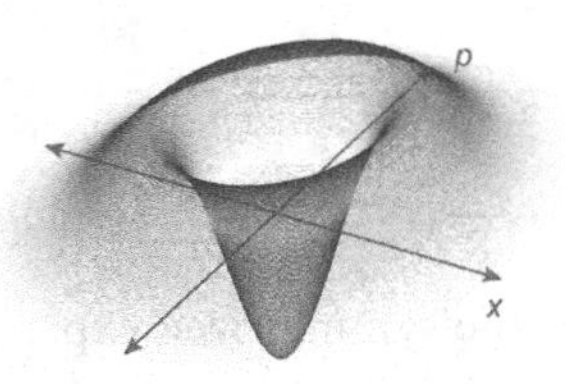

Fig. 9.2 Wigner function of a single photon state (seen from below). Its central part takes negative values.

9.1.4 Physical Content of the Wigner Function

The Wigner function has several credentials to be a joint probability density for simultaneous measurements of X and P, a situation that we studied in Chapter 5. But it does not have all of them. In particular, we have seen that X and P cannot be perfectly determined at the same time. So the notion itself of joint probability density, whatever its sign, loses most of its meaning in quantum physics. Using successive measurements is of no help, as the first measurement modifies the system in a random, uncontrolled way. Furthermore, there is no quantum state where these two parameters are perfectly defined: The Heisenberg inequality, which is a direct consequence of the existence of noncommuting observables, tells us that, whatever the quantum state, a joint probability density cannot look like a delta function in phase space.

We have seen that there exist generalized measurements that are effective in measuring X and P simultaneously, but these are necessarily imperfect, nonprojective measurements that add excess noise, or some "fuzziness" in phase space, which prevents us making these measurements perfectly accurate.

What is meaningful at the quantum level are Equations (9.22), (9.24), and (9.25), which give physically and mathematically sound probabilities that appear as marginals, i.e. as integrals over the Wigner function. In this point of view, the Wigner function appears as a very useful mathematical intermediate allowing us to "see" at one glance the marginal probability distributions of the quadratures X and P and of any rotated quadrature inbetween X_θ.

So, the fact that W can take negative values is not so surprising a feature, because W is not a joint probability density, just a mathematical tool. The problem lies at the level of a physical interpretation of the Wigner function: Quantum states characterized by a positive Wigner function can be considered as *semiclassical*, i.e. as states in which the quantum noise is a kind of classical noise of quantum origin. In such states, the fluctuations of the instantaneous value of the input quadratures $\hat{X}_{in}$ and $\hat{P}_{in}$ may be interpreted as the fluctuations of *local hidden variables*, which accounts for the randomness of quadrature measurements and can be mimicked by a Monte Carlo simulation. This convenient physical picture cannot be applied to states with Wigner functions with negative parts, which appear thereby as "more quantum" than the positive ones. In any case, great care should be taken when thinking of completely positive Wigner functions as semi-classical "probability distribution." Note that it is possible to put in evidence a violation by the CV EPR state (9.110) of a Bell inequality based on the parity operator (9.13). Therefore, one has direct evidence of non-locality on this state, though its Wigner function is everywhere positive.

9.1.5 Tomography of the Wigner Function

The previous sections have shown that the Wigner function of a quantum state is an illuminating tool to characterize it in a physical way. Actually, it is possible to have a direct experimental access to this function from data obtained by simple measurements. Let us call $\tilde{W}[u,v]$ the Fourier transform of the Wigner function $W(x,p)$,

$$\tilde{W}[u,v] = \frac{1}{2\pi}\int\int dxdp\, W(x,p)e^{-i(ux+vp)}. \tag{9.38}$$

This quantity is related by a Radon transformation to the Fourier transform

$$\tilde{P}_\theta[k] = \int dx_\theta P_\theta(x_\theta)e^{ikx_\theta} \tag{9.39}$$

of the probability distribution $P_\theta(x_\theta)$ of the rotated quadrature values $x_\theta = x\cos\theta + p\sin\theta$ by the formula

$$\tilde{W}[k\cos\theta, k\sin\theta] = \frac{1}{\sqrt{2\pi}}\tilde{P}_\theta(k). \tag{9.40}$$

According to Section 9.4, the probability distribution $P_\theta(x_\theta)$ can be measured by homodyne detection using a local oscillator of phase θ. All the experimental data necessary for determining the Wigner function are obtained by sampling the homodyne signal for a given value of θ. One then repeats the procedure for many values of θ covering the interval $[0,\pi]$. Note that the knowledge of the homodyne signal fluctuations for $\theta = 0$ and $\theta = \pi/2$, i.e. of $\hat{X}$ and $\hat{P}$, is far from being sufficient to completely determine the Wigner function. Also note that it is not possible to simultaneously measure the homodyne signal for different values of the LO phase unless one uses a generalized version of the double homodyne method of Section H3.2, using many beamsplitters and many LOs, which will introduce many units of quantum noise that will completely spoil the measurement.

The reconstruction of the Wigner function from experimental quadrature measurements is an example of *quantum state tomography* [117], enabling physicists to completely determine the quantum state of the system. One can directly use the formula (9.20) to perform it, but the result is affected by measurement and sampling noises. Experimentalists prefer maximum likelihood methods, which give the most probable Wigner function compatible with the measurements, a kind of generalized fit of the experimental data.

9.1.6 Other Phase Space Representations

The Wigner function is not the only function that can be used to calculate statistical moments and probability distributions in phase space. There are also other functions, like the Husimi Q-function, the Glauber–Sudarshan P-function and the Dirac D-function. The first three are closely linked phase-space functions F_s ($s = 0, 1, -1$) defined by the expression

$$F_s(\alpha) = Tr\left(\rho \hat{D}(\alpha)\right) \exp s|\alpha|^2/2, \tag{9.41}$$

where $s = 0$ for the Wigner function, $s = +1$ for the Husimi fonction, and $s = -1$ for the Glauber–Sudarshan function. $\hat{D}(\alpha) = \exp(\alpha\hat{a}^\dagger - \alpha^*\hat{a})$ is the displacement operator (Equation (F.16)). The three phase-space functions are Fourier transforms of characteristic functions $C(\xi)$, respectively, for the P-, W-, and Q-representations,

$$C_P(\xi) = Tr\rho e^{i\xi^*\hat{a}^\dagger} e^{i\xi\hat{a}}, \quad C_W(\xi) = Tr\rho e^{i(\xi\hat{a}+\xi^*\hat{a}^\dagger)}, \quad C_Q(\xi) = Tr\rho e^{i\xi\hat{a}} e^{i\xi^*\hat{a}^\dagger}. \tag{9.42}$$

One notes that, when developing the exponentials, in C_P the $\hat{a}^\dagger$ operators appear on the left and $\hat{a}$ on the right (normal ordering), that in C_Q the operators $\hat{a}^\dagger$ appear on the right and $\hat{a}$ on the left (anti-normal ordering), while in C_W the operators $\hat{a}^\dagger$ and $\hat{a}$ appear in a symmetrical way, invariant by the change $\hat{a} \leftrightarrow \hat{a}^\dagger$ (symmetric ordering).

Husimi *Q*-representation

This is more simply defined as

$$Q(\alpha) = \frac{1}{\pi}\langle\alpha|\rho|\alpha\rangle = \frac{1}{\pi}Tr\rho\hat{\Pi}_\alpha, \tag{9.43}$$

where $\hat{\Pi}_\alpha = |\alpha\rangle\langle\alpha|$. It has the form of a Born's rule probability (Equation (5.1)), and is therefore always positive. The POVM of the associated measurement is the complete set of operators $\hat{\Pi}_\alpha = |\alpha\rangle\langle\alpha|$ that we encountered in Section 5.7. In the case of quantum field quadratures it corresponds to a measurement performed with the help of a double homodyne detection (Section 9.5).

As stated in Chapter 5, the decomposition of a PO in terms of a PMO $\hat{M}_\alpha$ such that $\hat{\Pi}_\alpha = \hat{M}_\alpha^\dagger \hat{M}_\alpha$ is not unique. One can take, for example,

$$\hat{M}_\alpha = |\alpha\rangle\langle\psi|, \tag{9.44}$$

where $|\psi\rangle$ is any normalized wave function into which the initial state collapses, depending on the details of the actual experimental procedure.

The function Q is related to W by

$$Q(\alpha) = \frac{2}{\pi}\iint d^2\alpha' W_{\alpha'} e^{-2|\alpha-\alpha'|^2}. \tag{9.45}$$

It is a convolution product by a Gaussian that "smoothes" the Wigner function and washes out its negative parts.

Glauber–Sudarshan P-representation

This is more simply related to the density matrix by

$$\rho = \iint d^2\alpha P(\alpha)|\alpha\rangle\langle\alpha| \tag{9.46}$$

The values of P_ρ then appear as the coefficients of the diagonal representation of the density matrix on the coherent state basis. The direct expression of P is given by relation (9.41) with $s = -1$.

The Glauber–Sudarshan function for a coherent state $|\alpha_0\rangle$ is the delta function $P(\alpha) = \delta(\alpha - \alpha_0)$. It is positive everywhere. P can take negative values for other states, for example for squeezed states. It can be also a highly singular function of α, involving derivatives of the delta function, for example, for Fock states.

One defines a *nonclassical state* as a state that cannot be written as a statistical mixture of coherent states. If $P(\alpha)$ is negative for some value of α in expression (9.46), it means that this relation cannot be regarded as a statistical mixture. The negativity of P is therefore a witness of its "non-classicality."

The functions W and P are related by

$$W(\alpha) = \frac{2}{\pi} \iint d^2\alpha' P_{\alpha'} e^{-2|\alpha-\alpha'|^2}. \tag{9.47}$$

It is the same convolution product by a Gaussian as between Q and W, which washes out the irregularities of the P function.

Dirac Quasi-Probability Distribution

Let us briefly mention here another phase-space distribution, $D_\rho(x,p)$, known as the Dirac quasiprobability distribution [58]. Its expression can be written in terms of probability operators for a joint measurement of p and x, $\hat{\Pi}_p = |p\rangle\langle p|$ and $\hat{\Pi}_x = |x\rangle\langle x|$:

$$D_\rho(x,p) = Tr\rho\hat{\Pi}_p\hat{\Pi}_x = Tr\rho\delta(\hat{P} - p)\delta(\hat{X} - x). \tag{9.48}$$

One can write it more precisely as:

$$D_\rho(x,p) = Tr\rho|p\rangle\langle p|x\rangle\langle x| = \langle p|x\rangle\langle x|\rho|p\rangle. \tag{9.49}$$

It cannot be written with a projector as the PO, and is therefore not a von Neumann measurement of p and x.

The Dirac quasiprobability distribution, as a mean value of a non-Hermitian operator, is a complex quantity, which makes it in convenient for visualizing the state. From Equation (9.49) it is easy to calculate the marginals using the closure relations

$$\int dx D_\rho(x,p) = \langle p|\rho|p\rangle = Proba(p|\rho),$$
$$\int dp D_\rho(x,p) = \langle x|\rho|x\rangle = Proba(x|\rho). \tag{9.50}$$

They give, as for the Wigner function, the probability distributions for position and momentum, or for the quadrature operators, but not for the rotated quadratures x_θ.

For a pure state, the expression of D_ρ is very simple:

$$D_\rho(x,p) = e^{ipx}\psi(x)\psi^*(p). \tag{9.51}$$

For a given value of p, it is directly, within a phase factor, the wavefunction $\psi(x)$, with an analog property for the other quadrature. This distribution also allows us to calculate the mean value of polynomial functions of x and p, provided the classical expressions are written in the p antinormal ordering (p on the left, x on the right). This distribution has been used in various configurations, for example to simultaneously measure the distribution of both the transverse position and the transverse wavevector of a photon in a mixed state, using the "weak measurement" technique [12], which introduces a minimum excess noise in the measurement [4].

Comparison between the Phase-Space Functions *W*, *Q*, and *P*

The three phase-space functions can be used to calculate the mean value of an operator $\hat{A}_s$ equal to the product of m operators $\hat{a}^\dagger$ and n operators $\hat{a}$. The result depends on the choice of ordering of these operators as the operators do not commute, whereas the corresponding phase-space coordinates α and α^* do commute. More precisely, one has

$$\langle \hat{A}_s \rangle = \iint d^2\alpha\, C_s(\alpha)\alpha^{*m}\alpha^n. \tag{9.52}$$

The choice of C_s for $s = -1, 0, +1$ corresponds to different orderings in $\hat{A}_s$:

- The "normal" ordering $\hat{a}^{\dagger,m}\hat{a}^n$ (creation operators on the left, annihilation on the right) when one uses the P-function ($C_{s=-1} = P$);
- The "anti- normal" ordering $\hat{a}^n\hat{a}^{\dagger,m}$ (creation operators on the right, annihilation on the left) when one uses the Q-function ($C_{s=+1} = Q$);
- The symmetric ordering of creation and annihilation operators, defined earlier in this section, when one uses the W-function ($C_{s=0} = W$).

The different phase-space functions are interesting for treating different physical situations:

- Operators written in normal order, and therefore the P-function, have a zero mean value in vacuum state. The P-function is therefore well suited to treat photodetection signals and photon correlations;
- The antinormal order, and therefore the Q-function, gives a nonzero mean value in a vacuum state. It is well suited to the treatment of vacuum fluctuations effects, like spontaneous processes, but also, as explained in Section 9.5, to take into account the excess noise arising in generalized measurements like the double homodyne detection;
- The symmetric order of creation and annihilation operators is the only one to give marginals that are the probability distributions of quadrature

operators. So the Wigner function is the quantity to consider when one uses homodyne detection.

9.1.7 Evolution of the Wigner Function

The Wigner function is indeed an outstanding tool to visualize a given quantum state and its evolution. But it is more than that: It can be regarded as one of the main objects of interest for theoreticians, as it provides the most transparent comparison with classical physics. Formulating quantum mechanics fully in phase space bridges the mathematical gap separating the abstract Hilbert space from the underlying classical correspondence. To clarify the comparison with classical physics, in the next section we will restore the factor $\hbar$ in the commutator (Equation (9.1)) and in all the expressions that are derived from it.

General Considerations

When the evolution is Hamiltonian, the von Neumann equation $\partial_t \rho = \frac{1}{i\hbar}[\hat{H}, \hat{\rho}]$ transforms into the so-called Moyal equation,

$$\partial_t W = \frac{1}{i\hbar}(H \star W - W \star H) = \{\!\{H, W\}\!\}, \tag{9.53}$$

where H is the Wigner transform of the Hamiltonian operator, $H = \mathfrak{W}(\hat{H})$. We have introduced the Groenwold star product $f \star g$, the Wigner transform of a product of operators, $f \star g = \mathfrak{W}(\hat{F}\hat{G})$. We have also introduced the Moyal bracket $\{\!\{f, g\}\!\} = \frac{1}{i\hbar}(f \star g - g \star f)$.

The star product can be expressed in terms of x and p derivatives:

$$f \star g \equiv f \exp\left(\frac{i\hbar}{2}\left(\overleftarrow{\partial}_x \overrightarrow{\partial}_p - \overleftarrow{\partial}_p \overrightarrow{\partial}_x\right)\right) g = fg + \frac{i\hbar}{2}\{f, g\} + \cdots, \tag{9.54}$$

where the arrows indicate the direction of the derivatives $f \overleftarrow{\partial}_i \overrightarrow{\partial}_j g = (\partial_i f)(\partial_j g)$.

We see that the equation of motion for the Wigner function is identical to the classical Liouville equation for the classical phase-space distribution, plus quantum corrections: $\partial_t W = \{H, W\} + \mathcal{O}(\hbar^2)$, where $\{H, W\}$ is the classical Poisson bracket.

For a short-memory open quantum system, at the Markov approximation the system evolution is ruled by a differential equation of the Lindblad form (Equation (4.56)). It transforms into a phase space equation for the Wigner function. Taking as an example the case of a damped cavity, for the density matrix one has (Equation (4.65)),

$$\frac{d\rho}{dt} = \frac{1}{i\hbar}\left[\hat{H}, \rho\right] - \kappa(\hat{a}^\dagger \hat{a} \rho + \rho \hat{a}^\dagger \hat{a}) + 2\kappa \hat{a} \rho \hat{a}^\dagger, \tag{9.55}$$

where κ is the decay constant and $\hat{a}$ is the jump operator. The evolution equation for the Wigner function, using Equation.(9.53), involves the Wigner transforms of operators $\hat{a}^\dagger \hat{a} \rho$, $\rho \hat{a}^\dagger \hat{a}$, and $\hat{a} \rho \hat{a}^\dagger$. Using the McCoy procedure, one can express these different Wigner functions in terms of differential operations. One gets [47]:

$$\frac{d}{dt}W(\alpha,\alpha^*) = \{\{H,W\}\} + \frac{\kappa}{2}\left(\frac{\partial^2}{\partial\alpha\partial\alpha^*} + \frac{\partial}{\partial\alpha}\alpha + \frac{\partial}{\partial\alpha^*}\alpha^*\right)W(\alpha,\alpha^*), \quad (9.56)$$

where α and α^* are considered to be independent variables.

Finally, in the most general case, the evolution, or propagation, of the quantum state is expressed as a quantum map linking an initial density matrix to a final density matrix (Section 4.2). This transformation involves Kraus operators $\hat{K}_\ell$ (Equation (4.17)) and can be written as a Kraus sum (Equation (4.18)). The linearity of the Wigner transform (Equation (9.5)) implies that W obeys a similar relation, namely,

$$W_f = \sum_{\ell=1}^{N} \mathfrak{W}\left(\hat{K}_\ell \rho_i \hat{K}_\ell^\dagger\right). \quad (9.57)$$

Phase-Space Evolution in a Damped Harmonic Oscillator

Let us consider in more detail the case of a damped harmonic oscillator, studied in Section 4.6 and recalled in the previous paragraph, which features a single jump operator, the annihilation operator $\hat{A}_j = \hat{a}$. In the frame rotating at the oscillator frequency, the commutator term (4.65) in the evolution equation cancels, so that

$$\frac{d}{dt}W = \frac{\kappa}{2}\left(\frac{\partial^2}{\partial\alpha\partial\alpha^*} + \frac{\partial}{\partial\alpha}\alpha + \frac{\partial}{\partial\alpha^*}\alpha^*\right)W. \quad (9.58)$$

The evolution equation for a probability distribution is known in classical physics as a Fokker–Planck equation. Its two last terms describe the time evolution of the mean value of the probability distribution (drift). The first one describes the time evolution of its width (diffusion).

Let's first look for the stationary solution of Equation (9.58). One finds that

$$W_{stat} = e^{-\alpha\alpha^*} = e^{-(x^2+p^2)}. \quad (9.59)$$

This is the Wigner function of the vacuum state. This solution does not depend on the initial state: Whatever the initial state, the cavity is finally empty. Let us now consider the mean extension of the Wigner function around the origin $\hat{\sigma}^2(t) = \langle\alpha\alpha^*\rangle = \int d\alpha d\alpha^* \alpha\alpha^* W(\alpha,\alpha^*)$, which is also the mean energy. With the help of several integrations by parts, one finds that it obeys:

$$\frac{d}{dt}\hat{\sigma}^2 = \frac{\kappa}{2}\left(1 - 2\hat{\sigma}^2\right), \quad (9.60)$$

which has the solution $\hat{\sigma}^2(t) = \frac{1}{2} + \hat{\sigma}^2(0)e^{-\kappa t}$. We find that the extension $\hat{\sigma}^2$ of the Wigner function decays at a rate κ, until it reaches the value 1/2, that of the vacuum state. One also observes that the extension is time independent when $\hat{\sigma}^2(0) = 1/2$, i.e. when one starts from a coherent state.

Let us now consider the case where the initial state is the pure state $|\psi\rangle = |i\beta\rangle + |-i\beta\rangle$, a linear combination of coherent states of opposite phases, often named a "Schödinger kitten" as it superposes two nearly classical distinguishable states. The corresponding Wigner function, within a normalizing factor that is omitted, is

$$W_\beta(\alpha,t) = \frac{1}{\pi\left(1+\exp-2|\beta\exp-\kappa t|^2\right)} \times\left[e^{-2|\alpha-\beta\exp-\kappa t|^2}+\exp-2|\alpha+\beta\exp-\kappa t|^2 \pm 2\exp-2|\alpha|^2\exp-2|\beta|^2\left(1-\exp-\kappa t^2\right)\cos\left(4\Im m(\alpha)\beta\exp-\kappa t\right)\right]. \tag{9.61}$$

The interpretation of the three terms of the Wigner functions is illuminating: The first two terms are the positive Gaussian Wigner functions of the decaying coherent states considered separately, the centers of which, positioned at $\pm\beta e^{-\kappa t/2}$, decay to the origin with a decay constant $\kappa/2$. In the absence of the last term, they correspond to a situation where the initial state is a statistical superposition, not a coherent one, of the two Glauber states of opposite phases. The last term features non-Gaussian interference fringes close to the origin of phase space, $\cos(4\Im m(\alpha)\beta e^{-\kappa t/2})$. This last term has alternating positive and negative values in the central region of phase space. The amplitude of the fringes, witnessing the coherence of the quantum superposition, is proportional to $\exp(-2|\beta|^2(1-e^{-\kappa t}))$, which is close to $e^{-2|\beta|^2\kappa t}$: It decays with a rate $|\beta|^2\kappa$ proportional to the mean number of photons in the initial state. The corresponding decay time of the interferences, i.e. the decoherence time of the quantum superposition, is therefore much smaller than the cavity decay time.

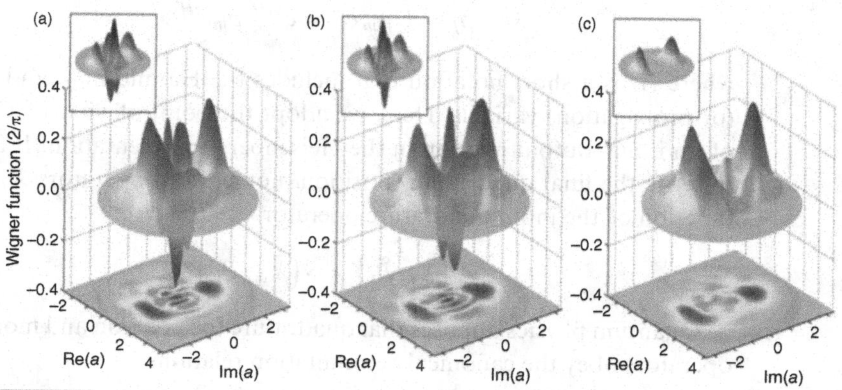

Fig. 9.3 (a,b) Experimentally reconstructed Wigner function of a decaying optical cavity for two initial even and odd cat states. The theoretical Wigner functions are given in the small insets. (c) Reconstructed Wigner function when the initial state of the atom is a statistical mixture of the two coherent states. From S. Deleglise et al. Reconstruction of nonclassical cavity field states with snapshots of their decoherence, *Nature*, 2008 [52].

Reference [52] describes an experiment with the Rydberg state QND detector of RF photons described in Section 3.2. The measured evolution of the reconstructed Wigner function is in close agreement with the theory (Figure 9.3).

Figure 9.3 shows the experimentally determined Wigner function of a decaying cavity when the initial state is a quantum superposition, with a + or − sign, of two opposite coherent states [52]. The last part of the figure does not exhibit this interference feature, as it describes a statistical superposition of the two initial coherent states.

9.2 Symplectic Quantum Maps

We now go back to the concept of quantum maps introduced in Chapter 4 and show that the introduction of the Wigner function allows us to write in a different and simple way the input–output relations. We will first consider *Hamiltonian evolutions*, corresponding to isolated systems, or systems submitted to external forces (like a classical pump field in quantum optics). We will then give an example of a non-unitary evolution of the Wigner function of an open system. Let us also recall that any quantum process can be seen as a Hamiltonian map by tracing the evolution over an ancilla supplementary mode (Section 4.2).

9.2.1 Definition and Mathematical Properties

A Hamiltonian quantum map is described by a unitary operator $\hat{U}$ acting on a density matrix,

$$\rho_f = \hat{U}\rho_{in}\hat{U}^\dagger = e^{-i\hat{H}'}\rho_{in}e^{i\hat{H}'}, \tag{9.62}$$

where $\hat{H}'$ is a short notation that includes the Hamiltonian and the time (or propagation) variable. Here, we adopt the point of view of Chapter 4 of an input–output relation in the Heisenberg representation. It allows us to write the final quadrature (position/momentum) operators $\hat{X}_f, \hat{P}_f$ as a function of the input quadrature operators:

$$\left(\hat{X}_f, \hat{P}_f\right) = \mathcal{S}\left(\hat{X}_{in}, \hat{P}_{in}\right). \tag{9.63}$$

Quantum physics imposes that quadrature (or position and momentum) operators obey the canonical commutation relations

$$\left[\hat{X}_f, \hat{P}_f\right] = \left[\hat{X}_{in}, \hat{P}_{in}\right] = i. \tag{9.64}$$

The transformation $\mathcal{S}$ can be of any form, provided it respects the canonical commutation relation (Equation (9.64)). In this case it is called a *canonical transformation*.

A symplectic transformation is a canonical transformation that is, in addition, a real linear combination of $\hat{X}_{in}$ and $\hat{P}_{in}$. It is therefore a special simple case of canonical transformation, which can be easily handled as one can use the powerful tools of linear algebra: *The transformation $\mathcal{S}$ is, in this simple case, fully characterized by the matrix* **S**,

$$\left(\hat{X}_f, \hat{P}_f\right)^T = \mathbf{S}\left(\hat{X}_{in}, \hat{P}_{in}\right)^T. \quad (9.65)$$

In this case, **S** is a 2×2 matrix. This definition of a symplectic transformation can easily be extended to the N mode case. $\mathcal{S}$ is then a $2N \times 2N$ real matrix **S** acting on the column vector $\hat{\mathbf{Q}} = (\hat{X}, \hat{P})^T$ of quadrature operators,

$$\hat{\mathbf{Q}}_f = \mathbf{S}\,\hat{\mathbf{Q}}_{in}. \quad (9.66)$$

These transformations form a group, called a symplectic group $Sp(2N, R)$, which has been extensively studied in mathematics and physics [61].

The commutation relations can be written in matrix form as

$$\left[\hat{\mathbf{Q}}, \hat{\mathbf{Q}}^T\right] = i\begin{pmatrix} 0 & 1 & \dots & \dots & 0 & 0 \\ -1 & 0 & \dots & \dots & 0 & 0 \\ \vdots & \vdots & \vdots & \vdots & \vdots & \vdots \\ \vdots & \vdots & \vdots & \vdots & \vdots & \vdots \\ 0 & 0 & \dots & \dots & 0 & 1 \\ 0 & 0 & \dots & \dots & -1 & 0 \end{pmatrix} = i\beta, \quad (9.67)$$

the matrix β being in the *symplectic form*: Its diagonal contains identical 2×2 matrices. The conservation of commutation relation by the transformation **S** implies that

$$\mathbf{S}\beta\mathbf{S}^T = \beta. \quad (9.68)$$

This relation can be taken as an operational definition of a symplectic transformation. Note that the same input–output relations can also be written in terms of the operators $\hat{a}$ and $\hat{a}^\dagger$, so that a symplectic transformation can also be defined as a linear transformation in $\hat{a}$ and $\hat{a}^\dagger$ that conserves the $[\hat{a}, \hat{a}^\dagger] = 1$ commutation relation. It turns out that quantum maps are often simpler when written in terms of annihilation and creation operators than in terms of quadratures.

One of the interesting things about symplectic transformations is that they have a very simple effect on Wigner functions. From Equations (9.17) and (9.66), one can find the important relation

$$W_f(\mathbf{Q}) = W_{in}(\mathbf{S}\,\mathbf{Q}). \quad (9.69)$$

The different values taken by the Wigner function are not changed by a symplectic quantum map: W_f is equal to W_{in}, but at a different point of the phase space. Symplectic transformations preserve, in particular, the negativity of the state. As with any Hamiltonian evolution, they leave invariant the purity as well.

Symplectic transformations describe many physical unitary processes acting on one or several modes. Their expressions can be obtained from the form of the evolution operator $\hat{U}$, and ultimately from the expression of the Hamiltonian $\hat{H}'$ that is responsible for the evolution/propagation of the quantum system. It is simple to show that symplectic, and therefore linear, input–output relations for the quadrature operators arise from a Hamiltonian that is a *quadratic form of the quadrature operators*, or a *quadratic form of the creation and annihilation operators*, like $\hat{a}^\dagger\hat{a}$ or $\hat{a}_1^\dagger\hat{a}_2 + \hat{a}_1\hat{a}_2^\dagger$.

9.2.2 Examples of Symplectic Maps

Here, we give examples of symplectic maps:

1. *Energy-conserving maps*
   These are described by orthogonal matrices **S**. They include the single mode free propagation matrix $\mathbf{S}_1$,

$$\hat{a}_f = e^{i\phi}\hat{a}_{in}, \quad \mathbf{S}_1 = \begin{pmatrix} \cos\phi & -\sin\phi \\ \sin\phi & \cos\phi \end{pmatrix}, \tag{9.70}$$

and the beamsplitter matrix $\mathbf{S}_2$,

$$\mathbf{S}_2 = \begin{pmatrix} \cos\alpha & 0 & 0 & -\sin\alpha \\ 0 & \cos\alpha & \sin\alpha & 0 \\ 0 & \sin\alpha & \cos\alpha & 0 \\ -\sin\alpha & 0 & 0 & \cos\alpha \end{pmatrix}, \tag{9.71}$$

which can be written in a much simpler way for the annihilation operators (Section H.3) as

$$\begin{pmatrix} \hat{a}_{1f} \\ \hat{a}_{2f} \end{pmatrix} = \begin{pmatrix} \cos\alpha & i\sin\alpha \\ i\sin\alpha & \cos\alpha \end{pmatrix} \begin{pmatrix} \hat{a}_{1i} \\ \hat{a}_{2i} \end{pmatrix}, \tag{9.72}$$

or, shifting the phase of mode 2 ($\hat{a}'_2 = i\hat{a}_2$) by $\pi/2$,

$$\begin{pmatrix} \hat{a}_{1f} \\ \hat{a}'_{2f} \end{pmatrix} = \begin{pmatrix} \cos\alpha & \sin\alpha \\ -\sin\alpha & \cos\alpha \end{pmatrix} \begin{pmatrix} \hat{a}_{1i} \\ \hat{a}'_{2i} \end{pmatrix}. \tag{9.73}$$

By concatenating such matrices, one can obtain the expression of any kind of multimode interferometric device.

The maps $\mathbf{S}_1$ and $\mathbf{S}_2$ are deduced from the unitary Hamiltonian evolutions $\hat{U}_1$ and $\hat{U}_2$ (Equation (9.62)):

$$\hat{U}_1 = e^{-i\phi\hat{a}^\dagger\hat{a}}, \quad \hat{U}_2 = e^{i\alpha(\hat{a}_1^\dagger\hat{a}_2 + \hat{a}_1\hat{a}_2^\dagger)}, \tag{9.74}$$

deriving from Hamiltonians proportional to $\hat{a}^\dagger\hat{a}$ and $\hat{a}_1^\dagger\hat{a}_2 + \hat{a}_1\hat{a}_2^\dagger$, respectively.

2. *Single mode attenuation or amplification*
   These maps are studied in Section H.2. They require the addition of an ancilla mode to be symplectic. For the attenuation case,
$$\hat{a}_f = \sqrt{A}\,\hat{a}_{in} + \sqrt{1-A}\,\hat{b}, \tag{9.75}$$
   and for the amplification case,
$$\hat{a}_f = \sqrt{G}\,\hat{a}_{in} + \sqrt{G-1}\,\hat{b}^\dagger. \tag{9.76}$$
3. *Propagation through a $\chi^{(2)}$ nonlinear medium*
   They are studied in Section I.2, and include the single mode squeezing transformation $\mathbf{S}_3$:
$$\hat{a}_f = \cosh\beta\hat{a}_{in} + \sinh\beta\hat{a}^\dagger_{in}, \quad \mathbf{S}_3 = \begin{pmatrix} e^\beta & 0 \\ 0 & e^{-\beta} \end{pmatrix}, \tag{9.77}$$
   and the two-mode parametric down conversion matrix $\mathbf{S}_4$:
$$\mathbf{S}_4 = \begin{pmatrix} \cosh\beta & 0 & 0 & 0 \\ 0 & \sinh\beta & 0 & 0 \\ 0 & 0 & \sinh\beta & 0 \\ 0 & 0 & 0 & -\cosh\beta \end{pmatrix} \tag{9.78}$$
   These maps are derived, respectively, from the unitary Hamiltonian evolutions
$$\hat{U}_3 = e^{i\beta((\hat{a}^\dagger)^2+(\hat{a})^2)}, \quad \hat{U}_4 = e^{i\beta(\hat{a}_1^\dagger\hat{a}_2^\dagger+\hat{a}_1\hat{a}_2)}, \tag{9.79}$$
   deriving from Hamiltonians proportional to $(\hat{a}^\dagger)^2 + (\hat{a})^2$ and $\hat{a}_1^\dagger\hat{a}_2^\dagger + \hat{a}_1\hat{a}_2$, respectively.
4. *Propagation in a weakly nonlinear medium*
   The Hamiltonian propagation in a nonlinear medium pumped by an intense laser includes, in the general case, terms that are the product of more than two annihilation and creation operators. They cannot be treated by the symplectic group approach. When the applied mean fields are large compared to their quantum fluctuations, it is possible to linearize the input–output equations with respect to their fluctuations, so that the corresponding quantum map becomes linear, and the transformation symplectic. This procedure can be applied to weak nonlinear effects in dielectrics, like in Kerr and four-wave mixing media, and all nonlinear optical process the Hamiltonian of which involves terms of order higher than 2.

Note that a change of mode basis in the Hilbert space describing the system, which is not a real physical process but a change of point of view, is described by an orthogonal transformation of the coordinates, and therefore by a symplectic transformation.

9.2.3 Bloch Messiah Decomposition

Let us here introduce two important sets of symplectic transformations:

- The compact group O of real orthogonal matrices, such as $OO^T = O^T O = \hat{1}$. These transformations conserve the energy that is proportional to the quantity $X^2 + P^2$. They include the unitary transformations studied in Chapter 4 as well as the basis changes.
- The set K of diagonal matrices D of the form

$$D = \begin{pmatrix} e^r & 0 \\ 0 & e^{-r} \end{pmatrix}. \tag{9.80}$$

They correspond to a squeezing transformation, studied in Section I.2, that reduces the quantum fluctuations below the vacuum level on one quadrature, and amplifies them on the other.

An interesting property of symplectic transformations is that they can be decomposed as a series of physically enlightening transformations. Let us mention in particular the Bloch Messiah decomposition (Euler for the mathematicians): One can show that all the symplectic transformations **S** can be written as the product

$$\mathbf{S} = \mathbf{O}_1 \mathbf{D} \mathbf{O}_2, \tag{9.81}$$

where $\mathbf{O}_1$ and $\mathbf{O}_2$ are orthogonal matrices and **D** is a diagonal matrix belonging to set K. This decomposition is analogous to the singular value decomposition of matrices, which requires the introduction of two different mode basis changes characterized by the orthogonal matrices $\mathbf{O}_1$ and $\mathbf{O}_2$ for the diagonalization of **S**.

9.2.4 Input–Output Relations for Quantum Fluctuations

We now have the mathematical tools that enable us to look at the evolution of the second moments, the variances, and the correlations in a quantum process. These moments are contained in the *covariance matrix* Γ given by (assuming for simplicity that the first moments, i.e. the mean values, are zero)

$$\Gamma = \begin{pmatrix} Var(X) & C(P,X) \\ C(P,X) & Var(P) \end{pmatrix}, \tag{9.82}$$

with $C(P, X) = (\langle \hat{P}\hat{X} \rangle + \langle \hat{X}\hat{P} \rangle)/2$ is the correlation. When the mean values of the quadratures are non-zero, one just needs to displace the Wigner function from the origin to the point of coordinates $\langle \hat{X} \rangle$, $\langle \hat{P} \rangle$. Relation (9.82) can be readily extended to the case of N modes, for which $\mathbf{Q}$ is the column vector of length $2N$ containing the quadratures $\mathbf{Q} = (X_1, X_2..., P_1, P_2...)^T$ and Γ a $2N \times 2N$ square covariance matrix.

The evolution of the covariance matrix under a symplectic process is straightforward and simple:

$$\Gamma_f = \mathbf{S}\Gamma_i \mathbf{S}^T. \tag{9.83}$$

Here, we give an example using the Bloch–Messiah decomposition: If the input state is the vacuum state, then $\Gamma_i = \hat{1}/2$ and

$$\Gamma_f = \mathbf{S}\mathbf{S}^T/2 = \mathbf{O}_1 \mathbf{D}\, \mathbf{O}_2 \mathbf{O}_2^T \mathbf{D} \mathbf{O}_1/2 = \mathbf{O}_1 \mathbf{D}^2 \mathbf{O}_1/2, \tag{9.84}$$

where

$$\mathbf{D}^2 = \frac{1}{4} diag(e^{2r_1}, e^{2r_2}, ..., e^{-2r_1}, e^{-2r_2}, ...). \tag{9.85}$$

This relation is valid for any number of modes. It shows that any symplectic quantum map can be considered as a multi-squeezing transformation $\mathbf{D}^2$ in the basis of modes generated from the initial mode basis by the transformation $\mathbf{O}_1$. In the initial basis, the squeezing is linearly distributed over many modes that are now entangled to each other, in a way analogous to the CV EPR state of Section 9.5.

Mathematically speaking, for any stochastic process the covariance matrix is symmetric and positive definite, which implies that, for all Γ,

$$\det \Gamma = Var(X)\,Var(P) - C(X,P)^2 \geq 0, \tag{9.86}$$

which is purely a mathematical constraint that includes the XP correlations. The existence of noncommuting variables in quantum physics imposes further constraints on the quantum noise, and therefore on the covariance matrix, that are not present for classical noise. More precisely, for quantum fluctuations one must have that

$$det\,\Gamma \geq \frac{1}{4}. \tag{9.87}$$

In the multimode case, one can write the quantum constraint in a compact way as

$$\Gamma \geq \frac{1}{2i}\beta, \tag{9.88}$$

meaning that the matrix $\Gamma + \frac{i}{2}\beta$ is a definite positive one, i.e. has all its eigenvalues positive or zero.

9.2.5 Williamson Reduction

Let us now introduce the Williamson theorem, which concerns a kind of symplectic diagonalization of covariance matrices. It states that for any covariance matrix Γ there is a symplectic transformation $\mathbf{S}$ such that

$$\mathbf{S}\Gamma\mathbf{S}^T = \mathbf{M} = diag(\nu_1, \nu_2 ... \nu_1, \nu_2, ...), \tag{9.89}$$

where the ν_is are N real positive numbers. The existence of noncommuting variables, and hence of a Heisenberg inequality for the second moments, implies a constraint, namely that all the ν_i must be greater than 1/2. Note that the Williamson reduction is not a regular diagonalization as it involves a symplectic matrix and not an orthogonal one, and that the *symplectic eigenvalues* ν_i are the same for the two quadratures X_i and P_i in each of the modes u_i. These symplectic eigenvalues can be determined as eigenvalues of the matrix $i\beta\Gamma$. Matrix $\mathbf{M}$ appears as a source of excess noise affecting in the same way the two quadratures of the different modes, which is in most cases a thermal noise. If the system is thermalized, all the ν values are equal to $n_T + 1/2$, where n_T is the number of thermal photons at temperature T, so that $\mathbf{M}$ is proportional to the identity.

Using the Bloch–Messiah decomposition of matrix **S**, one obtains

$$\Gamma_f = \left(n_T + \frac{1}{2}\right) \mathbf{O}_1 \mathbf{D}^2 \mathbf{O}_1. \tag{9.90}$$

The whole correlation matrix is equally affected by classical thermal fluctuations. The purity of the output state is related to the product of the eigenvalues by the relation $P_u = (2^N \nu_1 \nu_2, ..., \nu_N)^{-2}$.

9.3 Gaussian States

9.3.1 Definition

One calls Gaussian states quantum states which have a Gaussian-shaped Wigner function. They are more precisely of the form

$$W(x,p) = \frac{1}{\sqrt{(2\pi)^N det\Gamma}} e^{-\mathbf{Q}^T \Gamma^{-1} \mathbf{Q}}. \tag{9.91}$$

This expression is given in the simple case of zero mean values, where **Q** is the $2N$-element column vector containing the quadratures X, P. Γ is the covariance matrix introduced in Equation (9.82). Many quantum states are indeed Gaussian: Vacuum, coherent, and thermal states, as well as squeezed states and EPR states (see Section 9.5 and Appendix I), are Gaussian. They are simple quantum states because the Wigner function, and therefore all the properties of the Gaussian quantum state, are completely determined by the mean values and by the matrix Γ of second moments. This is, for example, the case of the purity of a Gaussian state, which is directly related to its covariance matrix,

$$P_u = \frac{1}{2^N \sqrt{det\Gamma}}, \tag{9.92}$$

with $det\Gamma = (1/4)^N$ for a pure state.

9.3.2 Hudson–Piquet Theorem

Gaussian states are concerned by an important mathematical result, the Hudson–Piquet theorem. It states that *for a pure state, single mode, or multimode, a Wigner function that is positive over the whole phase space is necessarily Gaussian*. So Wigner function positivity is a very strong requirement, as it constrains the mathematical shape of W at all points in the phase space. One can also say that, for a pure state, the existence of negative parts in the Wigner function implies that it is not Gaussian: For a pure state negativity means non-Gaussianity.

This theorem is only valid for pure states, so that there exist non-Gaussian mixed quantum states that have a Wigner function that is everywhere positive. For example, a statistical superposition of two coherent states, $\rho \propto |\alpha\rangle\langle\alpha| + |\alpha'\rangle\langle\alpha'|$, has a Wigner function that is the sum of of two Gaussians. It is at the same time non-Gaussian and positive.

According to Equation (9.69), symplectic transformations do not change the shape of the Wigner function: A Gaussian state, pure or mixed, is therefore transformed into another one. The covariance matrix Γ of the transformed Gaussian state is given by Equation (9.83). For this reason, symplectic transformations are called Gaussian maps or processes. In particular, any symplectic transformation S acting on a vacuum state will give rise to a Gaussian state. In Section 9.3.1, we gave a non-exhaustive list of physical symplectic transformations that all generate Gaussian, positive Wigner functions from vacuum. This is the case, in particular, for any nonlinear optical process that can be linearized because the quantum fluctuations are small compared to the means, which is an often-encountered situation. These Gaussian states are characterized by covariance matrices Γ given by

$$\Gamma = \frac{1}{2} S S^T. \tag{9.93}$$

When applied to Gaussian states, the Bloch–Messiah reduction (Equation (9.81)) implies that any pure Gaussian state can be seen as the result of a multimode squeezing transformation followed by a generalized interferometer, made of a network of beamsplitters and free space propagation. The Williamson theorem implies that any mixed Gaussian state can be seen as the result of the following sequence of operations acting on a multimode thermal state:

propagation in a generalized interferometer $\rightarrow$ multimode squeezing operation $\rightarrow$ propagation in a second generalized interferometer.

9.3.3 Non-Gaussian Character

It is not easy to define a proper measure of the *Wigner negativity* characterizing this higher level of quantum character. The minimum value of W is often used. The "negative volume" $\frac{1}{2}\iint dpdx(|W| - W)$ is another quantity [5] that characterizes the "strength" of Wigner negativity.

Suppose that the probability of measuring the x-quadrature is zero for some value x_0, $Proba(x = x_0) = 0$. According to Equation (9.24), this means that $\int W(x_0, p)dp = 0$, which cannot be obtained with a Wigner function that is everywhere positive: A zero in one of the marginal probability distributions is a witness of negativity of the Wigner function.

More generally, one can use the property that the Wigner function of a single mode state has negative parts if and only if its Husimi Q-function has zeros [36] to define a "hierarchy" of non-Gaussian states based on the number of zeros of the Q-function, named the "stellar hierarchy."

9.4 Measurements in Continuous Variable Systems

9.4.1 Homodyne Detection

Let us consider the following measurement scheme (Figure 9.4): The beam 1, which we want to measure, is mixed with an LO on a 50% beamsplitter. One detects the number of photons on the two beams (1 and 2) transmitted and reflected by the beamsplitter, and one records the difference between the two photocurrents. Such a measurement is associated with the observable $\hat{N}_- = \hat{N}_1 - \hat{N}_2$. Using the input–output relations (9.72) of the beamsplitter, one obtains

$$\hat{N}_- = \hat{a}_1^{in\dagger}\hat{a}_2^{in} + \hat{a}_1^{in}\hat{a}_2^{in\dagger}. \tag{9.94}$$

One sends an input state vector $|\Psi\rangle$ that one wants to characterize on beam 1, and a single mode coherent state of light $\left|\alpha e^{i\phi}\right\rangle$ where α is real on input beam 2, and the LO phase ϕ can be scanned by varying the LO propagation distance. The input state is therefore the factorized state $|\Psi\rangle \otimes |\alpha e^{i\phi}\rangle$. The mean value of the signal $\langle \hat{N}_- \rangle$ is therefore

$$\left\langle \Psi; \alpha e^{i\phi} \left| \hat{N}_- \right| \Psi; \alpha e^{i\phi} \right\rangle = \alpha\sqrt{2}\left\langle \hat{X}_\phi \right\rangle, \tag{9.95}$$

where $\hat{X}_\phi = (\hat{a}_1^{in\dagger} e^{i\phi} + \hat{a}_1^{in} e^{-i\phi})/\sqrt{2}$, which is equal to $\hat{X}$ for $\phi = 0$, and to $\hat{P}$ for $\phi = \pi/2$. The second moment is:

$$\left\langle \hat{N}_-^2 \right\rangle = \left(\alpha\sqrt{2}\right)^2 \left\langle \Psi \left| \hat{X}_\phi^2 \right| \Psi \right\rangle + \alpha^2 \left\langle \Psi \left| \hat{N}_1 \right| \Psi \right\rangle. \tag{9.96}$$

This equation shows that when $\alpha^2 \gg \langle \Psi | \hat{N}_2 | \Psi \rangle / 4$, the homodyne detection set-up measures the second moment of the rotated quadrature operator $\hat{X}_\phi$. The demonstration can be extended to all higher moments:

$$\left\langle \hat{N}_-^n \right\rangle \simeq \left(\alpha\sqrt{2}\right)^n \left\langle \Psi \left| \hat{X}_\phi^n \right| \Psi \right\rangle. \tag{9.97}$$

It relies on the relation

$$\hat{a}^\dagger |\alpha\rangle = \alpha^* |\alpha\rangle + |\phi_0\rangle, \quad \langle \phi_0 | \phi_0 \rangle = 1, \tag{9.98}$$

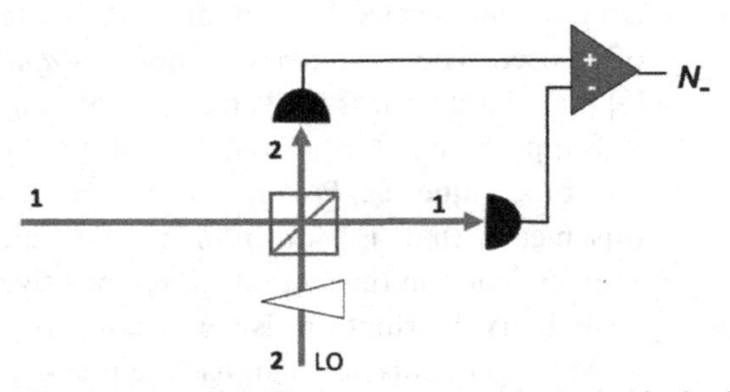

Fig. 9.4 Diagram of a homodyne detection set-up. The state 1 to measure is mixed with a local oscillator (LO) 2 on a beamsplitter. One records the current resulting from the difference of the two output photocurrents as a function of the LO phase.

which shows that, when $\alpha \gg 1$, a coherent state, in addition to being an eigenstate of $\hat{a}$, is "almost" an eigenstate of the creation operator, and therefore of the quadrature operators. The homodyne signal $\hat{N}_-$ therefore gives access, when $\alpha \gg 1$, to the complete probability distribution of quadrature operators values, which is important if one wants to study non-Gaussian states, for example.

In addition, the signal is multiplied by the LO field amplitude α, which can be a very large number when the LO is a single mode laser beam ($\alpha \simeq 10^7$ for a 1 mW beam in a 10 ms measurement time). Homodyne detection provides, by a very simple interferometric technique, an amplified and easily measurable signal that is proportional to the instantaneous value of the quadrature operator fluctuations, even when these fluctuations are of the order of the vacuum fluctuations. This simplicity accessing the instantaneous quantum noise is an important characteristic of quantum optics experimental studies. In the case of a source of successive identical input states, for example continuous wave light states, homodyne detection is very often followed by a spectral Fourier analysis of the fluctuating signal, which allows us to measure the noise spectrum $S(\Omega)$ of the homodyne signal fluctuations. For example, when the LO phase ϕ is zero, the noise spectral density $S(\Omega)$ is equal to $S(\Omega) = \langle \delta \hat{X}_\Omega^2 \rangle$, where

$$\delta \hat{X}_\Omega = \frac{1}{\sqrt{2\pi}} \int dt\, e^{i\Omega t} \delta \hat{X}(t). \tag{9.99}$$

When one uses photodetectors of quantum efficiency one, the homodyne detection with an intense LO is a projective, von Neumann measurement of a given quadrature. For example, for a measurement of the X quadrature giving as a result the value x, the PO is the projector

$$\hat{\Pi}_X(x) = |x\rangle\langle x|, \tag{9.100}$$

where $|x\rangle$ is the eigenstate of operator $\hat{X}$, of eigenvalue x, considered in Appendix I. Let's recall that the probability distribution of the quadrature is given by the marginal of the Wigner function,

$$Proba(x|\rho) = Tr\rho\hat{\Pi}_X = \int W_\rho(x,p)dp. \tag{9.101}$$

9.4.2 Double Homodyne Detection

By scanning the LO phase ϕ, homodyne detection allows us to access the two quadrature components of the field subject to the measurement successively. It would be highly desirable to be able to detect them simultaneously, so as to have access in a single shot to the whole field quantum fluctuations. Of course, the two quadratures correspond to two noncommuting operators, and we know that it is not possible to make these two measurements in a perfectly accurate way. In Section 5.7 we evaluated the excess noise of a specific POVM, allowing us to simultaneous measure position and momentum of a single particle. We now show that it is possible to exhibit a specific measurement set-up, called double homodyne detection, that allows us to measure the two field quadratures simultaneously.

The measurement set-up is sketched in Figure 9.5. A 50% beamsplitter BS divides the input beam to be measured, associated with the annihilation operator $\hat{a}$, in two output beams, $\hat{a}_1^{out}$ and $\hat{a}_2^{out}$, that are sent to two homodyne measurement devices, one measuring the X_1 quadrature and the other the P_2 quadrature. The map of the beamsplitter BS transformation is

$$\hat{a}_1^{out} = \frac{1}{\sqrt{2}}(\hat{a} + \hat{a}'), \quad \hat{a}_2^{out} = \frac{1}{\sqrt{2}}(-\hat{a} + \hat{a}'), \tag{9.102}$$

with $\hat{a}'$ being the second input port of beamsplitter BS. This symplectic transformation can be written in terms of Wigner functions, according to Equation (9.69),

$$W_{out}(\alpha_1^{out}, \alpha_2^{out}) = W_{in}\left(\frac{\alpha + \alpha'}{\sqrt{2}}, \frac{-\alpha + \alpha'}{\sqrt{2}}\right). \tag{9.103}$$

The splitting of the input LO by beamsplitter BS, which is in a coherent state, gives rise at its output to two uncorrelated coherent states that are used as Local Oscillators in the two homodyne detections. The variance of the quadrature operator $\hat{X}_1^{out}$, derived from the beamsplitter equation (9.102), is equal to $Var(X)/2 + 1/4$, while the variance of the quadrature operator $\hat{P}_2^{out}$ is equal to $Var(P)/2 + 1/4$. We see that the two measurements are affected by vacuum fluctuations entering through the second input port of the beamsplitter BS, so that the instantaneous values of the quantum fluctuations measured by the two homodyne detections are not the ones of the input beam we want to measure.

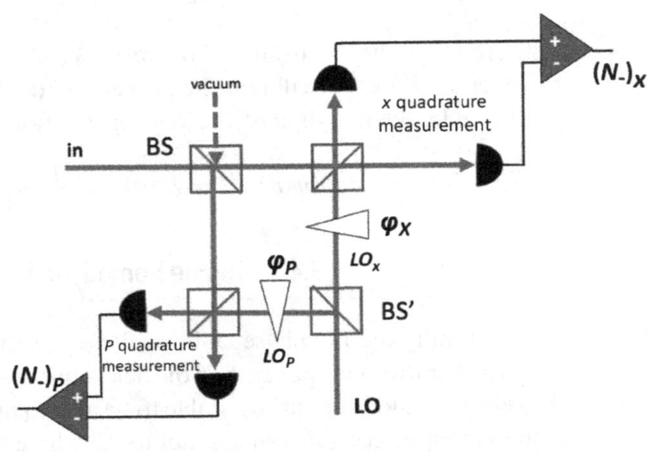

Fig. 9.5 Diagram of a double homodyne detection set-up. The light state to measure is split by a beamsplitter BS, and the two outputs are submitted to homodyne detections tuned respectively to the X and P quadratures. The unused input port of the beamsplitter BS introduces spurious vacuum noise in the detection that spoils the quality of the joint measurement.

Let us now consider the problem in more detail. The joint probability of measuring a value x on homodyne detection 1 and p on homodyne detection 2 is the product of the two separated probabilities

$$Proba(x,p|\rho) = \int dp_1\, W_1(x,p_1) \int dx_2\, W_2(x_2,p), \tag{9.104}$$

where W_1 and W_2 are the Wigner functions of the light states entering the two homodyne detectors. The Wigner function of the input factorized state is the product of the Wigner functions of the two states. More precisely, it is the product of the Wigner function W_ρ of the state of interest by the Wigner function of vacuum (relation (9.34)), so that

$$Proba(x,p|\rho) = \iint dx_2 dp_1\, W_\rho\left(\frac{1}{\sqrt{2}}(x+x_2), \frac{1}{\sqrt{2}}(p_1+p)\right) e^{-(x-x_2)^2} e^{-(p-p_1)^2}. \tag{9.105}$$

Let us change variables in the integral: $x' = (x+x_2)/2, p' = (p_1+p)/2$. One obtains

$$Proba(x,p|\rho) = \iint dx'dp'\, W_\rho\left(\sqrt{2}x', \sqrt{2}p'\right) e^{-2(x-x')^2} e^{-2(p-p')^2}. \tag{9.106}$$

We find that the joint probability is the convolution product of the Wigner function of the input state by a Gaussian. The input vacuum fluctuations have in some way introduced some amount of "fuzziness" in the measurement of the quadratures.

Note that Equation (9.106) is simply the expression of the Q-function in terms of the Wigner function (Equation (9.45)). The conclusion is that *the double homodyne detection is ultimately related to the Husimi Q-function,*

$$Proba(x,p|\rho) = Q_\rho\left(\sqrt{2}x', \sqrt{2}p'\right) = Tr\rho|\alpha\rangle\langle\alpha|, \tag{9.107}$$

with $\alpha = \sqrt{2}(x+ip)$. The corresponding PO is the projector on the coherent state,

$$\hat{\Pi}_{x,p} = |\alpha\rangle\langle\alpha|, \tag{9.108}$$

and a possible PMO is $\hat{M}_{x,p} = |\alpha\rangle\langle\phi_0|$, where $|\phi_0\rangle$ is the collapsed state.

9.4.3 Measurement of Correlations

The measurement of correlations of CV quantities A and B was considered in Section 5.7. When these quantities commute and can be measured independently, the simplest way is to simultaneously measure in successive time bins the instantaneous values of the fluctuations of A and of B, to multiply them electronically, and average this signal.

Another common technique is to record the instantaneous value of the difference between A and B. One uses the relation

$$Var(A-B) = Var(A) + Var(B) - 2C(A,B) \tag{9.109}$$

to deduce the correlation function. The relation simplifies further to $Var(A-B) = 2Var(A)(1-c)$, where c is the normalized correlation function (Equation (5.45)), if the two quantities have the same variance.

This technique can be extended to the case of any of the linear combinations of A and B considered at the end of Section 5.7, and to measure the variances of any combinations of A and B. Note that in the case when $A = X$, $B = P$, these combinations are proportional to the rotated quadratures X_θ that are easily measurable by homodyne detection.

9.5 Entanglement in Continuous Variable Systems

9.5.1 The EPR State

As mentioned in Chapter 7, the importance, uniqueness, and strangeness of entanglement in quantum mechanics was officially recognized for the first time in 1935 in the famous "EPR" paper [62], which did not mention spin or polarization DV entanglement, but position-momentum CV entanglement. Schrödinger further analyzed this situation and introduced the word "entanglement" [158]. Later, Bohm extended the discussion to spin entanglement and DV variables [21]. Here, we will expose the EPR situation in the the CV configuration and the "paradox" that it reveals.

Let us consider two one-dimensional particles A and B, of positions and momenta x_A, x_B, p_A, p_B. We assume that the two particles have interacted in the past. At the time of consideration, they are far apart and not interacting. The EPR article considers the case where the two particles are described by a two-particle quantum state of wavefunction,

$$\psi_{EPR}(x_A, x_B) \propto \delta(x_A - x_B - x_0) = \frac{1}{2\pi}\int dp\, e^{ipx_A}\, e^{-ipx_B}\, e^{-ipx_0}, \qquad (9.110)$$

where x_0 is a fixed position. We will come back to this wavefunction at the end of this section. The second form of the state shows that it is an integral of products of single atom plane wave functions e^{ipx}. The wavefunction is not factorized into an A part and a B part, and can be shown to be not factorized in any other basis: it is indeed a CV entangled state. It is easy to show from the second form that in such a state the variances $Var(x_A)$, $Var(x_B)$, $Var(p_A)$, and $Var(p_B)$ are infinite, so that the two particles, when considered individually, are completely delocalized and their individual velocities are completely random.

Let us now show that $\psi_{EPR}(x_A, x_B)$ is an eigenstate of the operator $\hat{X}_A - \hat{X}_B$ with eigenvalue x_0, and an eigenstate of operator $\hat{P}_A + \hat{P}_B$ with eigenvalue 0:

$$(\hat{X}_A - \hat{X}_B)\psi_{EPR} = (x_A - x_B)\psi_{EPR} = x_0\, \psi_{EPR}, \tag{9.111}$$

$$(\hat{P}_A + \hat{P}_B) = \frac{\hbar}{i}\left(\frac{\partial}{\partial x_A} + \frac{\partial}{\partial x_B}\right)\delta(x_A - x_B - x_0) = 0. \tag{9.112}$$

This implies that $Var(x_A - x_B) = 0$ and $Var(p_A + p_B) = 0$: *There is a perfect position correlation and perfect momentum anticorrelation between the two particles.* When Bob measures a given, random value x for the position of his particle, he knows for sure that Alice's particle has the position $x + x_0$. If Bob measures a given, random value p for the momentum of his particle, he knows for sure that Alice's particle has the opposite momentum $-p$.

The two-particle correlation can be written in terms of the conditional variances of Chapter 4

$$Var(x_A|x_B) = 0, \quad Var(p_A|p_B) = 0, \tag{9.113}$$

and therefore $Var(x_A|x_B)\,Var(p_A|p_B) = 0$.

The fact that variances involving position and momenta are simultaneously zero is due to the existence of the commutation relation

$$\left[\hat{X}_A - \hat{X}_B, \hat{P}_A + \hat{P}_B\right] = 0. \tag{9.114}$$

This situation reveals a "paradoxical" puzzling situation because it shows that quantum mechanics and, more precisely, the presence of entanglement does not forbid us to know perfectly either the position, or the momentum, in a nondestructive way and without any interaction with this particle. In their paper, EPR went further and claimed that this means that both the position and momentum of a particle are objective elements of reality, so that quantum mechanics, which considers them as "complementary," antinomic, self-excluding quantities, is not complete, i.e. does not describe the whole physical reality of the particle.

9.5.2 Implementation of the EPR State

The entangled position-momentum EPR state can be generated by a two-particle bound state that "explodes" at some time and emits its two components in opposite directions.

In the case of quantum electrodynamics, a pair of single mode squeezed states, studied in Appendix I, provides a convenient experimental implementation of a CV EPR state. They are generated from vacuum by degenerate parametric down conversion. The corresponding map for the $\hat{X}$ and $\hat{P}$ operators is given by Equation (I.19), which is clearly a symplectic (though not orthogonal) transformation. As a result, the corresponding Wigner function W_s is Gaussian and everywhere positive. For example,

$$W_s(x_1, p_1) = \frac{1}{\pi}e^{-\left(x_1^2 e^{-2S} + p_1^2 e^{2S}\right)} \tag{9.115}$$

is an X_1 quadrature squeezed light beam if $\beta > 0$. When $\beta \to +\infty$, the Gaussian function is more and more elongated along the P coordinate, and tends to a delta function of p_1.

Let us now go to a two-mode system, for example the two polarization modes of the light beam. It can also consist of two beams of different directions of propagation. We assume that the two modes are in vacuum squeezed states, one on the X_1 quadrature and the other on the P_2 quadrature, with the same squeezing parameter S. The Wigner function of this system is the factorized function

$$W_{in}(x_1, x_2, p_1, p_2) = \frac{1}{\pi} e^{-\left(x_1^2 e^{-2S} - p_1^2 e^{2S}\right)} \times e^{-\left(x_2^2 e^{2S} - p_2^2 e^{-2S}\right)}, \tag{9.116}$$

which has, when $S \to \infty$, a very narrow distribution of x_1 and p_2 quadratures.

When the polarizer is rotated by $\pi/4$, the two modes are mixed by the symplectic and orthogonal transformation Equation (9.72) with $\alpha = \pi/4$. The input–output transformation is the same if one makes the two beams cross on a 50% beamsplitter. We now use the fundamental relation (9.45) to write the Wigner function for the output quadratures X_1', X_2', P_1', P_2' of the two output beams,

$$W_{out}(\mathbf{Q}) = W_{in}(S\,\mathbf{Q}). \tag{9.117}$$

So that, finally,

$$\begin{aligned} W_{out} = \frac{1}{\pi} & \exp\left[-2\left((x_1 - x_2)^2 e^{-2S} - (p_1 + p_2)^2 e^{2S}\right)\right] \\ & \times \exp\left[-2\left((x_1 + x_2)^2 e^{2S} - (p_1 - p_2)^2 e^{-2S}\right)\right]. \end{aligned} \tag{9.118}$$

At large squeezing coefficient S, this Wigner function has a very narrow distribution in variables $(x_1 - x_2)$ and $(p_1 + p_2)$, describing a perfect X-correlation and P-anticorrelation. More precisely,

$$Var(\hat{X}_1 - \hat{X}_2) = e^{2S}, \quad Var(\hat{P}_1 + \hat{P}_2) = e^{-2S}, \tag{9.119}$$

$$Var(X_1) = Var(X_2) = Var(P_1) = Var(P_2) = \cosh S. \tag{9.120}$$

Using Equation (5.48), the conditional variance of X_1 knowing X_2, and of P_1 knowing P_2, are given by

$$Var\left(\hat{X}_1|\hat{X}_2\right) = Var\left(\hat{X}_1 - \hat{X}_2\right)\left(1 - \frac{Var(\hat{X}_1 - \hat{X}_2)}{4Var(\hat{X}_1)}\right), \tag{9.121}$$

$$Var\left(\hat{P}_1|\hat{P}_2\right) = Var\left(\hat{P}_1 + \hat{P}_2\right)\left(1 - \frac{Var(\hat{P}_1 + \hat{P}_2)}{4Var(\hat{P}_1)}\right). \tag{9.122}$$

They tend to zero when S increases, though the fluctuations of Alice's and of Bob's quadratures increase without limit: While it is not possible to accurately measure anything in each beam because of this excess noise, a post-selection on a given value of Bob's X_2 or P_2 quadrature measurements yields a conditional state that has a vanishing conditional variance on X_1 or P_1 when $S \to \infty$. Bob has perfect knowledge of Alice's X or P quadrature even if she is very far away. The EPR state studied in this section is the experimentally implementable version of the EPR paper quantum state [137]. It has been generated in several experiments, and physicists have measured conditional variances of X_1 and P_1 well below 1/2, a proof that one can measure accurately the two quadratures of the Alice's light beam without interacting with Alice's system.[1]

9.6 Nonseparability

The PPT criterion of Section 7.2 for characterizing nonseparable systems, Gaussian or non-Gaussian, has a particularly simple form when it is formulated in terms of the Wigner function. This is due to the fact that the transpose map has a very simple form in phase space: One can show that it transforms $W(x, p)$ into $W(x, -p)$ and therefore has the same action, in the case of particles, as *time reversal* [164]. So to assess the nonseparability of a bipartite quantum state described by its Wigner function $W(x_1, x_2, p_1, p_2)$, one calculates the symplectic eigenvalues of the partially transposed Wigner function $W^{PPT} = W(x_1, x_2, -p_1, p_2)$ by diagonalizing $i\beta\Gamma$. If one of these eigenvalues is below 1/2 this phase-space function is not the Wigner function of a quantum state, and therefore the initial quantum state is not separable. Be careful not to make the confusion between the positivity of the density matrix (a mathematical requirement for any physical quantum state) and the positivity of the Wigner function, which characterizes a subclass of quantum states.

The PPT analysis is possible only if one knows the whole Wigner function of the quantum state exactly, either theoretically or experimentally. One can prove that having a physical PPT state is a necessary and sufficient criterion of separability when Alice's Hilbert spaces is single mode (or single particle), and Bob's space comprises an arbitrary number of modes.

Turning to the particular case of Gaussian states, it can also be shown that, in the case of multimode mixed states, it is always possible to "*disentangle*" a nonseparable state by an appropriate change of the mode basis, i.e. by mixing Alice and Bob modes [174].

As a Gaussian state is completely determined by its covariance matrix Γ, nonseparability criteria depend only on second-order statistical moments.

[1] Note that the EPR state Wigner function is everywhere positive. It can be mimicked by a classical Monte-Carlo simulation. So the situation considered in the EPR paper is indeed puzzling and nonclassical, but it can be accounted for by introducing supplementary local variables, namely the instantaneous value of the quadrature fluctuations on the input modes.

The most used ones are the so-called Duan [60] and Duan–Mancini criterion [120]. The simpler criterion is the last one, which relies on the determination of the quantity

$$S_{Mancini} = Var(X_1 + X_2)\,Var(P_1 - P_2). \qquad (9.123)$$

If the state is separable, then $S_{Mancini} \geq 1/4$. This is not to be confused with the conditions that are necessary for the generation of an EPR state in terms of conditional variances $Var(X_1|X_2) < 1/2$ and $Var(P_1|P_2) < 1/2$ that imply $S_{Mancini} < 1/4$. This means that one needs more than nonseparability to get EPR correlations.

9.7 Unconditional Teleportation

Quantum teleportation can also be implemented in CV systems: In this case the information that one wants to transfer with a minimum distortion is the instantaneous value of the complex electric field, i.e. the quantum fluctuations of the two quadratures X_A and P_A of Alice's input light field, a task that cannot be perfectly achieved for a quantum system as the two field quadratures cannot be measured and left unspoiled at same time, as we saw in Section 5.7. We will consider here, not the quantum states themselves, but the fluctuating quadrature values in phase space, manipulated here as classical stochastic variables. The reasoning and, therefore, the choice of Alice's input quantum state, is limited to teleportation of positive Gaussian states

A sketch of the experiment [71] is presented in Figure 9.6. On Alice's left-hand side the input beam to teleport is sent into a double homodyne measurement device (Section 9.4) comprising two homodyne detectors tuned to detect the X and P quadrature fluctuations. The information about the measured X and P quadratures is contained in two photocurrents I_X and I_P that are sent to Bob with the help of classical channels like internet or telephone lines.

The second port of Alice's beamsplitter (dotted line) is not left free, which would introduce spurious vacuum noise in the double homodyne measurement. It is instead occupied by beam 1 of an EPR state, while beam 2 of the EPR state is sent to Bob on the right-hand side, who mixes it on a beamsplitter of transmission and reflection coefficients t and r with an auxiliary beam of quadratures X_{aux} and P_{aux} that is displaced in phase space, with the help of phase and amplitude modulators M_X and M_P, by quantities gX and gP where g is the gain of the electronics.

So what is measured by Alice's detectors and sent to Bob, are the instantaneous values of $(X_A + X_1^{EPR})\sqrt{2}$ and $(P_A - P_1^{EPR})/\sqrt{2}$. The two quadratures of Bob's output beam, X_B and P_B, are then:

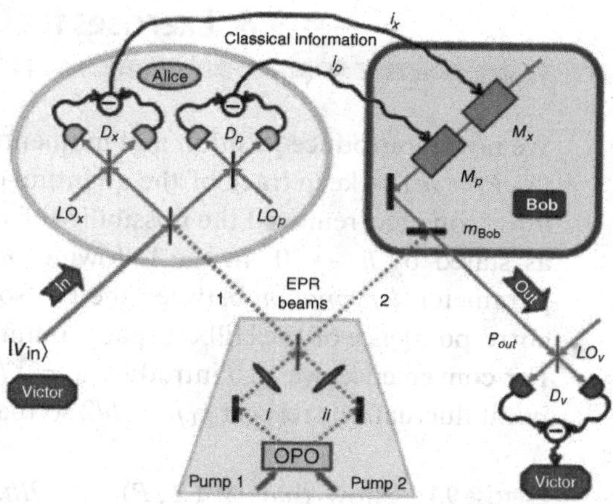

Fig. 9.6 Sketch of an unconditional teleportation device in which the input state (on the left) is mixed with one of the two EPR beams generated by the OPO, while the other EPR beam has its quadrature modified, by phase and intensity modulators, according to the results of the measurement on Alice's side transmitted by a classical communication channel. The recreated quantum state on Bob's side is, if everything is perfect, an exact copy of the input beam. From A. Furusawa et al. Unconditional quantum teleportation, *Science*, 1998 [71].

$$X_B = t\left(X_{aux} + \frac{g}{\sqrt{2}}\left(X_A + X_1^{EPR}\right)\right) + rX_2^{EPR}, \tag{9.124}$$

$$P_B = t\left(P_{aux} + \frac{g}{\sqrt{2}}\left(P_A - P_1^{EPR}\right)\right) - rP_2^{EPR}. \tag{9.125}$$

For the sake of simplicity we assume that the EPR state comes from two perfectly squeezed beams, meaning that the quadrature of the first EPR beam X_1^{EPR} is strictly equal to X_2^{EPR}, and P_1^{EPR} is strictly equal to P_2^{EPR}. One has that

$$X_B = tX_{aux} + \frac{gt}{\sqrt{2}}X_A + \left(\frac{gt}{\sqrt{2}} + r\right)X_1^{EPR}, \tag{9.126}$$

$$P_B = tP_{aux} + \frac{gt}{\sqrt{2}}P_A - \left(\frac{gt}{\sqrt{2}} + r\right)P_1^{EPR}. \tag{9.127}$$

If the amplitude reflection coefficient is taken as $r \simeq -1$, so that $t \ll 1$, and if the gain of the modulators is tuned so that $gt/\sqrt{2} \simeq 1$, the contributions of the auxiliary beam and of the EPR beam fluctuations cancel, and one has

$$X_B \simeq X_A, \quad P_B \simeq P_A. \tag{9.128}$$

The two quadratures have been transmitted from Alice's beam to Bob's beam without introducing additional noise. Note that this technique does not rely on any post-selection process. It works in a deterministic, unconditional way.

9.8 Exercises for Chapter 9

We now reintroduce position and momentum variables $\hat{X}$ and $\hat{P}$ such that $[\hat{X}, \hat{P}] = i\hbar\hat{1}$ to keep track of the quantum corrections. Note that by setting $\hbar = 1$, one has removed the possibility of understanding the classical limit as stated by $\hbar \to 0$. In the following, $\hbar$ is to be considered the single parameter *deformation* between the Poisson brackets and the phase-space correspondence of the Hilbert space commutators, the Moyal brackets [47]. For convenience, we also introduce $\hat{x} = \hat{X}/x_0$ and $\hat{p} = \hat{P}/p_0$, where the zero point fluctuations respect $x_0 p_0 = \hbar/2$ so that $[\hat{x}, \hat{p}] = i$.

Exercise 9.1 Show that $|W(X,P)| \leq 2/h$. Discuss the relationship with classical phase-space densities $w(X,P)$ and the uncertainty principle.

Exercise 9.2 Compute the Wigner function for the Fock state $|n\rangle$. Note that it is rotationally symmetric, $W_n(x,p) = W_n(x^2 + p^2)$.

Exercise 9.3 Compute the Wigner function for $|\psi\rangle = \alpha_n|n\rangle + \beta_m|m\rangle$.

(a) Show that it can be written as $W_\psi = W_n + W_m + W_I$, where W_I stands for the interference term. Observe that the interference of two rotational symmetric Wigner functions it is not rotation symmetric.

(b) Show that the Wigner function for the mixed state is simply $W_M = W_n + W_m$.

Exercise 9.4 Show that the Wigner transformation from ψ to W is invertible. Also show that all symmetric wave functions $\psi(x) = \psi(-x)$ attain the maximum values for $W(x,p)$ at the origin of coordinates. Show that odd wave functions $\psi(x) = -\psi(-x)$ achieve the minimum (negative) value. Consider the case of a Schrödinger-cat state equal to the sum of two coherent states of opposite values.

Exercise 9.5 Consider the ground state of harmonic oscillator and the Hamiltonian function $H = \left(\frac{p^2}{2m} + \frac{m\omega^2 x^2}{2}\right)$. Compute the mean value of the energy as $\int dxdp\, WH$. Convince yourself that, even for a case in which $W > 0$, it cannot be interpreted as a phase-space classical probability. Show that the energy variance is not expected to be zero classically for this phase space distribution. Obtain the correct result for the variance by using the Wigner–Weyl transform of $\hat{H}^2\left(\hat{H}^2 \to H^2 - \frac{(\hbar\omega)^2}{4}\right)$. See [95].

Exercise 9.6 Show that the Wigner function undergoes an evolution equation that is, up to $\mathcal{O}(\hbar^2)$, identical to the Liouville classical equation for phase-space density. Do this by considering the time derivative of $W(x,p) = \frac{1}{\hbar}\int e^{-ipy/\hbar}\psi^*(x - y/2)\psi(x + y/2)dy$ and use the Schrödinger equation for ψ. See [34, 95].

Exercise 9.7 Evolution of the Wigner function.

(a) The Groenwold star product is defined by

$$f \star g = \sum_{n=0}^{\infty}\sum_{k=0}^{n} \frac{(-1)^k}{n!}\left(\frac{i\hbar}{2}\right)^n \binom{n}{k} \partial_P^k \partial_X^{n-k} f \times \partial_P^{n-k}\partial_X^k g.$$

Convince yourself that it can be written as

$$f \star g = f \exp\left(\frac{i\hbar}{2}\left(\overleftarrow{\partial}_X \overrightarrow{\partial}_P - \overleftarrow{\partial}_P \overrightarrow{\partial}_X\right)\right) g,$$

where $f\overleftarrow{\partial}_X g = (\partial_X f)g$ and $f\overrightarrow{\partial}_X g = f(\partial_X g)$.

(b) Show that $f \star g = fg + \frac{i\hbar}{2}\{f, g\} + \mathcal{O}(\hbar^2)$ where the single brackets stands for the classical Poisson brakets. Show that the Moyal bracket defined by $\{\{f, g\}\} = \frac{1}{i\hbar}(f \star g - g \star f)$ equates to the Poisson bracket plus a quantum correction proportional to powers of $\hbar$. Justify the name "sine bracket."

(c) Show that in Fourier space the star product (which has a "complicated look" in phase space) is transformed into a simple phase factor.

(d) Discuss the relation between "the quantum condition" $\frac{1}{i\hbar}[\hat{x}, \hat{p}] = 1$ and the classical canonicity condition $\{x, p\} = 1$.

(e) Compute and compare $\{x^n, p^m\}$, $\{\{x^n, p^m\}\}$, and $\frac{1}{i\hbar}[\hat{x}^n, \hat{p}^m]$. Discuss the $\hbar \to 0$ limit.

Exercise 9.8 Since the star product is an exponential operator it can be shown to act as a "displacement" operator generating "Bopp shifts" [47] defined by

$$f(x, p) \star g(x, p) = f\left(x + \frac{i\hbar}{2}\vec{\partial}_p, p - \frac{i\hbar}{2}\vec{\partial}_x\right) g(x, p).$$

Use this fact to compute the Wigner eigenfunctions of the harmonic oscillator. Derive and use the star eigenequation of eigenenergy E,

$$H \star W = EW.$$

10 Experiment: Parameter Estimation

Here, we briefly describe a few examples of very precise or very sensitive measurements. We will come back to them in Chapter 11 to analyze the way they can be optimized.

10.1 Measurement of Light Intensity: Concentration of Atomic Species

In order to determine the concentration n of molecules (for example of a lethal gas) contained in a given sample of length L, one can make the very simple following experiment, sketched in Figure 10.1: A beam of light of tunable frequency f is sent through the sample, and the transmitted intensity I is measured by a photodetector. Let I_0 be the intensity of the impinging beam in photons per second. When the frequency of the light is resonant with an absorption frequency f_0 of the atom, the transmitted intensity drops to the value $I = I_0 e^{-n\hat{\sigma}L} \simeq I_0(1 - n\hat{\sigma}L)$, where $\hat{\sigma}$ is the absorption cross section of the atom. One then gets n from the formula $n = (I_0 - I)/I_0\hat{\sigma}L$.

10.2 Measurement of Light Phase: Length Difference

One wants to determine a variation δx of an optical path, as caused by a gravitational wave, for example. For this purpose, one uses an interferometric set-up, like a Mach–Zehnder interferometer (Figure 10.2), with a light of wavelength λ and intensity I_0 (in photons per second). One records the variation with δx of the difference between the two output intensities,

$$I_1 - I_2 = I_0 \left(\sin^2 \pi\delta x_0/\lambda - \cos^2 \pi\delta x_0/\lambda\right) = -I_0 \cos 2\pi\delta x_0/\lambda, \qquad (10.1)$$

where δx_0 is the path difference between the two arms. Hence, a small variation δx around a mean path difference of $\lambda/4$ (dark fringe) is related to the measured signal by

$$\delta x = (\lambda/2\pi)(I_1 - I_2)/I_0. \qquad (10.2)$$

Figure 10.3 shows a sketch of one of the gravitational interferometers that are currently measuring bursts of gravitational waves [122]. It is an

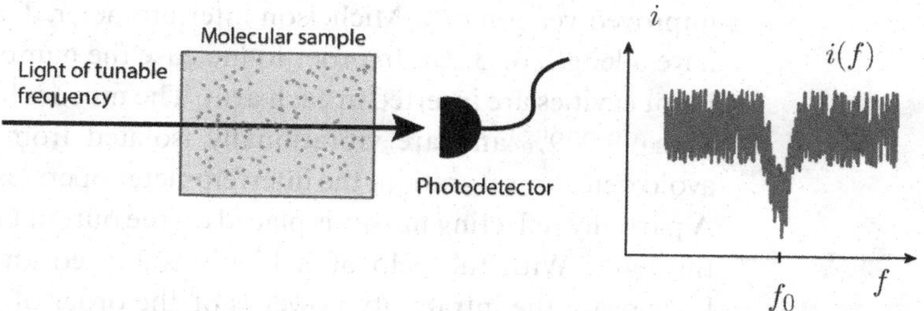

Fig. 10.1 Diagram of an absorption measurement performed on a weakly absorbing sample. The frequency of the probe beam is swept around the absorbing dip.

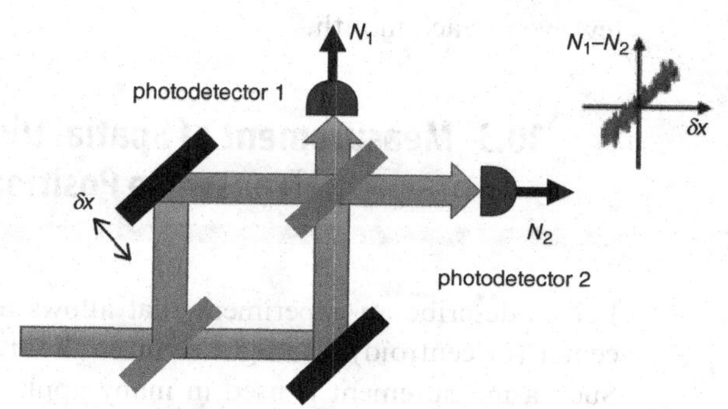

Fig. 10.2 Diagram of an interferometric phase shift measurement, based on a Mach–Zehnder interferometer containing a device able to change the phase of the field, for example by slightly varying the optical length of one of its arms. The difference between the intensities of the two output ports of the interferometer provides the estimation of the phase.

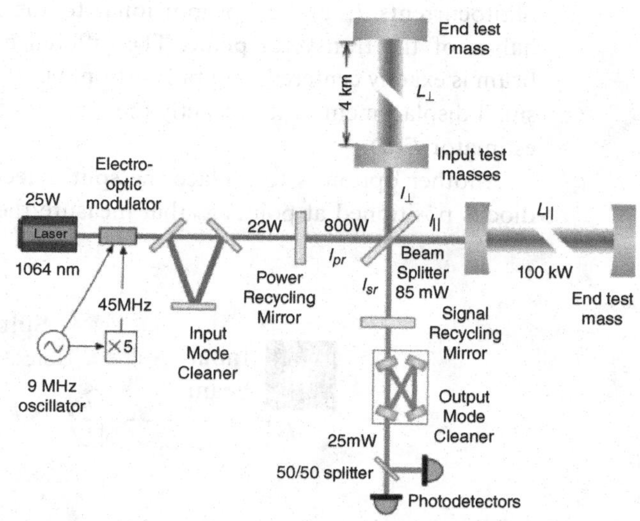

Fig. 10.3 Interferometric device allowing physicists to detect gravitational waves, based on a Michelson interferometer with kilometer-long arms. From D. Martynov et al. Sensitivity of the Advanced LIGO detectors at the beginning of gravitational wave astronomy, *Phys. Rev. D*, 2016 [122].

improved version of a Michelson interferometer. The interferometer arms have a length of 3 km. In order to increase the number of photons, Fabry–Perot cavities are inserted in each arm. The mirrors have reflectivities larger than 99.999% and are vibrationally isolated from the environment. To avoid detector saturation, the interferometer operates close to a dark fringe. A partially reflecting mirror is placed on the output beam in order to recycle the light. With the help of a highly stabilized and mode filtered input laser beam the intracavity power is of the order of 200 kW! This kind of set-up is able to detect a relative difference between the two arm lengths $\delta L/L$ of 10^{-21}, corresponding to one thousandth of the proton size. Several interferometers of this kind are currently deployed on earth, and detect a few events each month.

10.3 Measurement of Spatial Distribution of Light: Transverse Positioning

Let us describe an experiment that allows us to accurately position the center (or centroïd) of a light beam in the transverse plane (Figure 10.4). Such a measurement is used in many applications, such as, for example, in scanning tunneling microscopes, in which the very small deflection of the cantilever that scans the sample surface is often detected by sending on it a weak laser beam in the TEM_{00} mode and monitoring the deviation of the reflected beam. The displacement is usually measured by using a *split detector*, which is an analog double photodetector generating two photocurrents I_+ and I_- proportional to the light intensity in the two halves of the transverse plane. The difference $I_+ - I_-$ is zero when the beam is exactly centered, and proportional to the beam displacement d for small displacements – it is exactly the quantity required to provide a good estimator $E(d)$.

Another option is to replace the split detector by an array of photodiodes positioned at points x_i that measure the local intensity $\hat{I}(x_i)$. The

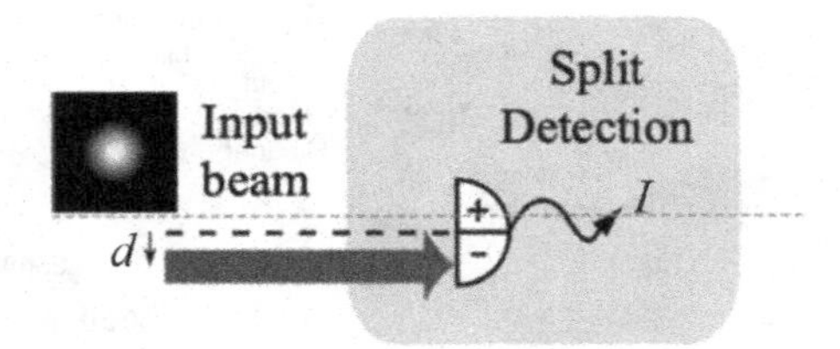

Fig. 10.4 Measurement of the transverse displacement of a light beam with the help of a split detector. One records the difference I between the intensities measured on the two halves of the split detector. If the beam is well centered, the signal I is zero, and a small transverse displacement d induces a nonzero signal. It therefore provides a good estimator of the displacement [51].

centroid position $\langle\hat{x}\rangle = \sum_i x_i \langle\hat{I}(x_i)\rangle$ is then calculated from these data. The method can be extended to the photon counting regime, replacing the analog detectors by single photon detectors positioned at points x_i that measure the local photon number $\hat{N}(x_i)$. The centroïd position $\langle\hat{x}\rangle = \sum_i x_i \langle\hat{N}(x_i)\rangle$ is then calculated from these data and averaged over many incoming photons N.

10.4 Measurement of Light Frequency: Metrology of Atomic Energy Levels

Here, we briefly describe the experiment that led to the value (11.1) for the Bohr frequency of the transition between the $1S_{1/2}$ and $2S_{1/2}$ levels of hydrogen (Figure 10.5) [68]. A frequency doubled diode laser at 243 nm excites by a two-photon transition the $2S_{1/2}$ level of hydrogen (at 121.5 nm) inside a linear enhancement resonant cavity. Hydrogen atoms are created by a RF discharge and, in order to reduce the doppler broadening, are cooled to 5K by collisions with the walls of a nozzle. The excited atoms then decay to the ground state via two-photon spontaneous decay, the regular single-photon one being forbidden by the selection rules. The $2S_{1/2}$ level lifetime is therefore exceptionally long (0.12 s), giving a linewidth of 2 Hz for the transition, one of the best transition Q-factors $(\nu/\Delta\nu) \simeq 10^8$ in atomic physics.

The frequency of the excitation laser at an exact resonance must then be determined as a function of the official time/frequency standard: the

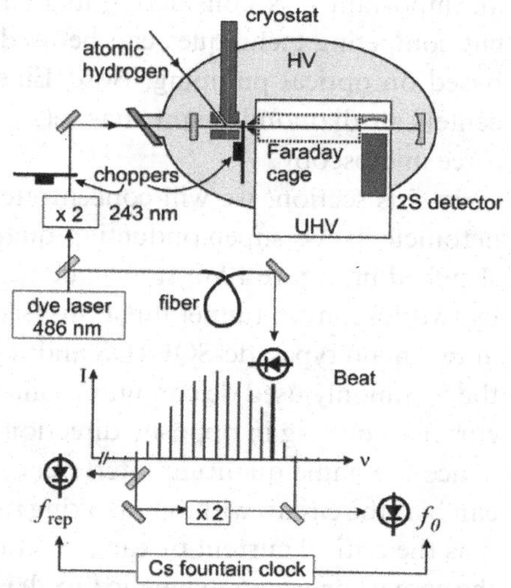

Fig. 10.5 Measurement of the 1S-2S transition in hydrogen, with the help of a frequency comb spanning more than one octave (bottom part of the figure), that allows us to precisely measure any frequency in the optical domain. From M. Fischer et al. New limits on the drift of fundamental constants, *Phys. Rev. Letters*, 2004 [68].

cesium fountain clock, the frequency of which is in the microwave domain, not the optical domain. To bridge the 10^5 gap in frequency, one uses a *laser frequency comb*, which emits coherent light made of many equidistant frequencies $\omega_0 \pm n\omega_c$, n being an integer, ω_0 the central frequency (in the optical domain), and ω_c the inter-mode frequency (in the microwave domain). This technique permits us to accurately determine the frequencies of all the teeth of the comb, which finally provide the ticks of a kind of frequency ruler spanning a large part of the optical domain, including the frequency of the excitation laser.

The excitation laser at 486 nm and the reference frequency comb are then mixed on a beamsplitter and the beat note frequency of lowest value is recorded. Knowing the teeth frequencies and the beat note frequency, the authors have all the information necessary to determine the excitation frequency of hydrogen with high accuracy, of the order of 10^{-15} for the relative accuracy.

Such impressive precision allows physicists, by comparing the present measurement to another one performed in 1999, to determine an upper limit for the relative temporal variation of the fine structure "constant" α, $\delta\alpha/\alpha$, equal to $(-0.9 \pm 2.9) \times 10^{-15}$ per year, a value compatible with zero.

10.5 Measurement of a Magnetic Field: The Superconducting Quantum Interference Device (SQUID)

Performing high sensitivity estimations of a small dc magnetic field B is an important task considering its numerous possible applications. Different competing techniques can be used [128]: For example, magnetometry based on optical pumping, Bose–Einstein condensates, nitrogen vacancy centers in diamond, giant magneto-resistance, the Hall effect, magnetic force microscopy.

In this section, we will concentrate on one of the most sensitive magnetometers: the superconducting quantum interference device (SQUID), sketched in Figure 10.6. It is made of a superconducting loop interrupted by two Josephson tunnel-junctions (see Appendix K). SQUIDs are divided in two main types, dc-SQUIDS and rf-SQUIDS. Here, we will only discuss the commonly used dc configuration. The loop hosts two superconducting currents rotating in opposite directions that interfere at the quantum level, hence the name quantum interference device. The maximal current I that can bias the circuit without breaking its superconductivity is $I_c = 2I_0$, where I_0 is the critical current of each junction, here considered identical. This is the case when a zero magnetic flux Φ is threaded by the loop. In the case of nonzero flux the critical current I_c of the SQUID depends on Φ [38]:

$$I_c(\Phi) = 2I_0 \left|\cos\left(\pi\frac{\Phi}{\Phi_0}\right)\right|, \tag{10.3}$$

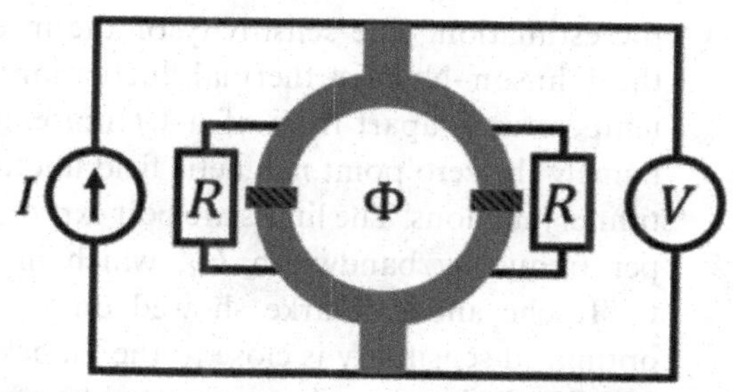

Fig. 10.6 Diagram of a SQUID, which is based on the Aharonov–Bohm interferometric effect. A superconducting loop is interrupted by two Josephson junctions that are shunted by resistances R. A constant current feeds the SQUID, and the generated voltage V is used to estimate the flux Φ threading the SQUID loop. From C. Tesche and J. Clarke. dc SQUID: noise and optimization, *J. Low Temperature Physics*, 1977 [168].

where $\Phi_0 = h/2e$ is the magnetic field quantum. This modulation is caused by a Mach–Zehnder-like interference: The wave-function of the superconducting fluid is recombined after being divided in the arms of the loop. The effective length of the interferometer arms is controlled by the flux due the Aharonov–Bohm effect [143]. By measuring the critical current, one can measure with great sensitivity the magnetic field normal to the SQUID surface. This is done by ramping up the external current bias I until the superconductivity of the Josephson junctions is broken and a finite voltage is developed across the device. The phenomenon suggests another mode of operation: By biasing the dc-SQUID slightly above I_c, a voltage develops across the broken tunnel junctions. This voltage is given by

$$V = \frac{R}{2}\sqrt{I^2 - I_c^2(\Phi)}, \tag{10.4}$$

where R is the finite resistivity of the junctions [38]. Operated in this regime, the dc-SQUID acts as a direct flux-to-voltage transducer producing up to a few millivolts signal over a single flux quantum. Using Equation (10.4) we find the sensitivity of the estimation to be

$$\frac{\partial V}{\partial \Phi} = -\pi \frac{I_0 R}{\Phi_0} \frac{I_0 \sin 2\pi \frac{\Phi}{\Phi_0}}{\sqrt{I^2 - I_c^2}}. \tag{10.5}$$

It is, in principle, infinite when $I = I_c$, but experimental factors contribute to re-normalize the divergence in Eq. (10.5). The device remains performant regardless [38].

A SQUID can sense a change in magnetic field much smaller than the magnetic flux quantum Φ_0. To give an order of magnitude, Φ_0 is close to the value of the magnetic flux generated by earth's magnetic field over the nucleus of a biological cell [39].

Reference [168] provides a detailed analysis of the sources of noise that limit the sensitivity of SQUIDs, and [128] contains a comparison with the other experimental techniques and a discussion of how to optimize the estimation. The sensitivity of the measurement is mainly limited by the Johnson–Nyquist thermal fluctuations in the resistances. At very low temperatures, apart from of a $1/f$ noise, it is limited by quantum effects, namely the zero point magnetic field fluctuations and the shot noise in the tunnel junctions. The limits are best expressed in terms of energy resolution per frequency bandwidth E_R, which has the dimension of an action. C. Tesche and J. Clarke showed on a specific SQUID model that the optimized sensitivity is close to the Planck constant $\hbar$ [168]. Experimental SQUID devices have been reported to have an experimental sensitivity as low as $2\hbar$ [128].

10.6 Exercises for Chapter 10

Exercise 10.1 Compute the density of molecules if the transmitted intensity across the sample of length $L = 1$ mm is $I/I_0 = 0.1$ and the scattering cross-section is $\sigma = (10 \text{ nm})^2$.

Exercise 10.2 Consider a cloud of two-level atoms of length L interacting with a monochromatic field of wavelength λ. The index of refraction can be computed from the Bloch equations to be $\mathcal{N} = 1 + \frac{\sigma n \lambda}{4\pi}\left(\frac{i-\delta}{1+\delta^2}\right)$. Here, $\delta = (\omega - \omega_0)/2$ and ω and ω_0 are the probe and atomic frequencies, respectively. The transmission coefficient t is such that $t^2 = I/I_0$. Use $\mathcal{N}$ to compute t.

Exercise 10.3 Say you have a laser beam of about 10 μm in diameter and a much larger photodetector. Explain how would you use a micrometric screw and a knife edge to measure the shape of your beam.

Exercise 10.4 Atomic spectroscopy can be modeled using a classical harmonically bound electron under dissipation (Lorentz model). Show that from this model one expects asymmetric line shapes and that the Lorentzian line shape is only an approximation valid around the summit of the line.

Exercise 10.5 What is the magnetic flux generated by earth's magnetic field though a red blood cell? What fraction of the critical current will that correspond to in a SQUID?

Exercise 10.6 A detectable gravitational wave produces a distortion in the space-time fabric inducing a modulation in the refraction index of a vacuum of about $\Delta n = \pm 10^{-22}$. The modulation is of opposite sign in orthogonal directions making a Michelson-type interference ideal for their detection. Compute the intensity at the output of the interferometer as a function of the optical path in each arm. Consider the case in which the laser has frequency noise and explain why the measurement is insensitive to it. Show that the minimal change in refraction index detectable is $\Delta n \sim \lambda/\pi L\sqrt{2\langle N\rangle}$. Here, λ is the laser wavelength, $\langle N\rangle$ is the mean number of photons detected, and

$L_1 \sim L_2 \sim L$ is the length of both arms of the Michelson interferometer. Consider a 100 W and $\lambda = 1$ μm laser: estimate from this the arm length L needed?

Exercise 10.7 How would you use a Fabry–Perot interferometer to do a better job? Compute the reflection coefficient R by summing coherently all the reflections in the mirrors (of reflectivity r_1 and $r_2 = 1$, respectively). Show that under the pertinent limits the phase shift the light acquires in the cavity before exiting is largely enhanced.

11 Theory: Parameter Estimation

11.1 Quantum Metrology

This final chapter is devoted to the characterization and optimization of parameter estimation based on quantum measurements. We will then evaluate the perturbation, or "back-action," that these measurements bring to the system. We will ultimately give other examples of quantum non demolition measurements.

The probabilistic character of quantum mechanics, and the existence of the Heisenberg inequality, often called "indeterminacy principle," seems to imply that the quantum world is undetermined, blurred, and "veiled," and that the physical parameters of microscopic objects cannot be precisely measurable. It is easy to find many counter-examples to this statement. The most spectacular is the $1S_{1/2} \rightarrow 2S_{1/2}$ transition frequency in hydrogen, undoubtedly a quantum object of importance, as it can be precisely compared with the predictions of quantum theory. The experiment is briefly described in Section 10.5. It leads to the following value of the transition frequency:

$$f_{1S\rightarrow 2S} = (2{,}466{,}061{,}413{,}187{,}035 \pm 10)\text{Hz} \qquad (11.1)$$

With a relative uncertainty of 4×10^{-15} it is one of the best-known experimentally determined physical quantities in the whole domain of physics!

Other physical parameters of quantum particles have been measured with great accuracy; for example, the magnetic moment of $g\mu_B$ of the electron, where μ_B is the Bohr magneton. The dimensionless gyromagnetic factor g has been measured to be:

$$g = 2.0023193043617 \pm 15 \times 10^{-13}. \qquad (11.2)$$

The relative uncertainty is here 7×10^{-13}.

So it is better to state that, even though some couples of parameters of microscopic objects cannot be simultaneously measured with perfect accuracy, such as the position/momentum of a particle, or the amplitude/phase of an oscillating field, there is no a priori restriction concerning the accuracy of a single parameter estimation. Note that the transition frequencies of an atom, the magnetic moment of a particle, and the dimensionless fine structure constant are intrinsic physical properties with immutable values: this is the domain of fundamental constants metrology, whereas position and momentum, for example, are contingent properties. This remark does not imply that determining the position and velocity of an object with minimum uncertainty is not an interesting and useful task for physicists.

"Quantum metrology," or quantum parameter estimation, is the domain of quantum physics, which studies these problems theoretically and experimentally and looks for the possible quantum limits in the estimation and the ways to reduce them. In this final chapter we also introduce the domain.

11.2 The Heisenberg Inequality

11.2.1 Reminder

The Heisenberg inequality was rigorously proved in Refs. [102, 154]. It is a direct mathematical consequence of the noncommutativity of some observables, which we will name $\hat{A}$ and $\hat{B}$. It states that

$$\forall\, |\psi\rangle, \forall\, \hat{A}, \hat{B}\,\text{Hermitian} \quad Var(A)\,Var(B) \geq \frac{1}{4}\left|\langle\psi\left|[\hat{A},\hat{B}]\right|\psi\rangle\right|^2. \quad (11.3)$$

It has the following meaning: Given N identically prepared successive copies of the state $|\psi\rangle$, one successively measures the physical quantity A on $N/2$ copies, and the physical quantity B on the other $N/2$ copies. One gets fluctuating results, from which one calculates the variances $Var(A)$ and $Var(B)$. These quantities fulfill the Heisenberg inequality (11.3). Repeated measurements of these quantities on any quantum state are affected by unavoidable fluctuations around the mean, often called "quantum noise," and the smaller the quantum noise affecting one of these quantities, the bigger the other one. Of course, when $\langle[\hat{A},\hat{B}]\rangle \neq 0$, both $Var(A)$ and $Var(B)$ can be equal to zero and the two quantities simultaneously determined with perfect accuracy.

Note that the inequality (11.3) is just a mathematical constraint. It has nothing to do with the perturbation of a quantum state by the measurement process we studied in Chapter 5. We will study the relation between these two different features of quantum measurements in detail in Section 11.4.

11.2.2 Generalized Heisenberg Inequality

Here, we give the demonstration in the pure state case, which can be straightforwardly extended to mixed states. Let us consider a quantum state $|\psi\rangle$ and two observables $\hat{A}$ and $\hat{B}$. One introduces the "fluctuation operators" of zero mean,

$$\delta\hat{A} = \hat{A} - \langle\hat{A}\rangle, \quad \delta\hat{B} = \hat{B} - \langle\hat{B}\rangle, \quad (11.4)$$

and write the Cauchy–Schwarz inequality $\langle\alpha|\alpha\rangle\langle\beta|\beta\rangle \geq |\langle\alpha|\beta\rangle|^2$ with $|\alpha\rangle = \delta\hat{A}|\psi\rangle$ and $|\beta\rangle = \delta\hat{B}|\psi\rangle$. One obtains that

$$Var(A)\,Var(B) \geq \left|\left\langle\delta\hat{A}\,\delta\hat{B}\right\rangle\right|^2. \tag{11.5}$$

We have

$$\delta\hat{A}\,\delta\hat{B} = \frac{1}{2}\left(\left[\delta\hat{A}\,\delta\hat{B}\right] + \left\{\delta\hat{A}\,\delta\hat{B}\right\}\right), \tag{11.6}$$

where $\{\delta\hat{A}\,\delta\hat{B}\}$ is the anticommutator $\delta\hat{A}\delta\hat{B} + \delta\hat{B}\delta\hat{A}$. It is easy to show that the commutator is anti-Hermitian and that the anticommutator is Hermitian, implying that the mean of the former is imaginary, and that of the latter real. Finally, one gets an inequality that is "tighter" than the usual one:

$$Var(A)\,Var(B) \geq \frac{1}{4}\left|\left\langle\psi\left|\left[\hat{A},\hat{B}\right]\right|\psi\right\rangle\right|^2 + \mathbf{C}^2_{A,B}, \tag{11.7}$$

where $\mathbf{C}_{A,B}$ is the correlation function introduced in Chapter 5.

11.2.3 Consequences

Equation (11.7) implies different inequalities:

1. $\mathbf{C}^2_{A,B}/(Var(A)\,Var(B)) \leq 1$. This implies that the symmetrized version of the normalized correlation function $c(A, B)$ defined in Equation (5.45) lies between -1 and 1. This inequality is a mathematical one, valid for any correlated couple of stochastic variables, classical or quantum.
2. $Var(A)\,Var(B) \geq \frac{1}{4}|\langle\psi|[\hat{A},\hat{B}]|\psi\rangle|^2$, the usual Heisenberg inequality. The saturation of the usual Heisenberg inequality implies that $\mathbf{C}_{A,B} = 0$, i.e. that there are no correlations between A and B in a minimum state.
3. The normalized correlation function obeys the inequality

$$c(A,B)^2 \leq 1 - \frac{\left|\left\langle\psi\left|\left[\hat{A},\hat{B}\right]\right|\psi\right\rangle\right|^2}{4\,Var(A)\,Var(B)}. \tag{11.8}$$

   Two quantities that do not commute cannot be perfectly correlated;
4. The conditional variance obeys the inequality

$$Var(A)\,Var(B)\left(1 - c(A,B)^2\right) \geq \left|\left\langle\psi\left|\left[\hat{A},\hat{B}\right]\right|\psi\right\rangle\right|^2/4. \tag{11.9}$$

   In the first term, one recognizes the conditional variance (5.48), at least when the statistics are Gaussian, so that one can write in this particular case an interesting Heisenberg inequality in the case of conditional measurements:

$$Var(A)\,Var(B|A) \geq \frac{1}{4}\left|\left\langle\psi\left|\left[\hat{A},\hat{B}\right]\right|\psi\right\rangle\right|^2, \tag{11.10}$$

   or equivalently (Figure 11.1),

$$Var(B)\,Var(A|B) \geq \frac{1}{4}\left|\left\langle\psi\left|\left[\hat{A},\hat{B}\right]\right|\psi\right\rangle\right|^2. \tag{11.11}$$

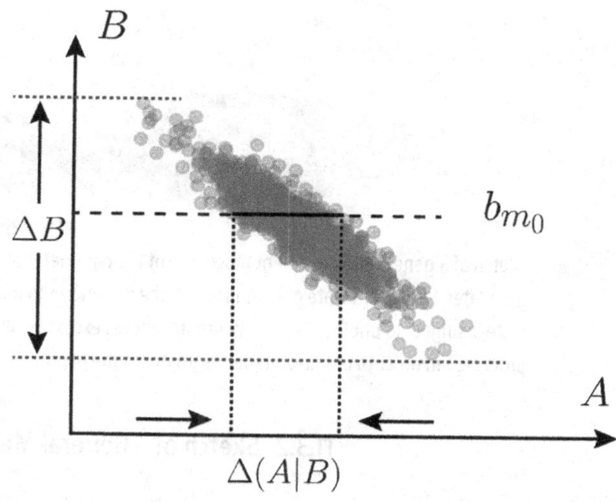

Fig. 11.1 The Heisenberg inequality $Var(B)Var(A|B) \geq \frac{1}{4}\left|\langle\psi\left|\left[\hat{A},\hat{B}\right]\right|\psi\rangle\right|^2$ sets constraints on the product of the variance on quantity B and of the conditional variance of the conjugate quantity A given the value b_{m_0} of the measurement result on B.

11.3 Quantum Cramér–Rao Bound

11.3.1 Position of Problem

One of the objectives in physics is to determine as precisely as possible the value of different physical parameters. We will call p a generic parameter, assumed to be real. In Chapter 10, we gave four examples of parameter estimation that we will use in our discussion. These were chosen from the domain of optics, a widely used technique for performing very accurate measurements. In these examples, from the measurement data and the knowledge of the physical properties of the measurement device, one gets a mathematical formula (or a data processing protocol) giving the parameter p of interest as a function of the measured quantities. This p-dependent formula is called an *estimator* $E(p)$ of p. The unavoidable imperfections of the device are responsible for a *technical noise* that limits the accuracy of the determination of p. This technical noise can be reduced, or even canceled, by improving the measurement device. But we also know that light is affected by a *quantum noise* of fundamental nature that cannot be completely eliminated. It gives rise to a *quantum limit* in the estimation of the parameter, which we would like to characterize and reduce as much as possible by optimizing the measurement protocol.[1]

[1] Note that measurements in physics are very often performed with the help of electrical signals instead of optical ones. We will not discuss them in this short textbook.

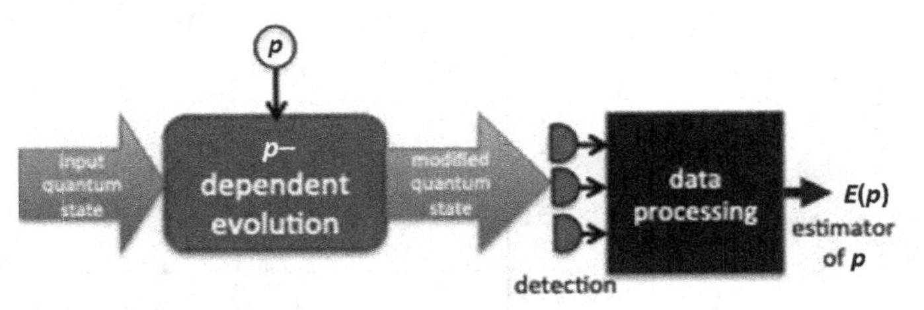

Fig. 11.2 Sketch of a generic measurement procedure of the parameter p: The experimental set-up generates a quantum state of light in which the parameter to measure has been imprinted. This state is subject to photodetection, which introduces excess noise, and the measurement results are processed in order to obtain a proper estimator of the parameter from the experimental data.

11.3.2 Sketch of a General Measurement

From these examples, one can draw a general diagram for the experimental determination of a physical parameter p (Figure 11.2). A physical device (the measurement set-up) transforms the initial quantum state $|\psi_0\rangle$ of the system into an output quantum state $|\psi(p)\rangle$ that depends on the parameter of interest p. Let us define the unitary operator $\hat{U}(p)$ and the generator $\hat{G}(p)$ of translations in p by

$$|\psi(p)\rangle = \hat{U}(p)|\psi_0\rangle = e^{ip\hat{G}}|\psi_0\rangle. \tag{11.12}$$

The state $|\psi(p)\rangle$ is then submitted to a quantum measurement that records one or several quantities (for the examples of Chapter 1, the intensity of a single beam is recorded for a concentration measurement, two intensities for an interferometric measurement, and many intensities for transverse positioning), and that can be repeated n times. The experimental data are then processed in order to generate the estimator $E(p)$, from which the experimental value of the parameter p is deduced. One assumes that all the technical imperfections of the device have been eliminated so that, in particular, the state of the measurement device is a pure state. The only source of uncertainty is the quantum fluctuations due to the measurement process itself. $E(p)$ is then a fluctuating quantity, the mean value of which is assumed to be the searched value p (unbiased estimator: Mean$(E(p)) = p$). Its variance, $Var(E(p))$, gives a value characterizing the accuracy of the measurement. It is, in particular, equal to the minimum variation δp_{min} of the parameter that gives rise to a measurable change in the measured data.

In this process, the experimentally determined data are not uncertain because they have been measured to values that can be recorded in the memory of the computer (Figure 11.3). What is not known is the value of the parameter p that leads to these recorded values of the data. One returns to the Bayesian approach of the probability: Each new measurement adds information which changes the conditional probability. The measurement device is characterized by the conditional probabilities for

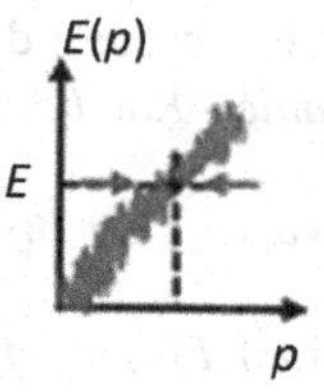

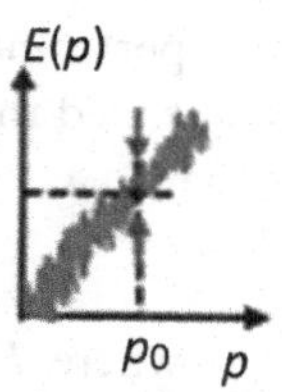

Fig. 11.3 Accuracy in the estimation of a parameter p. (Left) The horizontal arrow gives the uncertainty on the parameter p given that a certain experimental result $E(p)$ has been obtained. (Right) The vertical arrow gives the dispersion of the measured values $E(p)$ for a known value p_0 of the parameter. Both are linked by the Bayes formula.

the value of the parameter p knowing the result $E(p)$ of the measurement: $Proba(p|E(p))$. This probability is related through Bayes' theorem to the reverse conditional probability $Proba(E(p)|p)$, a quantity that one can determine provided one knows the details of the experimental set-up and the probability distribution of the quantum fluctuations of the measured quantities.

11.3.3 Quantum Cramér–Rao Bound

There are different ways to improve the accuracy of a measurement:

1. The simplest is to repeat the measurement n times, which improves the sensitivity by a factor of $\sqrt{n}$. This is not always possible such as, for example, if one wants to measure a parameter that is changing quickly with time.
2. Another way is to improve the data processing technique used to deduce the estimator from the recorded measurement results. Information theory [152] tells us that there is a lower limit for $Var(p_{est})$ known as the *classical Cramér–Rao bound*, B_{CR}, a quantity minimized over all possible data processing protocols, and equal to

$$Var(p_{est}) \geq B_{CR} = \frac{1}{nF(p)}, \tag{11.13}$$

where F is the *Fisher information*, a very important concept for parameter estimation. It is defined as

$$F(p) = \int dE(p)\ Proba(E(p)|p) \left| \frac{\partial(\ln(Proba(E(p)|p)))}{\partial p} \right|^2 . \tag{11.14}$$

This limit is very general and depends only on the conditional probability distribution $Proba(E(p)|p)$. It does not tell us the way to reach it. Note that the minimum measurable variation of p, equal to the square root of the variance, decreases with n as $1/\sqrt{n}$.
3. It is also possible to optimize the measurement procedure. One then reaches a new bound, optimized over all possible POVMs likely to be

performed on the physical system described by the wavevector $|\psi(p)\rangle$, called the *quantum Cramér–Rao bound* B_{QCR} [65, 92, 93],

$$Var(p_{est}) \geq B_{QCR} = \frac{1}{nF_Q(p)}, \tag{11.15}$$

where F_Q is the *quantum Fisher information*. One can show that it is equal to

$$F_Q(p) = 4\,Var\left(\hat{G}(p)\right), \tag{11.16}$$

where $Var(\hat{G})$ is the variance in state $|\psi_0\rangle$ of the operator $\hat{G}(p)$ defined in Equation (11.12) [65].

For example, if the parameter p that one wants to measure is the spatial displacement x of a quantum particle, then $\hat{U}(x)$ is the spatial translation operator $\hat{U}(x) = e^{i\hat{p}_x x/\hbar}$, and $\hat{G}(x) = \hat{p}_x/\hbar$ is proportional to the conjugate momentum operator $\hat{p}_x$, so that the minimum measurable displacement δx_{min} is

$$\delta x_{min} = \frac{\hbar}{2\sqrt{n\,Var(\hat{p}_x)}}. \tag{11.17}$$

Actually the minimum measurable displacement δx_{min} can be much smaller than the dispersion of x values in the initial quantum state: Suppose that the initial state fulfills $Var(\hat{p}_x)\,Var(\hat{x}) = M^2\hbar^2/4$ where M is a large number. In this case one has

$$\delta x_{min} = \frac{\sqrt{Var(\hat{x})}}{M\sqrt{n}}. \tag{11.18}$$

A precise measurement is obtained by multiplying the measurements and by using a quantum state well above the Heisenberg inequality limit on the conjugate quantity.

Note that, in the general case and for a single run ($n = 1$), one can write Equation (11.15) in the form of a Heisenberg-like inequality,

$$Var(p_{est})\;Var(\hbar\hat{G}) \geq \frac{\hbar^2}{4}, \tag{11.19}$$

where $\hbar\hat{G}$ appears as a kind of *complementary, or conjugate, quantity* of the measured quantity p, even though the context is, of course, different.

4. Another possibility is to optimize the initial quantum state $|\psi_0\rangle$ on which one imprints the information about p. So far, there is no simple and general expression for the lower limit of the variance $Var(p_{est})$ optimized over all possible initial quantum states, pure or mixed. So the determination of a state allowing experimentalists to estimate a parameter with a minimum minimorum uncertainty is not yet solved in the general case, though in the literature one can find many expressions of lower limits in particular cases. Also, note that if there is an observable $\hat{P}$ such that $\hat{P}|p\rangle = p|p\rangle$ with eigenvalues belonging to a continuous interval of values, then there are no quantum fluctuations when one repeatedly measures the parameter p on this state.

5. An optimization process that we have not taken into account so far concerns the shape and number of the *modes* of the electromagnetic field in which the quantum states "live." Let us introduce the following mode, which plays a central role in the present discussion, called the "detection mode," $u_{det}(\mathbf{r}, t)$:

$$u_{det}(\mathbf{r}, t) = p_0 \frac{\partial}{\partial p} u_{mean}(\mathbf{r}, t, p)|_{p=0}, \tag{11.20}$$

where $u_{mean}(\mathbf{r}, t, p)$ is the mean field mode,

$$u_{mean}(\mathbf{r}, t, p) = \frac{1}{E_\ell \sqrt{N}} \left\langle \psi(p) \left| \hat{E}^{(+)}(\mathbf{r}, t) \right| \psi(p) \right\rangle, \tag{11.21}$$

where E_ℓ is the single photon field, N the mean photon number, p_0 is the scaling factor necessary to normalize to 1 the mode u_{det}, and u_{det} can be used as the first mode of a new mode basis. For the sake of simplicity, we will restrict our discussion to Gaussian pure states that have p-dependent covariance matrices Γ. One can show that in this case the quantum Cramér–Rao bound has the following expression [144]:

$$B_{QCR} = \frac{p_0^2}{2N} \mathcal{N}_{detection}. \tag{11.22}$$

The noise factor $\mathcal{N}_{detection}$ depends on the covariance matrix Γ. It is equal to $(1/\Gamma^{-1})_{u_{det}}$. More precisely, it is the diagonal element of matrix Γ^{-1} in the detection mode. If the noise in this mode is not correlated with the fluctuations in the other modes, $\mathcal{N}_{detection}$ is simply *the variance of the quantum noise in the detection mode.* So, the noise perturbing the measurement of parameter p is the quantum noise of the detection mode, hence its name. The fluctuations of the other modes, quantum or classical, do not affect the measurement of the parameter p. $\mathcal{N}_{detection}$ is equal to the vacuum noise variance 1/2 for a coherent state. To reduce further the quantum Cramér–Rao bound, one needs to inject a squeezed state in the proper quadrature of the detection mode.

In addition, the quantum Cramér–Rao limit for the variance is inversely proportional to the total mean number N of photons in the light beam. In the family of Gaussian states for which the present analysis is valid, there are also the coherent states generated by intense lasers, for which N can reach values of the order of 10^{15} for measurement times in milliseconds. B_{QCR} can then be much lower than the limit that can be achieved using non-Gaussian states such as *NOON* states, which we will define below, for which the $1/N^2$-scaling for the variance is more favorable than in Gaussian states, but which are only available for small N values.

11.3.4 Examples

Intensity Measurement

We go back to the example of Section 10.1, in which one uses an intensity measurement to determine the concentration of some molecular species. The parameter to measure is here the photon number N, associated with the number operator $\hat{N}$. We examine different possible states of light likely to be used in the measurement:

1. *Number states*. They are the eigenstates of the measurement observable, and have no intensity fluctuations. One would think that such states would provide a perfectly accurate measurement. However, the spectrum of the number operator is discrete and not continuous. Hence, the minimum measurable intensity variation when one uses the number state $|n\rangle$ corresponds to the absorption of a single photon $|n\rangle \to |n-1\rangle$. This implies that the minimum detectable density of molecules is given by

$$n_{min} = \frac{1}{IT\hat{\sigma}L} = \frac{1}{N\hat{\sigma}L}. \tag{11.23}$$

   This limit, scaling as $1/N$, is an example of the so-called *Heisenberg limit*.
2. *Coherent states*. In practical situations, one uses a laser as a source, which generates a light described at the quantum level by a coherent state $|\alpha\rangle$, having Poissonian photon statistics, for which $Var(N) = |\alpha|^2 = \langle N\rangle$. The minimum detectable concentration is then

$$n_{min} = \frac{1}{\sqrt{N}\hat{\sigma}L}. \tag{11.24}$$

   This limit, scaling as $1/\sqrt{N}$, is called the *standard quantum noise limit*, or the shot noise limit.[2] It can be reduced by increasing the laser intensity or the measurement time.

   Up to now, Fock states have been generated experimentally only for small N values, around a few tens, whereas coherent states generated by lasers may have mean photon numbers of the order of 10^{15}. Practically speaking, in the present situation it is preferable to use lasers instead of number states.
3. *Sub-Poissonian light*. This has a photon number variance $Var(n)$ smaller than the Poisson value $\langle N\rangle$, hence its name. Such nonclassical states of light are generated by, for example, high quantum efficiency diode lasers [167, 184] that emit a photon for each transition of an electron from the valence band to the conduction band. This implies that the statistics of emitted photons is identical to the statistics of electrons in the electrical current traversing the junction. This statistics are governed by Nyquist thermal electrical current fluctuations of variance $4k_B T\Delta f/R$ (where Δf

[2] There is another shot noise limit, which is obtained when one uses photodetectors of low quantum efficiency. The photocurrent statistics in this case are dominated by the random occurrence of quantum jumps in the detector junction, and do not depend on the statistics of arrival of photons.

is the bandwidth), so it is possible to reduce them below the shot noise variance $2qI\Delta f$ (where q is the electron charge) by using an electrical resistance R of high enough value in the circuit.

4. *Bright squeezed light.* When the mean field is large compared to its fluctuations ("bright beam"), the fluctuations $\delta\hat{N}$ of the photon number are given by

$$\delta\hat{N} = \delta(\hat{a}^\dagger\hat{a}) \simeq \delta\hat{a}^\dagger\langle\hat{a}\rangle + \delta\hat{a}\langle\hat{a}\rangle^* = \sqrt{2}\langle\hat{a}\rangle\delta\hat{X} = \sqrt{2\langle N\rangle}\delta\hat{X} \quad (11.25)$$

when $\langle\hat{a}\rangle$ is real. This means that the photon number variance is, in this case, proportional to the X quadrature variance, which can be reduced below the standard quantum noise limit by a factor e^{-2S} (Section I.2). The minimum signal in this case is given by

$$n_{min} = \frac{1}{\sqrt{N}\hat{\sigma}L}e^{-2S}. \quad (11.26)$$

It is reduced by the squeezing factor and varies like $\frac{1}{\sqrt{N}}$.

5. *Twin beams.* They are generated by parametric down-conversion in a doubly resonant cavity resonant for the signal and idler modes, a device called optical parametric oscillator [90], mentioned in Section I.3. Above an oscillation threshold for the pump intensity, it delivers "twin beams," i.e. two spatially coherent, laser-like signal and idler fields having the same intensity fluctuations, $Var(\hat{a}_s^\dagger\hat{a}_s - \hat{a}_i^\dagger\hat{a}_i) = 0$. The set-up is then very simple [165]. One inserts the sample on the signal beam and records the signal-idler intensity difference. When using twin beams, this quantity has no noise. Here, we can make the same remark as for direct detection: The minimum variation of this quantity is one photon, so that the minimum detectable concentration is

$$n_{min} = \frac{1}{N_s\hat{\sigma}L}, \quad (11.27)$$

where N_s is the number of photons in the signal beam. It is a Heisenberg-like limit.

In practice, the signal-idler difference is not zero because of imperfections in the device. The residual noise is $Var(N_s - N_i) = \frac{1}{2}\sqrt{N_s}e^{-2S}$, to be compared with the value obtained using the two output beams 1 and 2 of a 50% beamsplitter, $Var(N_1 - N_2) = \frac{1}{2}\sqrt{N_1}$, so that,

$$n_{min} = e^{-2S}\frac{1}{2\sqrt{N_s}\hat{\sigma}L}. \quad (11.28)$$

Phase Shift Measurement

This example is important, as many ultrasensitive measurements are performed using interferometric techniques. The phase-shift operator $\hat{U}_\phi$ is in this case equal to $\hat{U}_\phi = e^{-i\phi\hat{a}^\dagger\hat{a}}$. It looks like the evolution operator $\hat{U}(t) = e^{-i\omega t\hat{a}^\dagger\hat{a}}$ because phase shifts induced by spatial propagation or time evolution are closely related for monochromatic light. The $\hat{G}$ operator introduced in Equation (11.16), equal to $\hat{G} = \hat{a}^\dagger \hat{a}$, is the photon number operator. One therefore has the quantum Cramér–Rao bound for a single measurement ($n = 1$):

$$\delta\phi_{min} = \frac{1}{2\sqrt{Var(N)}}. \tag{11.29}$$

So the larger the photon number dispersion, the lower the quantum Cramér–Rao bound. Different techniques can be implemented to reach it, and improve it:

1. "*ON*" *state*. The first technique one can think of is to measure a phase shift of a single mode beam. In this case, the operator $\hat{U}_\phi$ has the form given above. The best quantum state, having the largest photon number dispersion while keeping the same mean $\langle N \rangle$, is in this case the "*ON*" state,

$$|ON\rangle = (|0\rangle + |2N\rangle)/\sqrt{2}. \tag{11.30}$$

The minimum detectable phase shift is then,

$$\delta\phi_{min} = \frac{4}{N} = \frac{4}{IT}. \tag{11.31}$$

It corresponds to the Heisenberg limit for such a measurement.

2. "*NOON*" *state*. Direct phase measurements are hampered by an inevitable classical phase noise of the single mode source. It is therefore better to use a set-up with two paths, and therefore two modes, for example a Michelson or a Mach–Zehnder interferometer (Figure 10.2), in which the common mode phase noise can be eliminated by a differential intensity measurement between the two output beams of the interferometer. For this configuration, one still needs a maximum spread of the the photon distribution, and the best state turns out to be the "NOON state":

$$|NOON\rangle = (|2N, 0\rangle + |0, 2N\rangle)/\sqrt{2}, \tag{11.32}$$

for which the minimum detectable phase shift is the same as in Equation (11.31).

When it propagates through the interferometer, the NOON state evolves into

$$|\Psi_\phi\rangle = \frac{1}{\sqrt{2}} e^{-i\phi\hat{a}_1^\dagger\hat{a}_1}(|2N, 0\rangle + |0, 2N\rangle) = \frac{1}{\sqrt{2}}\left(e^{-i2N\phi}|2N, 0\rangle + |0, 2N\rangle\right). \tag{11.33}$$

Note that the sensitivity of this scheme is increased because the phase coefficient $e^{-i2N\phi}$ evolves $2N$ times faster than in a single photon state. In the present case, $Var(N) = 4N^2$, so that $\delta\phi_{min} = 1/4N$. The NOON state also allows us to reach the Heisenberg limit.

3. *Gaussian pure states.* Let us now restrict ourselves to the "practical" states, which can be produced with very high photon numbers, i.e. the Gaussian pure states, including coherent and squeezed states. It is possible to calculate the quantum Cramér–Rao bound in this case. One finds the same limit as in Equation (11.29). For intense Gaussian states of real mean value, $Var(N)$ is equal to $2\langle N\rangle Var(\hat{X})$, and the quantum Cramér–Rao is

$$\delta\phi_{min}^{Cramr\text{–}Rao} = \frac{1}{2\sqrt{2NVar(X)}}, \tag{11.34}$$

where N is the total number of photons used in the measurement. For a coherent state, $Var(X) = 1/2$, so that

$$\delta\phi_{min}^{Cramr\text{–}Rao,coherent} = \frac{1}{2\sqrt{N}}. \tag{11.35}$$

The $1/\sqrt{N}$ dependence is the signature of a "standard quantum noise" or shot noise limit. For a pure P-squeezed state of squeezing factor e^{-2S}, which is also anti-squeezed by the same factor on the X quadrature, the limit is:

$$\delta\phi_{\rm min}^{\rm Cramer\text{–}Rao,squeezed} = \frac{e^{-S}}{2\sqrt{N}}. \tag{11.36}$$

In Chapter 10, we presented an experimental set-up, namely the Mach–Zehnder interferometer, allowing us to measure a phase shift. It is possible, using the well-known beamsplitter formula (H.28), to find the expression of the measured signal $\hat{N}_{1,out} - \hat{N}_{2,out}$ in terms of two input modes $\hat{a}_{1,in}$ and $\hat{a}_{2,in}$ of the first Mach–Zehnder beamsplitter, and in the approximation of the linearized quantum fluctuations, when one shines an intense beam of real mean field $\langle a_{1,in}\rangle = \alpha_1$ on input port 1 and a field of zero mean on input port 2, one finds that

$$\hat{N}_{1,out} - \hat{N}_{2,out} = \left(\alpha_1^2 + \alpha_1\sqrt{2}\hat{X}_{1,in}\right)\cos 2\phi + \alpha_1\sqrt{2}\hat{X}_{2,in}\sin 2\phi. \tag{11.37}$$

We consider a small variation $\delta\phi$ of the phase around the balance point $\phi = \pi/4 + \delta\phi$. The variance of the noise on the signal in absence of a phase shift ($\delta\phi = 0$) has the expression

$$Var(N_{1,out} - N_{2,out}) = 2\alpha_1^2 Var(X_{2,in}). \tag{11.38}$$

It is equal, in a way similar to the balanced homodyne signal (Equation (9.96)), to the product of the mean photon number $N_1 = \alpha_1^2$ entering through port 1 and the fluctuations of the X quadrature entering through port 2 [35]. The signal in the absence of noise when $\delta\phi \neq 0$ is simply $2\alpha_1^2\delta\phi$. Equating the signal squared to the noise variance gives the minimum detectable phase shift measured by this precise apparatus gives

$$\delta\phi_{min}^{Mach\text{–}Zehnder} = \frac{\sqrt{Var(X_{2,in})}}{\sqrt{2N_{1,in}}}, \tag{11.39}$$

equal to $1/(2\sqrt{N_{1,in}})$, or $e^{-S}/(2\sqrt{N_{1,in}})$, if vacuum, or squeezed vacuum, enters through port 2. It is identical to the Cramér–Rao bound for coherent and squeezed states (Equations (11.35) and (11.36)).

We have therefore found a practical way to reach the quantum Cramér–Rao bound in an interferometer. This optimum noise reduction scheme with the help of squeezed states is currently implemented in the giant interferometers that have been built in several locations to detect gravitational waves [2]. In the years 2020s it increases the signal-to-noise ratio by roughly 50%.

Measurement in the Transverse Plane

A subject that is very important because of its potential applications to microscopy concerns *imaging*. In optical instruments made of lenses the minimum size of a detail in an image is due to the diffraction of light through the instrument and is given by the transverse extension of the image of a point-like object, called *point spread function* (PSF). It is the Rayleigh criterion. Actually, one can do much better using appropriate data processing techniques [91], for example by numerical deconvolution of the PSF. One then reaches a limit related to the quantum noise of the measurement. To simplify this discussion, we will restrict ourselves to the estimation of a single parameter p in the transverse plane: that of the lateral position of a TEM_{00} light beam center along the direction Ox of the transverse plane. The mean field mode is the displaced Gaussian mode $u_{mean}(x,t,p) = (2/\pi w_0^2)^{1/4} e^{-(x-p)^2/w_0^2} e^{-i\omega t}$, w_0 being the beam waist size. The detection mode, defined by Equation (11.20) is $u_{det}(x,t,) = p_0 \partial u_{mean}/\partial p$ with $p_0 = w_0/2$: it is the TEM_{10} Hermite–Gauss mode. The minimum measurement displacement for a Gaussian pure state in mode TEM_{00} is the quantum Cramér–Rao bound given by Equation (11.17)

$$\delta p_{min}^{QCR} = \frac{w_0}{2\sqrt{N}} \sqrt{\mathcal{N}_{detection}} \tag{11.40}$$

Using a TEM_{00} laser beam in an intense coherent state, N can be very large, and therefore the minimum displacement $w_0/(2\sqrt{N})$ can be orders of magnitude smaller than the light spot size, provided one uses intense enough light and/or long exposure times. The minimum value of w_0 for an optical microscope is of the order of the wavelength, i.e. $\simeq 1$ μm, and the minimum measurable displacement of such a spot can be as small as 10^{-14} m using a focused beam of total power 1 mW and an exposure time of 1 s. The quantum limit can be even further reduced by injecting a squeezed vacuum state in the detection mode [51]. As explained in Section 10.3, one can use the split detector method for such a measurement: The minimum displacement measurable by this precise method (signal-to-noise ratio equal to 1) is then calculated to be $w_0/(\pi\sqrt{N})$, which is 22% above the quantum Cramér–Rao bound. This implies that there are more sensitive methods for measuring the displacement: The QCR limit can be reached, for instance,

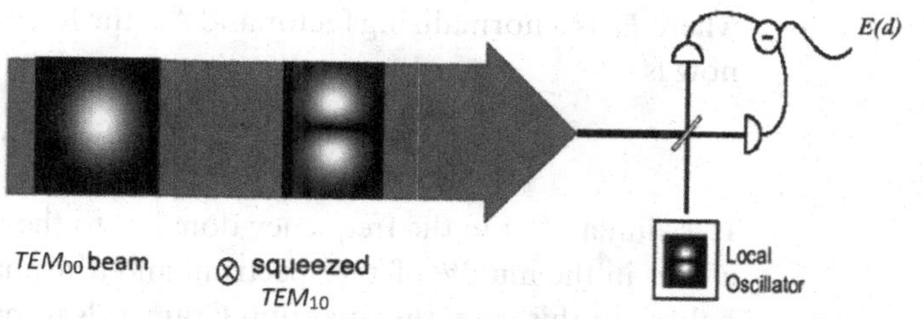

Fig. 11.4 Optimized measurement of the transverse displacement of a TEM_{00} mode using homodyne detection with the detection mode as a local oscillator, which is, in the present case, the TEM_{10} mode. The measurement can be improved further by adding a squeezed state in the TEM_{10} detection mode [51].

by using balanced homodyne detection with the detection mode as a local oscillator, which is in the present case a TEM_{10} beam centered at $x = 0$ (Figure 11.4).

One knows that the regular homodyne detection, detailed in Section 9.4, gives successive access to the two field quadratures $\hat{X}$ and $\hat{P}$, which are quantum conjugate observables, simply by shifting by $\pi/4$ the phase of the local oscillator. One can do the same for the homodyne detection with a TEM_{10} local oscillator: By shifting by $\pi/4$ the LO, one goes from the estimation of the beam displacement to the estimation of the conjugate quantity. It turns out that it is the *tilt* of the beam with respect the Oz axis.

The QCR bound can also be reached by using detector arrays, either analog or digital, followed by data processing. As explained in Chapter 10, the optimized transverse position estimator is then the centroid position $\langle \hat{x} \rangle = \sum_i x_i \langle \hat{I}(x_i) \rangle$. The bound can be lowered by using nonclassical states of light: vacuum squeezed states in the detection mode TEM_{10} for the homodyne detection configuration, and non-Gaussian states such as NOON states for the centroid measurement configuration.

One can extend this discussion by considering more practical imaging situations, for instance the estimation of the distance between two incoherent Gaussian-shaped sources of light. It requires the determination of the quantum Cramér–Rao bound for mixed states, which is out of the scope of this text.

Frequency Measurement

As shown in Section 10.4, frequency measurements are performed by mixing on a beamplitter a light beam of unknown frequency ω_0 and linewidth $\Delta\omega$, supposed to be in transverse mode $u_{TEM_{00}}$, with a reference beam, often a laser frequency comb. A low-pass filter selects the beat note of lowest frequency ω_{ref}, which is actually the parameter one wants to measure. It lies in the radiofrequency domain, so that one can measure the oscillating light field u_{beat} itself and not only its intensity,

$$u_{beat}(\omega) = E_0 e^{-i\omega t} e^{-(\omega-\omega_{ref})^2/4\Delta\omega^2}, \tag{11.41}$$

where E_0 is a normalizing factor and $\Delta\omega$ the linewidth. The detection mode now is

$$u_{det}(\omega) = \frac{\omega - \omega_{ref}}{\Delta\omega} u_{beat}(\omega). \tag{11.42}$$

It is similar, but in the frequency domain, to the $u_{TEM_{00}}$ spatial mode, with a zero in the middle of the spectrum and a π phase shift between the two halves. In this case, the quantum Cramér–Rao bound for a coherent state illumination can be shown to be

$$B_{QCR} = \frac{\Delta\omega}{\sqrt{N}} = \frac{\Delta\omega}{\sqrt{IT}}. \tag{11.43}$$

As for the other examples in this section, the QCR limit for coherent states is much smaller than the linewidth, by a factor $\sqrt{N}$, which is very large in the microwave domain and with long interrogation times. This limit is very difficult to attain in metrology experiments. However, it is possible to increase the signal-to-noise ratio using a homodyne detection with the detection mode of spectral shape $u_{det}(\omega)$ as a local oscillator. This has been done in a demonstration experiment [31].

11.4 Measurement-Induced Perturbation of a Quantum State

11.4.1 The Heisenberg Microscope

As we know from Chapter 5, any measurement performed on a quantum state either destroys it or perturbs it. When the state is not destroyed, the question is how to minimize the perturbation while keeping a good accuracy in the measurement. This kind of issue was considered qualitatively by Heisenberg in his famous example of the "Heisenberg microscope" that we recall here in a modernized context. The experimental set-up is displayed in Figure 11.5.

In order to precisely measure the position of a particle along the x axis, assumed to be initially at rest, one uses a light beam of wavelength λ propagating along Oz. The light that is scattered by the particle is collected by a microscope objective that images on the image plane the particle situated on the object plane. As we saw in the Section 11.3, in the case of illumination by a quantum coherent state, the accuracy of the positioning is given by the size of the image of the particle divided by the square root of the photon number. The image size is limited by diffraction to the value $\Delta x_{meas} \simeq \lambda/NA$, $NA = \sin(\phi_M)$ being the numerical aperture of the lens, where ϕ_M is the half-angle of the light cone focused by the microscope. It can be increased by using short wavelengths (UV, X-rays) or large numerical apertures.

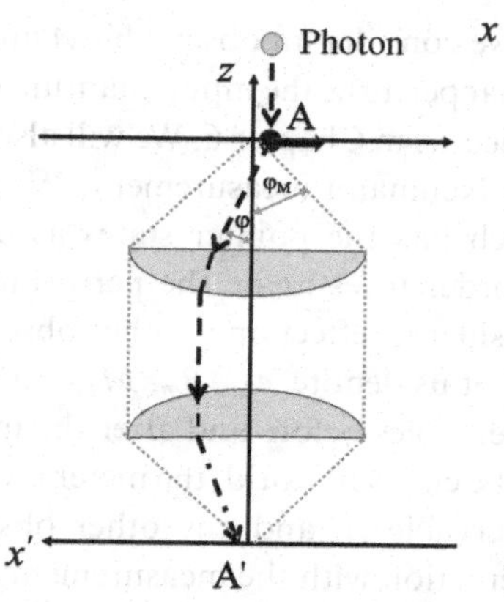

Fig. 11.5 The "Heisenberg microscope": The dashed arrows represent the trajectory of a photon incident along the z-direction and scattered by the particle located at point A, the transverse position x of which one wants to accurately measure; the bold arrow on the x-axis shows the inevitable random recoil effect induced by the scattering of the photon on the particle.

This light inevitably perturbs the particle, because there is a momentum transfer between the light and the particle that pushes the particle in a random direction. This perturbation can be reduced by lowering the light intensity, and the minimum perturbation will be obtained when a single photon is scattered. The momentum gained by the particle along x and due to the scattering of the photon in direction ϕ with respect to Oz is $p_x = \hbar k \sin\phi = h \sin\phi/\lambda$. The photon is scattered by the particle in a random direction inside a cone of collected light of half-angle ϕ_M, so that the uncertainty on the acquired momentum is $\Delta p_{x,after} \simeq h \sin\phi_M/\lambda$. Finally, one obtains

$$Var(x_{meas})\, Var(p_{x,after}) \simeq h^2. \tag{11.44}$$

This relation looks like the Heisenberg inequality (11.3) and it is easy to confuse the two relations: The Heisenberg microscope is often presented in textbooks as an illustration of the Heisenberg inequality. But actually it is very different, as it deals with the effect of a measurement on the quantum state of the system after it has been submitted to a measurement process, whereas the Heisenberg inequality is a general mathematical property of quantum states and operators. It does not depend on the measurement postulate (M3) of quantum mechanics (see Appendix A) giving the quantum state after a measurement.

11.4.2 Noise-Disturbance Inequality

A rigorous noise-disturbance relation has been derived by Ozawa [138]. To demonstrate it, we must use the Heisenberg representation, in which the observables evolve in the quantum measurement process instead of the quantum states.

We consider an observable $\hat{A}$ and a measurement device of the quantity A that perturbs the input quantum state and that we treat here as a quantum device, as in Chapter 6. We will also restrict ourselves to the case of an ideal von Neumann measurement. We call $\hat{M}$ the meter quantum observable, which has the pointer states as eigenstates, as described in Section 5.1. In order to estimate the perturbing effect of the measurement of A, we consider its effect on another observable $\hat{B}$.

Let us denote $\hat{A}_{in}$, $\hat{B}_{in}$, $\hat{M}_{in}$, and $\hat{A}_{out}$, $\hat{B}_{out}$, $\hat{M}_{out}$ to be the values of the observables before and after the interaction with the measurement device. In the classical world, the meter gives exact information about the measured observable $\hat{A}$, and any other observable like $\hat{B}$ is not perturbed by the interaction with the measurement set-up, so that we should have

$$\hat{M}_{out} = \hat{A}_{in}, \quad \hat{B}_{out} = \hat{B}_{in}. \tag{11.45}$$

This is not the case in quantum mechanics. As with the quantum amplifier (Section H.2), we must introduce additional "noise terms" in the previous relations:

$$\hat{M}_{out} = \hat{A}_{in} + \hat{N}_A, \quad \hat{B}_{out} = \hat{B}_{in} + \hat{D}_B, \tag{11.46}$$

where $\hat{N}_A$ represents the added quantum noise that prevents a perfect knowledge of the measured quantity A. $\hat{D}_B$ represents the *disturbance* in the nonmeasured quantity B arising from the interaction of the physical system with the measurement device. First, we use the Heisenberg inequality in Section 11.2 three times:

$$\begin{aligned}
&\Delta N_A \Delta D_B + \Delta A_{in} \Delta D_B + \Delta N_A \Delta B_{in} \\
&\quad \geq \frac{1}{2}\left(\left|\left\langle\left[\hat{N}_A, \hat{D}_B\right]\right\rangle\right| + \left|\left\langle\left[\hat{A}_{in}, \hat{D}_B\right]\right\rangle\right| + \left|\left\langle\left[\hat{N}_A, \hat{B}_{in}\right]\right\rangle\right|\right) \\
&\quad \geq \frac{1}{2}\left|\left\langle\left[\hat{N}_A, \hat{D}_B\right]\right\rangle + \left\langle\left[\hat{A}_{in}, \hat{D}_B\right]\right\rangle + \left\langle\left[\hat{N}_A, \hat{B}_{in}\right]\right\rangle\right|,
\end{aligned} \tag{11.47}$$

where $\hat{M}_{out}$ and $\hat{B}_{out}$ are operators acting on different Hilbert spaces, so that $[\hat{M}_{out}, \hat{B}_{out}] = 0$, implying that

$$\left[\hat{N}_A, \hat{D}_B\right] + \left[\hat{A}_{in}, \hat{D}_B\right] + \left[\hat{N}_A, \hat{B}_{in}\right] = -\left[\hat{A}_{in}, \hat{B}_{in}\right]. \tag{11.48}$$

Finally, one obtains

$$\Delta N_A \Delta D_B + \Delta A_{in} \Delta D_B + \Delta N_A \Delta B_{in} \geq \frac{1}{2}\left|\left\langle\left[\hat{A}_{in}, \hat{B}_{in}\right]\right\rangle\right|. \tag{11.49}$$

This inequality, named the Heisenberg–Ozawa inequality, puts strong constraints on the different variances when the two input observables $\hat{A}_{in}$ and $\hat{B}_{in}$ do not commute.

To be more specific, we will focus on the long-debated case of position and momentum measurements,

$$\Delta N_x \Delta D_p + \Delta x \Delta D_p + \Delta N_x \Delta p \geq \frac{\hbar}{2}. \tag{11.50}$$

This inequality implies that:

- $\Delta N_x = \Delta D_p = 0$ is forbidden: There is always some added noise and/or disturbance in a measurement process. The disturbance of the momentum p due to a measurement of x is called the "back-action" of the position measurement device.
- $\Delta N_x = 0$ is possible even for a finite momentum disturbance ΔD_p, but then $\Delta x \Delta D_p \geq \frac{\hbar}{2}$. A perfect measurement of the position is possible, the associated back-action on momentum being limited by the inverse of the dispersion of x values in the initial state.
- A nonmomentum-disturbing, called "back-action evading" or "quantum nondemolition" (QND) position measurement ($\Delta D_p = 0$) is possible, but this measurement is not perfect. One has the constraint $\Delta N_x \Delta p \geq \frac{\hbar}{2}$. The noise added in the measurement is limited by the inverse of the dispersion of p values in the initial state.
- The Heisenberg–Ozawa inequality does not forbid quantum states and measurement devices such that $\Delta N_x \Delta D_p < \frac{\hbar}{2}$. There exist physical situations where a (nonexisting but sometimes invoked) noise-disturbance Heisenberg inequality is violated. This is possible when the initial dispersions Δx and Δp are large enough. Recent experiments have indeed put in evidence this apparent "violation" [156].

11.5 Examples of Quantum Nondemolition (QND) Measurements

11.5.1 General Discussion

So far, we have mainly been concerned with destructive measurements, in which the system no longer exists after the measurement. But often one is interested in the precise estimation of a time-dependent parameter. This is, for example, the case in gravitational wave (GW) detection, where the exact shape of the GW induced signal carries a lot of information about its physical origin. A good measurement in this context is one that repeatedly gives a faithful value of the time-dependent parameter without destroying the system under study and with a minimum measurement induced back-action. These sequential measurements, which are called "quantum non demolition" (QND) measurements, allow physicists to perform multiple "quantum tapping" on the physical system, i.e. to sequentially estimate the parameter without adding the measurement-induced spurious noise that has been studied in the previous section. A QND measurement, as with any measurement, involves a system coupled to a meter. It is characterized by several criteria [79]:

- The meter should provide a precise measurement of the system;
- The system should not be degraded by the measurement;

- The system output values should be quantum correlated with the meter measurement results.

In order to be precise about the concepts, let us take the example of quantum optics and consider the simplest tapping device, where on the input signal field $\hat{E}_s^{in}$ that one wants to measure, one inserts a beamsplitter of intensity transmission coefficient T, giving rise to an output signal field $\hat{E}_s^{out}$ and a reflected meter field $\hat{E}_m^{out}$, with the familiar relations for the quadrature $\hat{X}$ (the relation being the same for the $\hat{P}$ quadrature):

$$\hat{X}_s^{out} = \sqrt{T}\hat{X}_s^{in} + \sqrt{1-T}\hat{X}_m^{in}, \quad \hat{X}_m^{out} = \sqrt{T}\hat{X}_m^{in} - \sqrt{1-T}\hat{X}_s^{in}. \qquad (11.51)$$

The $\hat{X}_m^{in}$ term brings unwanted vacuum fluctuations into the system. For the correlations in the case of a coherent input state, one obtains

$$C_m = C\left(\hat{X}_s^{in}\hat{X}_s^{out}\right) = T; \quad C_{out} = C\left(\hat{X}_s^{in}\hat{X}_m^{out}\right) = 1 - T. \qquad (11.52)$$

The quality of the measurement device as a quantum state preparation can be assessed by the conditional variance of the output signal knowing the measurement

$$Var_{s|m} = Var\left(\hat{X}_s^{out}\middle|\hat{X}_m^{out}\right) = Var\left(\hat{X}_s^{out}\right)\left(1 - C^2\left(\hat{X}_m^{out}\hat{X}_s^{out}\right)\right). \qquad (11.53)$$

The requirements for a good QND device are self-contradictory in this classical situation: The absence of "demolition" is obtained for $C_m \simeq 1$, i.e. $T \simeq 1$, while good measurement fidelity requires $C_{out} \simeq 1$, i.e. $T \simeq 0$. Let us now inject a squeezed vacuum (of squeezing factor e^{-S} on quadrature $\hat{X}$) in the second port of the beamsplitter. The correlations and the conditional variance are now

$$C_m = \frac{1-T}{1-T+Te^{-S}}, \quad C_{out} = \frac{T}{T+Te^{-S}}, \quad Var_{s|m} = Te^{-S}. \qquad (11.54)$$

When $S \to \infty$, both C_m and C_{out} tend to one, and $Var_{s|m}$ tends to zero, so that the $\hat{X}$ quadrature measurement is perfectly nondestructive. Perfect multiple tapping can be achieved with such "back-action evading" measurement techniques, which makes it interesting for applications in telecommunications.

Correlations of the quadrature $\hat{P}$ are given by Equation (11.54) with S replaced by $-S$. When $S \to \infty$, both C_m and C_{out} tend to zero: The "back-action" induced by the measurement of $\hat{X}$ has been completely transferred to the conjugate variable $\hat{P}$.

11.5.2 Examples

QND Light Intensity Measurement by Nonlinear Beam Coupling

A frequent scheme for a QND measurement is that of two-beam coupling (Figure 11.6): The signal light beam S to measure is coupled to an ancilla meter beam M, the quadrature of which is destructively measured. We

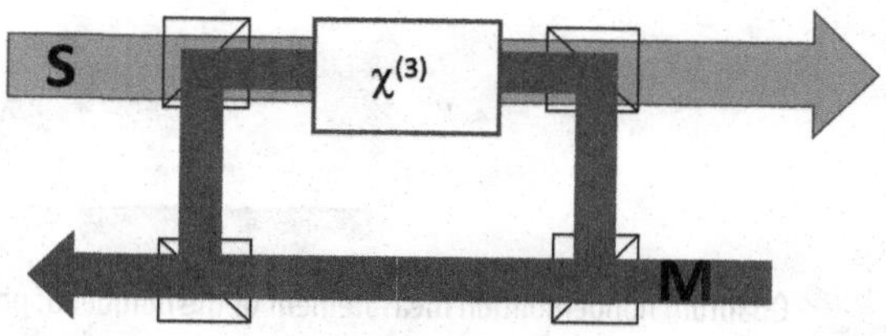

Fig. 11.6 QND quadrature measurement by cross-Kerr effect. A probe beam M is inserted in a Mach–Zehnder interferometer, and the beam S to measure overlaps the probe M inside one arm of the interferometer where a crossed-Kerr nonlinear medium has been inserted. This medium induces a phase on beam M that is proportional to the intensity of beam S. This phase shift is read by the interferometric phase measurement technique of Section 10.2

saw in the Section 11.5.1 that linear coupling of two beams with the help of a beamsplitter does not provide a QND measurement. The coupling with a nonlinear optical medium is required. The cross-Kerr effect involves two beams, a signal beam and a meter beam, propagating in a third-order nonlinear medium. The interaction Hamiltonian has a simple form:

$$\hat{H} = \hbar\omega_s \hat{a}_s^\dagger \hat{a}_s + \hbar\omega_m \hat{a}_m^\dagger \hat{a}_m + \hbar\chi \hat{a}_s^\dagger \hat{a}_s \hat{a}_m^\dagger \hat{a}_m. \tag{11.55}$$

It is not a symplectic Hamiltonian, because it involves a product of more than two annihilation or creation operators. It commutes with the photon number of both modes, which are therefore two constants of motion, a good prerequisite to be QND measurable quantities. If the signal mode is in the Fock state $|n_s\rangle$, the Hamiltonian for the meter mode is

$$\hat{H}_m = \hbar\omega_m \hat{a}_m^\dagger \hat{a}_m + \hbar\chi n_s \hat{a}_m^\dagger \hat{a}_m. \tag{11.56}$$

Its energy depends on the signal mode photon number, which, for the output signal field, transforms into a phase shift that is proportional to the signal mode photon number: The index of refraction for the meter beam depends on the signal intensity, whereas in the single mode Kerr effect, the index of refraction for the mode depends on its own intensity.

In the QND device, the signal mode intensity fluctuations induce a change of the phase of the output meter, which is monitored by inserting the meter beam in a Mach–Zehnder interferometric device (Figure 11.6). In the cross-Kerr medium described by the Hamiltonian (11.56), which is symmetric with respect to the signal and meter modes, the symmetric effect also takes place: The index of refraction of the signal beam depends on the intensity fluctuations of the meter beam, so that the back action of the QND measurement affects the output phase of the signal beam, but not its intensity. This cross-Kerr technique for a QND measurement is used in quantum optics experiments and also in circuit QED (Appendix K).

QND Photon Number Measurement by Optomechanics

Figure 11.7 gives one possible implementation of a QND measurement of the number of photons in a cavity by optomechanical means, taken

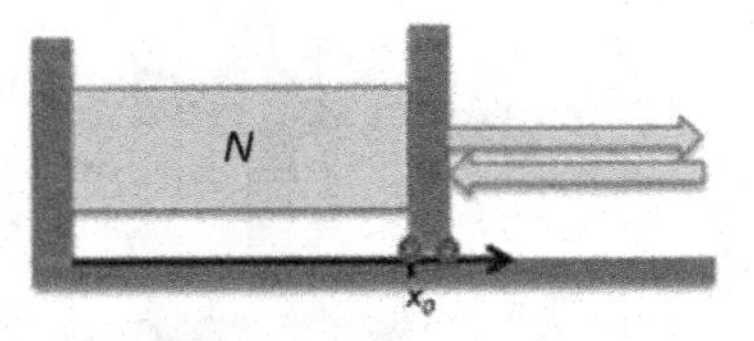

Fig. 11.7 Quantum nondemolition measurement of the number of photons inside a resonant optical cavity with a moving mirror. The radiation pressure of the intracavity field pushes the cavity mirror out of the position of exact resonance, which induces a detectable change in the reflected light.

from [88]. It consists of an optical cavity containing the N photons to be measured, one mirror of which, of mass M, is free to move from its initial position x_0 from time $t = 0$. A weak ancilla external light beam of frequency ω is reflected by the moving mirror. One measures its frequency after reflection, which is shifted by the Doppler effect, and therefore one knows the mirror velocity v and momentum $p = Mv$ at a given time T.

The system (i.e. the intracavity field) and the meter (the moving mirror) are coupled by the momentum exchange between the bouncing photons and the moving mirror, i.e. by the radiation pressure force $F = N\hbar\omega/x_0$. Classically, the mirror momentum at time T is $p(T) = N\hbar\omega T/x_0$, and therefore the photon number is derived from p by $N = px_0/(\hbar\omega T)$. The quantity $N_{est} = p(T)x_0/(\hbar\omega T)$ is therefore an *estimator* of the photon number.

From a quantum point of view, the set (system + meter) is described by the Hamiltonian

$$\hat{H} = \hbar\omega\hat{N} + \hat{p}^2/2M - \hbar\omega\hat{n}\hat{p}/x_0, \tag{11.57}$$

which commutes with $\hat{N}$: The photon number is thus a constant of motion even in presence of the meter. It is not modified by the measurement

$$\hat{N}_{out} = \hat{N}_{in}. \tag{11.58}$$

Working in the Heisenberg picture, at time T one has

$$\hat{p}(T) = \hat{p}(0) + \frac{\hbar\omega T}{x_0}\hat{N} \tag{11.59}$$

and therefore

$$\hat{N}_{est} = \hat{N} - \frac{x_0}{\hbar\omega T}\hat{p}(0), \tag{11.60}$$

where $\hat{N}_{est} = \hat{p}(T)x_0/(\hbar\omega T)$ is the quantum estimator of the photon number. The second term in Equation (11.60) is responsible for introducing unwanted noise in the photon number estimation. Its variance is given by

$$Var(N_{est}) = Var(N) + \frac{x_0^2}{\hbar^2\omega^2 T^2}\Delta^2 p(0). \tag{11.61}$$

The excess noise, represented by the second term, is related to the initial unavoidable uncertainty of the mirror momentum. It decreases for long times.

Where is the "back-action" in such a device? It turns out that it affects the phase Φ of the intracavity field, which here plays the role of the conjugate quantity to the photon number. One has

$$\hat{\Phi}_{out} = \hat{\Phi}_{in} + \omega T. \quad (11.62)$$

Actually, ω varies with time because it undergoes a Doppler shift at each bounce of photons on the moving mirror. A simple calculation shows that variation of cavity length and variation of frequency are linked by $\delta\omega/\omega = -\delta x_0/x_0$. Let us note that in this device the intracavity field energy $E = N\hbar\omega$ is not conserved (the kinetic energy of the mirror increases), but the "adiabatic invariant" E/ω is indeed conserved. In addition, the frequency shift turns out to be just the one needed to keep the field resonant with the varying cavity.

Relation (11.62) implies, for the noise acquired by the phase, that

$$Var(\Phi_{out}) = Var(\Phi_{in}) + \Delta^2\omega\, T^2 = Var(\Phi_{in}) + \left(\omega^2 T^2/x_0^2\right)\Delta^2 x_0. \quad (11.63)$$

The second term is precisely the variance of the measurement induced disturbance $Var(D)$. The noise/disturbance relation in this specific case can be written as

$$Var(N)\, Var(D) = \frac{1}{\hbar^2} Var(x_0)\, Var(p(0)) \geq \frac{1}{4}. \quad (11.64)$$

11.6 Exercises for Chapter 11

Exercise 11.1 Prove the phase-number uncertainty relation, valid for all quantum states of small quantum fluctuations with respect to the mean values

$$Var(N)\, Var(\phi) > 1/4.$$

Exercise 11.2 Use the Heisenberg inequality

$$Var(A)\, Var(B) \geq \frac{1}{4}\left\|\langle[\hat{A},\hat{B}]\rangle\right\|^2 + \left\|\frac{1}{2}\langle\{\hat{A},\hat{B}\}\rangle - \langle\hat{A}\rangle\langle\hat{B}\rangle\right\|^2$$

to find a restriction to the possible values of a qubit Bloch vector $\vec{U}$. Find a simple expression for the purity $P_u = Tr[\hat{\rho}^2]$ in terms of $\|\vec{U}\|$.

Exercise 11.3 Compare the sensitivity of a coherent state to the sensitivity of a cat state to measure displacements. Are states with larger photon number more sensitive?

Exercise 11.4 Derive the Cramér–Rao bound

$$\Delta X \geq 1/\sqrt{NF(X)_{X=X_0}},$$

where ΔX is the standard deviation of the measurement over the parameter X, N is the number of repetitions of the experiment, X_0

is the true value of the parameter to be estimated, and $F(X) \equiv \int d\xi p(\xi \mid X) \left[\frac{\partial \ln p(\xi|X)}{\partial X}\right]^2$ is the Fisher information. Here, ξ is the measurement results. What is the condition to saturate the inequality?

Exercise 11.5 We will now analyze the Mach–Zehnder interferometer as a metrological tool to estimate small phase shifts ϕ. We will refer to the two input fields by $\hat{a}_{\text{in}}$ and $\hat{b}_{\text{in}}$ and use the subscript "out" for the outputs and to the two beam splitters as $BS_{1,2}$. Here, we closely follow Refs. [53, 183]. See also [49].

(a) Consider the Jordan–Schwinger operators

$$\hat{J}_x = \frac{1}{2}\left(\hat{a}^\dagger \hat{b} + \hat{b}^\dagger \hat{a}\right), \quad \hat{J}_y = \frac{i}{2}\left(\hat{b}^\dagger \hat{a} - \hat{a}^\dagger \hat{b}\right), \quad \hat{J}_z = \frac{1}{2}\left(\hat{a}^\dagger \hat{a} - \hat{b}^\dagger \hat{b}\right),$$

and show they form an angular momentum algebra. That is, show that $[\hat{J}_i, \hat{J}_j] = i\epsilon_{ijk}\hat{J}_k$, $\hat{J}^2 = \frac{\hat{N}}{2}\left(\frac{\hat{N}}{2}+1\right)$ with $\hat{N} = \hat{a}^\dagger \hat{a} + \hat{b}^\dagger \hat{b}$.

(b) Find the matrix expression for the conversion of the inputs in outputs as a function of the phase φ accumulated in the optical paths difference. Show that the transformation of $\hat{a}_{\text{in}}$ and $\hat{b}_{\text{in}}$ are given by rotations generated by the angular momentum operatorss $\hat{U} = \exp(-i\theta\hat{J} \cdot \hat{n})$.

(c) Show that a quantum state transforms as $|\psi\rangle_{\text{out}} = e^{-i\hat{J}_x\pi/2} e^{i\hat{J}_z\varphi} e^{-i\hat{J}_x\pi/2}|\psi\rangle_{\text{in}}$, which corresponds to a y rotation.

Exercise 11.6 Derive expressions for $\langle \hat{J}_z \rangle_{\text{out}}$ and $\Delta^2 \hat{J}_z|_{\text{out}}$ as a function of $\langle \hat{J}_i \rangle_{\text{in}}$ and $\Delta^2 \hat{J}_i|_{\text{in}}$. Find the sensitivity of the photon number difference $\hat{J}_z$ to the differential phase accumulation parameter φ in the interferometer.

Exercise 11.7 Compute the value of $\Delta\varphi$ as a function of the number of photons if the two input states of the Mach–Zehnder interferometer are:

(a) Two Fock states in $\hat{a}_{\text{in}}$ and $\hat{b}_{\text{in}}$: $|N, 0\rangle$. Verify that the precision of the estimation is given by the shot noise limit $1/\sqrt{N}$.

(b) A coherent state $|\alpha\rangle$ and vacuum. What value of $\langle \varphi \rangle$ provides the highest sensitivity to the arm-length variation? This is the optimal operation point.

(c) A coherent state $|\alpha\rangle$ and squeezed vacuum $|\xi\rangle$, with $|\xi\rangle = \exp[(\xi^* \hat{a}^2 - \xi \hat{a}^{\dagger 2})/2]|0\rangle$. State in which $\langle \hat{N} \rangle = \sinh^2 \xi$. Convince yourself that taking the same phase for the coherent states (say, both α and ξ real, for example) yields the optimal result. Analyze the limit of large photon numbers at the optimal operating point for φ and show that the measurement scheme beats the standard quantum limit.

(d) The state

$$|\psi\rangle_{\text{in}} = \frac{1}{\sqrt{2}}(|j, 0\rangle + |j, 1\rangle) = \frac{1}{\sqrt{2}}\left(\left|\frac{N}{2}\right\rangle\left|\frac{N}{2}\right\rangle + \left|\frac{N}{2}+1\right\rangle\left|\frac{N}{2}-1\right\rangle\right).$$

Show that the precision limit reaches the Heisenberg bound ($1/N$) in this case.

Exercise 11.8 Show that the quantum Fisher information over the parameter X defined by

$$F_Q(X) = \int d\xi \frac{1}{p(\xi \mid X)} \left[\frac{\partial p(\xi \mid X)}{\partial X} \right]^2$$

is bounded by above by $4\langle \Delta \hat{P}^2 \rangle_0$. Here, $\hat{P}$ is the generator of the unitary transformation $|\psi(X)\rangle = \hat{U}(X)|\psi(0)\rangle$, $\Delta \hat{P}^2$ its variance, and $p(\xi \mid X) = \langle \psi(X)|\hat{\Pi}(\xi)|\psi(X)\rangle$, where the set of operators $\hat{\Pi}(\xi)$ form a POVM (Hermitian, positive semi-definite, $\int d\xi \hat{\Pi}(\xi) = 1$, etc). As before, ξ should be interpreted as the measurement data.

Note that X and $\hat{P}$ need not be canonically conjugate (i.e. X can be a parameter without an associated observable operator "$\hat{X}$"). Give an example [49].

Exercise 11.9 Consider the preparation of N qubits in the state $|+\rangle = (|0\rangle + |1\rangle)/\sqrt{2}$ and that the evolution of each qubit is given by $|0\rangle \to |0\rangle$, $|1\rangle \to \exp(i\phi)|1\rangle$. The state $|+\rangle$ evolves into

$$|+\rangle \to |\phi\rangle \equiv \left(|0\rangle + e^{i\phi}|1\rangle\right)/\sqrt{2}.$$

Compute the Fisher information of $\hat{\sigma}_x$ measurements to estimate the phase ϕ. Compare it with the quantum Fisher information and show that it is the optimal measurement.

A Basic Postulates of Quantum Mechanics: a Reminder

A.1 Postulates

Here, we recall the well-known basic postulates of quantum mechanics, formulated in terms of state vectors.

(S) *Description of physical systems*
An isolated physical system, or a system submitted to fixed external forces, prepared in a totally controlled way, is described by a single mathematical object called a *state vector*, or ket, $|\psi\rangle$, belonging to a well-defined Hilbert space $\mathcal{H}$ of dimension N (finite or infinite). The state vector is normalized to 1:

$$|\langle\psi|\psi\rangle|^2 = 1. \tag{A.1}$$

(E) *Evolution*
The time evolution of the state vector $|\psi(t)\rangle$ is ruled by the Schrödinger equation,

$$i\hbar\frac{d}{dt}|\psi(t)\rangle = \hat{H}(t)|\psi(t)\rangle, \tag{A.2}$$

where $\hat{H}(t)$ is the Hamiltonian of the system, which is the Hermitian operator associated with the energy.
(M) *Measurement*
(M1) *Quantization*
The measurement of the physical quantity A can only give as a result one of the N eigenvalues a_n of the associated Hermitian operator. These operators are called "observables."
(M2) *Born's rule*
In general, the result of the measurement cannot be predicted with certainty. The conditional probability $Proba(a_n, |\psi\rangle)$ of obtaining the result a_n, given that the system is in the quantum state described by the state vector $|\psi\rangle$, is given by

$$Proba(a_n, |\psi\rangle) = \langle\psi|\hat{P}_n|\psi\rangle, \tag{A.3}$$

where $\hat{P}_n$ is the projector on the eigensubspace associated with the eigenvalue a_n.

(M3) *State collapse*

The conditional state of the system just after the measurement *and in the subset of cases where the measurement has given the result a_n* is

$$|\psi^{after|a_n}\rangle = \frac{\hat{P}_n|\psi\rangle}{\sqrt{Proba(a_n, |\psi\rangle)}}. \quad (A.4)$$

A.2 Comments

A.2.1 Description of the System

A physical system, whatever its complexity, is described by a single mathematical object $|\psi\rangle$, which is a normalized vector belonging to a well-defined Hilbert space $\mathcal{H}$. This apparently purely mathematical property implies a very important physical property of quantum systems, usually called the *superposition principle*:

If $|\Psi_1\rangle$ and $|\Psi_2\rangle$ describe two possible states of the physical system, then $|\Psi_3\rangle = \lambda_1|\Psi_1\rangle + \lambda_2|\Psi_2\rangle$, where λ_1 and λ_2 are any complex numbers (with $|\lambda_1|^2 + |\lambda_2|^2 = 1$), is another possible state of the physical system.

This property is not just a technical mathematical feature of the state vector. It is actually one of the deepest roots of the quantum world and originates from its wave-like character, since classical waves have the same property.

The superposition principle implies the possibility of *quantum interference*: A classical example is the double slit experiment for photons or massive particles. If $|\Psi_1\rangle$ describes an electron passing through slit one, and $|\Psi_2\rangle$ describes an electron passing through slit two, then $|\Psi_3\rangle = (|\Psi_1\rangle + |\Psi_2\rangle)/\sqrt{2}$ describes another possible state of the electron, namely the one in which the electron has the chance to pass through both slits.

Another example is an optical system that has a state $|\Psi_1\rangle$ that transmits light and a state $|\Psi_2\rangle$ that reflects it. The superposition $(|\Psi_1\rangle + |\Psi_2\rangle)/\sqrt{2}$ describes something that reflects and transmits light at the same time. In quantum optics, and in contrast with literature, a door can be at the same time open **and** shut!

Let us call $|a\, e^{i\phi}\rangle$ the quantum state describing a monochromatic light beam of amplitude a and phase ϕ (that we will detail in Appendix F). Then $\left|a\, e^{i(\phi+\pi)}\right\rangle$ is the quantum state describing a coherent light of amplitude a and phase $\phi + \pi$, i.e. having an electric field that is the opposite of the previous one at any time. Physicists have experimentally produced the states $|a\, e^{i\phi}\rangle \pm |a\, e^{i(\phi+\pi)}\rangle$ (properly normalized to one) that describe very strange objects, such as the nonzero superposition of an oscillating electric field and its opposite (considered in Section 9.1)!

When the two superposed states are close to classical objects and are distinguishable, physicists name the superposition "Schrödinger cats" or "Schrödinger kittens," an allusion to the rather unphysical but striking situation of a superposition of a dead and alive cat imagined by Schrödinger.

A.2.2 Temporal Evolution

The evolution postulate presented in this appendix is the so-called Schrödinger representation, in which mean values $\langle A\rangle = \langle\psi(t)|\hat{A}|\psi(t)\rangle$ are calculated by using a state vector that evolves with time. Let us introduce the unitary *evolution operator*, such that

$$|\psi(t)\rangle = \hat{U}(t)|\psi(0)\rangle, \tag{A.5}$$

where the evolution of $\hat{U}(t)$ is ruled by an operatorial Schrödinger equation,

$$i\hbar\frac{d}{dt}\hat{U}(t) = \hat{H}(t)\hat{U}(t). \tag{A.6}$$

One can also use the Heisenberg representation, defining new observables $\hat{A}_H$ associated with the physical quantity A by,

$$\hat{A}_H(t) = \hat{U}^\dagger(t)\hat{A}(t)\hat{U}(t). \tag{A.7}$$

It is easy to see that the mean value of A is now given by

$$\langle A\rangle = \langle\psi(0)|\hat{A}_H(t)|\psi(0)\rangle, \tag{A.8}$$

where the observable evolves with time, while the state vector keeps its initial value.

The temporal evolution of the observable in the Heisenberg representation is now given by

$$i\hbar\frac{d}{dt}\hat{A}_H(t) = i\hbar\hat{U}^\dagger(t)\frac{\partial\hat{A}(t)}{\partial t}\hat{U}(t) + \left[\hat{A}_H(t),\hat{H}(t)\right]. \tag{A.9}$$

A.2.3 Measurement

Any real physical quantity A, such as the energy, position, electric field, magnetic moment, etc., is associated with a Hermitian linear operator or "observable," $\hat{A}$, operating in the Hilbert space of state vectors $\mathcal{H}$. By definition, $\hat{A}^\dagger = \hat{A}$, so that $\hat{A}$ can be diagonalized,

$$\hat{A}|u_n\rangle = a_n|u_n\rangle, \tag{A.10}$$

the eigenvalues a_n being real, and the eigenvectors $|u_n\rangle$ forming an orthonormal basis of the Hilbert space $\mathcal{H}$. In the degenerate case, several orthogonal eigenvectors $|u_n^i\rangle$ are associated with the same eigenvalue a_n.

One defines *projectors* on eigenspaces $\hat{P}_n$ by

$$\hat{P}_n = \sum_i |u_n^i\rangle\langle u_n^i|. \tag{A.11}$$

They are such that

$$\hat{P}_n\hat{P}_{n'} = \delta_{n,n'}\hat{P}_n; \quad \sum_n \hat{P}_n = \hat{1}. \tag{A.12}$$

The projectors are therefore orthogonal to each other.

The observable $\hat{A}$ can also be written as

$$\hat{A} = \sum_n a_n\hat{P}_n. \tag{A.13}$$

Measurement rules M2 and M3 are valid only for "ideal measurements" where the measurement process is as perfect as possible. We will call them *von Neumann measurements*. Such measurements do not destroy the physical system. Just after the measurement, the system is in state $|\psi^{after|a_n}\rangle$, a state called a *conditional*, or *post-selected, state* because it is not the one that is always obtained after the measurement, but only the one obtained in the special and uncontrolled opportunities when the measurement gives the precise result a_n.

Von Neumann measurements are also called "projective measurements," because for the quantum state of the system after the measurement they yield the normalized projection of the initial state onto the eigensubspace corresponding to the measured value. Note that the squared norm of the un-normalized projected state is exactly the probability for that conditional preparation. They are also called orthogonal measurements because the different states after the measurement are orthogonal to each other.

Let us first note that the number of possible measurement results a_n is necessarily limited to the dimension of the Hilbert space $\mathcal{H}$. In the case of degenerate eigenvalues, it can be even smaller.

Let us now assume that the preparation process is such that it delivers many identical copies of the system in state $|\psi\rangle$. One can then make many measurements of the same physical quantity A, and the averaged quantities can be predicted from Born's law[1] with almost certainty. In particular, the mean value of A, which we will denote $\langle A\rangle$, is

$$\langle A\rangle = \sum_n a_n Proba(a_n, |\psi\rangle) = \langle\psi|\hat{A}|\psi\rangle \tag{A.14}$$

and the mean of any analytic function $f(A)$ of A, is

$$\langle f(A)\rangle = \langle\psi|f(\hat{A})|\psi\rangle. \tag{A.15}$$

Let us now consider two successive measurements of quantity A on the same system (which must not be confused with two measurements of A

[1] Excerpt from Max Born's 1926 paper about the interpretation of the wave function (in: *On the quantum theory of collisions*, Zeits. for Phys. **37**, 863 (1926)): ...the wavefunction $\Phi_{nm}(\alpha, \beta, \gamma)$ gives the probability for the electron to be thrown out in the direction designated by angles α, β, γ..
During proofreading Born added the following footnote: More careful consideration shows that the probability is proportional to the **square** of the quantity Φ_{nm}: a footnote worth a Nobel prize!

on identically prepared quantum states). The probability of successively obtaining results a_n and $a_{n'}$ is,

$$Proba(a_n, a_{n'}|\psi) = \frac{\langle\psi|\hat{P}_n\hat{P}_{n'}\hat{P}_n|\psi\rangle}{\langle\psi|\hat{P}_n|\psi\rangle}. \tag{A.16}$$

Because $\hat{P}_n\hat{P}_{n'} = \delta_{n,n'}\hat{P}_n$, the probability of getting a result $a_{n'}$ different from the first result a_n is zero, whereas the probability of getting for the second measurement the same result a_n as the first result a_n is one. In this special case the probabilities are certainties: After the first measurement we know with certainty the result of the measurement of A on the system. The first measurement acts as a *state preparation*: Quantum physics is characterized by intrinsic randomness, but is at least *repeatable*! We will use that important point in Chapter 5.

Let us also note that the state after the measurement resulting from the collapse is, in general, very different from the state before it. The measurement process results in a very strong modification of the state vector, so that most of the information about the initial state is lost. This is the reason why it is not possible to perfectly determine from successive measurements on the same system its exact initial state when its preparation is not known.

A.3 Time Reversibility

In addition, the evolution ruled by Equation (A.2) has an important property: that of being essentially *reversible in time*.

As an example, let us consider the simple case of a 1D particle submitted to a time-independent and *real* potential $V(x)$. The evolution of the wave function $\psi(x, t)$ obeys the Schrödinger equation

$$i\hbar\frac{d}{dt}\psi(x,t) = \hat{H}\psi(x,t), \tag{A.17}$$

where $\hat{H} = -(\hbar^2/2m)\partial^2/\partial x^2 + V(x)$. Let us now take the complex conjugate of this equation,

$$-i\hbar\frac{d}{dt}\psi^*(x,t) = i\hbar\frac{d}{d(-t)}\psi^*(x,t) = \hat{H}\psi^*(x,t). \tag{A.18}$$

Therefore, *the wave function $\psi^*(x,t)$ is a solution of the same Schrödinger equation, but with a reversed time $t \to -t$.*

If one starts at $t = 0$ from a given initial wave function $\psi(x,0)$, the Hamiltonian evolution will lead at time t_0 to the wave function $\psi(x,t_0) = \hat{U}(t_0)\psi(x,0)$. Let us then consider as a new initial condition at time t_0 the complex conjugate wave function $\psi^*(x,t_0)$. The Hamiltonian evolution from t_0 to $2t_0$ due to $\hat{H}$ will make the wave function evolve from $\psi^*(x,t_0)$ to $\psi^*(x,0)$, which has the same probability distribution of position as the initial one, $|\psi(x, 2t_0)|^2 = |\psi(x, 0)|^2$, and opposite momenta, $|\tilde{\psi}(p, 2t_0)|^2 = |\tilde{\psi}(-p, 0)|^2$, where $\tilde{\psi}(p)$ is the wavefunction in momentum space. Such an evolution is therefore *reversible*, like the corresponding one in classical mechanics.

More generally, one can show that any Hamiltonian evolution can be reversed. This means that, for any system characterized by a time independent Hamiltonian, there exists a *time reversal operator*, $\hat{T}$, such that, for any wave vector and any t_0,

$$\hat{U}(t_0)\hat{T}\hat{U}(t_0)|\psi(0)\rangle = \hat{T}|\psi(0)\rangle, \tag{A.19}$$

as a particle $\hat{T}$ is the complex conjugation operation of its wave function. For a spin 1/2 system, one can show that $\hat{T}$ is proportional to the Pauli matrix $\hat{\sigma}_y$: $\hat{T} = -i\hat{\sigma}_y$.

A.4 Exercises for Appendix A

Exercise A.1 Read Albert Einstein's 1905 paper concerning a "heuristic point of view" about radiation [63, 64]. Is it based on the photoelectric effect? What was "corpuscular" about light to him? What was the experimental evidence to support his claim?

Exercise A.2 Read Erwin Schrödinger's 1926 paper on the wave equation [159]. Discuss the parallel between ray optics and classical mechanics with wave optics and quantum mechanics. Discuss why his original wave equation involved a second derivative in time.

Exercise A.3 Read Max Born's 1954 Nobel lecture [23]. Discuss the historical origin of the statistical interpretation of quantum mechanics. Where did he learn about "matrices" and how did this help Heisenberg?

B Generalized Postulates of Quantum Mechanics

The generalized postulates of quantum mechanics are introduced and discussed in Chapters 2, 4, and 5.

(S) *Description of physical systems*
A physical system is described by a density matrix, ρ, which is a linear operator acting on a Hilbert space $\mathcal{H}$ of dimension N (finite or infinite) that is positive and normalized to 1:

$$\forall\, |\psi\rangle \in \mathcal{H} \quad \langle\psi|\rho|\psi\rangle \geq 0; \qquad Tr\rho = 1. \tag{B.1}$$

(E) *Process*
A quantum system initially in state ρ_{in} is, after the process, which can be an evolution in time or in space, in state ρ_f given by

$$\rho_f = \sum_{\ell=1}^{N_K} \hat{K}_\ell\, \rho_{in}\, \hat{K}_\ell^\dagger, \tag{B.2}$$

where the so-called Kraus operators $\hat{K}_\ell$ are N_K linear operators acting on the Hilbert space $\mathcal{H}$ of state vectors satisfying the condition

$$\sum_{\ell=1}^{N_K} \hat{K}_\ell^\dagger \hat{K}_\ell = \hat{1} \tag{B.3}$$

(M) *Measurement*
(M1) *Quantization* A measurement can only give as a result a value a_n belonging to a list $\{a_1, a_2, ..., a_{N_A}\}$ of discrete or continuous values, where N_A is not limited by the dimension N of the Hilbert space.
(M2) *Born's rule* In general, the result of a measurement cannot be predicted with certainty. The conditional probability $Proba(a_n|\rho)$ of obtaining the result a_n given that the system is in the quantum state described by the density matrix ρ is given by

$$Proba(a_n|\rho) = Tr\rho\hat{\Pi}_n, \tag{B.4}$$

where the $\hat{\Pi}_n$ are N_M positive operators (but not necessarily projectors) associated with the value a_n of the property, satisfying the condition $\sum_{n=1}^{N_M} \hat{\Pi}_n = \hat{1}$.
(M3) *State collapse* The conditional, or post-selected, state of the system just after a measurement that has given the result a_n, called $\rho^{after|a_n}$, is

$$\rho^{after|a_n} = \frac{1}{Tr\rho\hat{\Pi}_n} \hat{M}_n \rho \hat{M}_n^\dagger, \tag{B.5}$$

where the $\hat{M}_n$ are N_A linear operators (but not necessarily positive) acting on the Hilbert space $\mathcal{H}$ and satisfying the condition $\hat{M}_n^\dagger \hat{M}_n = \hat{\Pi}_n$.

The state after a measurement whose result is ignored and therefore not post-selected to a given value (unconditional evolution) is then obtained by averaging over the possible a_n, which yields

$$\rho^{after} = \sum_{n=1}^{N_A} \hat{M}_n \, \rho \, \hat{M}_n^\dagger,$$

which is identical to Equation (B.2).

C Description of Composite Systems

Very often, one is interested in the description not only of a single particle or physical object, but of several of them. Here, we will briefly show how such composite systems are described in quantum mechanics.

Let us stress that the situation considered in this appendix is very different from that in Appendix A, which considers the sum of two state vectors $|\psi_1\rangle$ and $|\psi_2\rangle$ that leads to an interference phenomenon. The former describes a *single particle with two possible trajectories*, whereas the latter describes *a state of two particles*.

Let us consider as a start the system formed by two massive 1D particles traveling along the Ox direction. It is characterized by a two-variable wave function $\psi(x_1, x_2)$ such that $|\psi(x_1, x_2)|^2$ gives the joint probability of detecting particle 1 at position x_1 *and* particle 2 at position x_2.

C.1 Tensor Product of Hilbert Spaces

C.1.1 Basis

If a particle labeled 1 is described by a state vector $|\psi_1\rangle$ belonging to a Hilbert space $\mathcal{H}_1$ of dimension d_1 spanned by a basis of vectors $|u_i\rangle$ ($i = 1, ..., d_1$), and particle 2 is described by a state vector $|\psi_2\rangle$ belonging to a Hilbert space $\mathcal{H}_2$ of dimension d_2 spanned by a basis of vectors $|v_j\rangle$ ($j = 1, ..., d_2$), then the system formed by the two particles is described by a state vector $|\psi\rangle$ belonging to the *tensor product* $\mathcal{H} = \mathcal{H}_1 \otimes \mathcal{H}_2$ of spaces $\mathcal{H}_1$ and $\mathcal{H}_2$. This space has dimension $d_1 \times d_2$). It is spanned by the basis of vectors $|u_i\rangle \otimes |v_j\rangle$ ($i = 1, ..., d_1, j = 1, ..., d_2$), which we will write $|u_i, v_j\rangle$ for simplicity. It has a Hilbert inner product defined on its vector basis by

$$\langle u_i, v_j | u_{i'}, v_{j'}\rangle = \delta_{i,i'}\delta_{j,j'}. \tag{C.1}$$

A state belonging to $\mathcal{H}$ therefore has the form

$$|\Psi\rangle = \sum_{i,j} \lambda_{i,j} |u_i, v_j\rangle. \tag{C.2}$$

Note that it cannot always be written in a factorized way, such as

$$|\Psi\rangle = \sum_{i,j} \lambda_i^1 \lambda_j^2 |u_i, v_j\rangle = \left(\sum_i \lambda_i^1 |u_i\rangle\right) \otimes \left(\sum_j \lambda_j^2 |v_j\rangle\right). \tag{C.3}$$

If such a factorization is not possible in any basis, the state is said to be *entangled* (Chapters 6 and 7).

In the case of two 1D particles considered in Appendix E, a factorized state has a wave function of the form $\psi(x_1, x_2) = \langle x_1, x_2|\Psi\rangle = \psi_1(x_1)\psi_2(x_2)$, whereas an entangled state is characterized by a wave function that cannot be written as a product, for example $\psi(x_1, x_2) = \psi_1(x_1)\psi_2(x_2) + \psi_2(x_1)\psi_1(x_2)$, where $\psi_1 \neq \psi_2$.

C.1.2 Operators

Let us consider an operator $\hat{A}_1$ operating on Hilbert space $\mathcal{H}_1$, and an operator $\hat{B}_2$ operating on Hilbert space $\mathcal{H}_2$. The operator $\hat{A}_1 \otimes \hat{B}_2$ operating on Hilbert space $\mathcal{H} = \mathcal{H}_1 \otimes \mathcal{H}_2$ is defined by

$$\hat{A}_1 \otimes \hat{B}_2|u_i, v_j\rangle = (\hat{A}_1|u_i\rangle) \otimes (\hat{B}_2|v_j\rangle). \tag{C.4}$$

In particular, one can extend an operator $\hat{A}_1$ acting on space $\mathcal{H}_1$ to an operator acting on tensor space $\mathcal{H}$ by

$$\hat{A}_1|u_i, v_j\rangle = (\hat{A}_1|u_i\rangle) \otimes |v_j\rangle. \tag{C.5}$$

In short, the extension of $\hat{A}_1$ is $\hat{A}_1 \otimes \hat{1}$. Note that for the commutators between operators acting on different subspaces, one always has

$$\left[\hat{A}_1, \hat{B}_2\right] = 0 \quad \forall\, \hat{A}_1,\, \hat{B}_2. \tag{C.6}$$

For example, in the case of two 1D particles, $[\hat{x}_1, \hat{p}_2] = 0$ whereas $[\hat{x}_1, \hat{p}_1] = i\hbar$.

C.1.3 Examples

Another example is provided by the system formed by two qubits, labeled 1 and 2, which is of dimension 4. A possible basis of the corresponding Hilbert space is the set $\{|1+, 2+\rangle, |1-, 2+\rangle, |1+, 2-\rangle, |1-, 2-\rangle\}$. Another useful basis is the set of entangled so-called Bell states $\{|\Psi_\pm\rangle, |\Phi_\pm\rangle\}$ defined by

$$|\Psi_\pm\rangle = \frac{1}{\sqrt{2}}(|1+, 2-\rangle \pm |1-, 2+\rangle), \tag{C.7}$$

$$|\Phi_\pm\rangle = \frac{1}{\sqrt{2}}(|1+, 2+\rangle \pm |1-, 2-\rangle). \tag{C.8}$$

In many instances, the two Hilbert spaces that are concatenated are of different natures. For example, the system formed by a qubit traveling along the Ox axis is described by a state vector $|\Psi\rangle$ belonging to the tensor product of a qubit and a 1D particle, spanned by the basis $|\pm, x\rangle$:

$$|\Psi\rangle = \sum_{\pm} \int dx \psi_\pm(x)|\pm, x\rangle, \tag{C.9}$$

which requires the introduction of two wave functions $\psi_+(x)$ and $\psi_-(x)$ such that $|\psi_+(x)|^2$ (resp. $|\psi_-(x)|^2$) gives the probability of detecting the qubit in state $|+\rangle$ (resp. $|-\rangle$) at position x.

C.2 No-Cloning Theorem

Is it possible to "clone" a single unknown input quantum state $|u_{in}\rangle$, i.e. to make a perfect copy of it? We will see that such a process is forbidden by the laws of quantum mechanics.

The demonstration is actually very simple. One can model a "cloning machine" as a device in which two physical systems associated with Hilbert spaces $\mathcal{H}_1$ and $\mathcal{H}_2$ of equal dimension interact and then separate from each other. At the input, one sends the factorized state $|\Psi_{in}\rangle = |u_{in}\rangle \otimes |v\rangle$, where $|v\rangle$ is a fixed blank quantum state on which the copy of u_{in} will be printed. At the output of the device, a perfect cloning machine is expected, after having interacted once with the quantum state u_{in}, to generate the factorized state $|\Psi_{out}\rangle = |u_{in}\rangle \otimes |u_{in}\rangle$ containing the desired identical copies.

Let us now consider two different arbitrary input states $|u_{in}\rangle$ and $|u'_{in}\rangle$ to be cloned. Successfully cloning the first implies having the state $|u_{in}\rangle \otimes |v\rangle$ at the input and the state $|u_{in}\rangle \otimes |u_{in}\rangle$ at the output. Successfully cloning the second implies having the state $|u'_{in}\rangle \otimes |v\rangle$ at the input and $|u'_{in}\rangle \otimes |u'_{in}\rangle$ at the output.

What happens if we inject the state $|\psi\rangle = (|u_{in}\rangle + |u'_{in}\rangle)/\sqrt{2}$ into the cloning machine? The quantum evolution in the cloning machine is a quantum map (studied in Chapter 4), and as such is linear with respect to the input states. Therefore, for the input $|\psi\rangle \otimes |v\rangle$ one gets as an output the entangled state $(|u_{in}\rangle \otimes |u_{in}\rangle + |u'_{in}\rangle \otimes |u'_{in}\rangle)/\sqrt{2}$. This state is different from the desired cloned state $|\psi\rangle \otimes |\psi\rangle$. The conclusion is that a "cloning machine" that can successfully clone a pair of states will not be able to clone their superpositions! Thus an ideal cloning machine for arbitrary unknown states cannot exist. Note that the argument relies only on the linearity of quantum mechanics.

One can end up with the same conclusion using a different argument: A perfect cloning machine yields two identical copies of a given single unknown input state. One then uses these two copies to make two *simultaneous von Neumann measurements* of position on one copy, AND of momentum on the other, and therefore measure the position of the particle without perturbing the measurement of its momentum, and measure its momentum without perturbing the measurement of its position. This would contradict the foundations of quantum mechanics (see Section 5.7).

Note that it is not forbidden is to make *successive von Neumann measurements* of x OR p if one has been able to design a well-controlled experimental set-up producing a row of identically prepared quantum states.

An instructive example is given by the double homodyne detection used on a light beam (Section 9.4), where the two output states, generated by a beamsplitter, are not perfectly identical because of the vacuum fluctuations entering by the unused input of the device: The no-cloning theorem implies that a perfect “state splitter” does not exist.

But an imperfect cloning machine can exist. It produces two output states that are not perfectly identical to the input unknown state, and/or not perfectly identical to each other. It has been shown that the overlap $|\langle u_{out}|u_{in}\rangle|^2$, also called “fidelity,” cannot be larger than $\sqrt{5/6}$ [30].

C.3 Exercises for Appendix C

Exercise C.1 What is the condition for performing experiments over a macroscopic sample of atoms and consider the measurements as produced by individual systems?

Exercise C.2 Show that if you could enforce a deterministic nonunitary evolution you would be able to clone states. Why is this not possible?

Exercise C.3 Show that the Hilbert space of a particle with 2^N levels (say a truncation of the hydrogen atom or a harmonic oscillator) and the Hilbert space of N qubits is mathematically identical. What is the role of entanglement in quantum information processing then?

D Qubits

The simplest quantum systems are described by a Hilbert space of dimension 2. They are of paramount importance in quantum technologies as they are the quantum extensions of the classical bits that are the basic tools of classical computers. They are the most promising candidates as basic bricks of quantum information, hence their name qubits. In contrast to the classical bit, which can be in one or the other of its two states, the qubit can be time in both at the same. More precisely, it can be prepared in a coherent linear superposition of the two states, featuring a quantum parallelism that, together with the possibility of entangling different qubits, is harnessed to speed up information processing and computing.

D.1 Bloch Vector

Qubits are ubiquitous in the quantum world: spin 1/2 particles, polarization states of photons, etc. Moreover, in many cases two of the energy levels of systems like atoms, ions, and quantum dots (usually, but not always, including the ground state) can be singled out. In these cases, the qubit space is a two-dimensional subspace of a larger Hilbert space. Here, we will adopt a general approach that is applicable to all these cases, introducing the *Bloch vector* (Figure D.1) as a convenient tool to make up simple physical pictures of the system.

A qubit has two orthonormal basis states that we will note $|\pm\rangle$, of energies $\pm\hbar\omega_0/2$. We introduce the three Hermitian operators $\hat{\sigma}_i$ ($i = x, y, z$), known as Pauli matrices,

$$\hat{\sigma}_x = \begin{pmatrix} 0 & 1 \\ 1 & 0 \end{pmatrix}, \quad \hat{\sigma}_y = \begin{pmatrix} 0 & -i \\ i & 0 \end{pmatrix}, \quad \hat{\sigma}_z = \begin{pmatrix} 1 & 0 \\ 0 & -1 \end{pmatrix}, \tag{D.1}$$

the first and second columns corresponding, respectively, to qubit basis vectors $|+\rangle$ and $|-\rangle$. These matrices have the following properties:

$$\hat{\sigma}_i^2 = \hat{1}, \quad \hat{\sigma}_h\hat{\sigma}_j = -\hat{\sigma}_j\hat{\sigma}_h, \quad [\hat{\sigma}_h, \hat{\sigma}_j] = 2i\epsilon_{hjk}\hat{\sigma}_k, \tag{D.2}$$

where ϵ_{hjk} is the Levi–Civita fully antisymmetric tensor. Many other properties of these matrices can be found in Exercise 2.2.

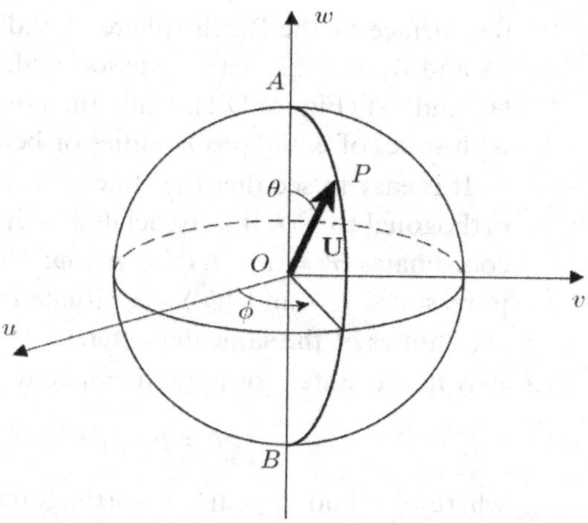

Fig. D.1 The Bloch vector $U(t)$ representing a qubit. The pure states are on the surface of the sphere, the mixed states inside. Points A and B correspond to the qubit energy eigenstates. The equatorial plane corresponds to the states of maximum coherence.

The following non-Hermitian operators are also useful:

$$\hat{\sigma}_+ = (\hat{\sigma}_x + i\hat{\sigma}_y)/2 = \begin{pmatrix} 0 & 1 \\ 0 & 0 \end{pmatrix} = |+\rangle\langle -|, \tag{D.3}$$

$$\hat{\sigma}_- = (\hat{\sigma}_x - i\hat{\sigma}_y)/2 = \begin{pmatrix} 0 & 0 \\ 1 & 0 \end{pmatrix} = |-\rangle\langle +|. \tag{D.4}$$

These operators are jump operators that transfer the qubit from one state to the other.

It is worth noting that the operators $\hat{\sigma}_x/2$, $\hat{\sigma}_y/2$, and $\hat{\sigma}_z/2$ are the three components of a spin 1/2 angular momentum operator, denoted $\overrightarrow{\hat{S}}$.

One then defines the Bloch vector $\overrightarrow{U}$ of a given two-level quantum state, pure or mixed, by

$$\overrightarrow{U} = \langle \overrightarrow{\hat{\sigma}} \rangle = 2\langle \overrightarrow{\hat{S}} \rangle. \tag{D.5}$$

The Bloch vector is a mathematical object living in an abstract three-dimensional space (of axes u, v, and w in Figure D.1). It allows us to visualize any quantum state of the qubit, pure or mixed, in the three-dimensional geometrical space:

- A pure state $|\psi\rangle$ can always be written as

$$|\psi\rangle = \cos\frac{\theta}{2}|+\rangle + \sin\frac{\theta}{2}e^{i\phi}|-\rangle. \tag{D.6}$$

The corresponding Bloch vector has coordinates (U_x, U_y, U_z) such that

$$U_x + iU_y = \sin\theta e^{i\phi}, \quad U_z = \cos\theta. \tag{D.7}$$

Therefore, θ and ϕ are the spherical coordinates of the direction of the Bloch vector. Its length is equal to 1 and its end point is situated on the surface of the Bloch sphere of radius 1. The north and south poles (A and B) of the sphere are associated, respectively, with the basis states $|+\rangle$ and $|-\rangle$ (Figure D.1), while the points on the equator are associated with states of equal probabilities of being in $|+\rangle$ and $|-\rangle$.

It is easy to see that the state $|\psi'\rangle = \cos\frac{\pi-\theta}{2}|+\rangle + \sin\frac{\pi-\theta}{2}e^{i(\phi+\pi)}|-\rangle$ is orthogonal to $|\psi\rangle$. It is associated with a Bloch vector that has spherical coordinates $\theta' = \pi - \theta$, $\phi' = \pi + \phi$: This means that the two orthogonal pure states $|\psi\rangle$ and $|\psi'\rangle$ are situated on the Bloch sphere at the two extremities of the same diameter.

- Any mixed state ρ of a qubit can be written as

$$\rho = p_1|\psi_1\rangle\langle\psi_1| + p_2|\psi_2\rangle\langle\psi_2|, \tag{D.8}$$

where $|\psi_1\rangle$ and $|\psi_2\rangle$ are the orthogonal eigenstates of ρ, and p_1 and p_2 the eigenvalues (with $p_1 + p_2 = 1$).

Finally, in the case where the qubit states are energy eigenstates differing by an energy $\hbar\omega_0$, the Hamiltonian of the system is $\hat{H} = \frac{\hbar\omega_0}{2}\hat{\sigma}_z$, so that the mean energy of a qubit is $\hbar\omega_0/2$ times the projection of the associated Bloch vector on the z axis.

D.2 From DV Projection Noise to CV Quantum Noise

The different experimental examples given in Chapter 1 showed that it is possible to measure the probability for a single qubit to be in the excited state. It is equal to $p_+ = Tr(\rho\hat{P}_+)$, where $\hat{P}_+ = |+\rangle\langle+|$ is the projector on the excited state. The variance of this quantity is

$$Var\left(\hat{P}_+\right) = Tr\left(\rho\hat{P}_+^2\right) - \left(Tr\left(\rho\hat{P}_+\right)\right)^2 = p_+(1-p_+) \tag{D.9}$$

because $\hat{P}_+$ is a projector. This "projection noise," which affects all population measurements on a qubit, depends only on the mean population and has a maximum value of 1/4 for a Bloch vector lying on the equator ($p_+ = 1/2$). It is, in general, rather large unless the state of the system is in one of its energy eigenstates. It leads to an uncertainty in energy measurements performed on single quantum systems.

Let us now consider N identical qubits that can be coherently coupled (for example, by interacting with the same field). They form a kind of "quantum register." We call $|+i\rangle$ and $|-i\rangle$ the quantum states of qubit i, and $\hat{\vec{S}}_i$ the associated angular momentum operator introduced in the previous section. The whole system is now characterized by a total angular momentum $\hat{\vec{S}} = \sum_{i=1}^{N}\hat{\vec{S}}_i$, with angular momentum quantum numbers S and M_S. It behaves as a single quantum object, described by the collective Bloch vector $\vec{U} = 2\langle\hat{\vec{S}}\rangle$. Being the sum of N spins 1/2, which are angular momentum states, using the sum rules of angular momenta we know that S

can take any value between 0 and N/2, with $-S \leq M_S \leq S$. The ground state of the system $|S = N/2, M_S = -N/2\rangle$ is the tensor product of all ground states $|-i\rangle$:

$$|S = N/2, M_S = -N/2\rangle = |N/2, -N/2\rangle = |-1\rangle \otimes |-2\rangle \cdots |-N\rangle. \quad \text{(D.10)}$$

Its first excited state $|\Phi_N\rangle$ is

$$\begin{aligned} |\Phi_N\rangle = \hat{S}_+|N/2, -N/2\rangle &= |N/2, -N/2+1\rangle \\ &= \sum_{i=1}^{N} \hat{S}_{+i}|N/2, -N/2\rangle \\ &= \frac{1}{\sqrt{N}} \sum_{i=1}^{N} |-1\rangle \otimes |-2\rangle \otimes \cdots |+i\rangle \cdots \otimes |-N\rangle. \end{aligned} \quad \text{(D.11)}$$

It is the coherent sum of N quantum states made of $N-1$ ground qubit states and one excited state $|+i\rangle$. When N is large, these collective states are macroscopic quantum states. They are often used as quantum registers.

In particular, macroscopic samples of atoms and ions, contained in a trap or in a glass cell, are described by a large collective Bloch vector $\vec{U}$ that is aligned with the Oz axis in the ground state. The first excitations of the sample do not modify the Oz component appreciably, so that $\langle \hat{S}_z \rangle$ remains close to $-N/2$. The Bloch vector end point is confined to a two-dimensional transverse plane. The well-known angular momentum commutation relation $[\hat{S}_x, \hat{S}_y] = 2i\hat{S}_z$ becomes

$$\left[\hat{S}_x, \hat{S}_y\right] \simeq -iN. \quad \text{(D.12)}$$

Within the scaling factor N, this relation is the usual commutation relation between position and momentum $\hat{P}$ and $\hat{X}$ of a particle. When they are measured, $\hat{S}_x$ and $\hat{S}_y$ are affected by quantum fluctuations. If N is big enough, so that the granularity of the sample is negligible, this quantum noise is continuously varying and obeys a Heisenberg inequality:

$$Var(S_x)\,Var(S_y) \geq \frac{N^2}{4}. \quad \text{(D.13)}$$

As stated in the introduction to Chapter 9, classical polarization states of light are characterized by the Stokes vector lying on the Poincaré sphere, which has a similar algebra to the Bloch vector. If the light contains a macroscopic number of photons, quantum fluctuations of polarization are quite similar to that of a collective Bloch vector [18].

D.3 Exercises for Appendix D

Exercise D.1 Read John Bell's 1964 paper (published in 1966) [16]. Can you derive a hidden variable theory to reproduce all the statistical predictions of quantum mechanics for a single qubit? Discuss von

Neumann's theorem about hidden variable theories and the hidden assumption that Bell uncovered.

Exercise D.2 Qubits are the unit of quantum information and they must be protected to process computations. Consider a qubit encoded in a two-level atom as $|\psi\rangle = \alpha|-\rangle + \beta|+\rangle$. Can a bit-flip error be corrected? Consider the qubit encoded in tree atoms as $|\psi\rangle = \alpha|-,-,-\rangle + \beta|+,+,+\rangle$. Can a single bit-flip be corrected? Is there any way to detect this error before it is too late? Describe the set of QND measurements and the feedback protocol required for this as a function of the test result. What are the Pauli operators in this three atom qubit? Can two flips be corrected?

Exercise D.3 Consider the cat-qubit spanned by the orthogonal basis $|\pm z\rangle = \frac{1}{\mathcal{N}}(|\alpha\rangle \pm |-\alpha\rangle)$ (superpositions of coherent states). Show that in the cat-qubit space the bosonic operators correspond to the cat-qubit Pauli operators as $\hat{a}$, $\hat{a}^\dagger \approx \hat{\sigma}_x$ ($\hat{\sigma}_x|\pm z\rangle = |\mp z\rangle$). Is there a way to detect this without destroying the quantum information?

Exercise D.4 Consider a two-level subsystem spanned by the qubit states $|0_+\rangle = \frac{1}{\mathcal{N}}(|\alpha\rangle + |-\alpha\rangle)$ and $|1_+\rangle = \frac{1}{\mathcal{N}}(|i\alpha\rangle + |-i\alpha\rangle)$, their mean photon number being $\approx |\alpha|^2$. Show that if the systems loses a photon (application of the operator $\hat{a}$) the information about an arbitrary superposition is not lost, but the encoding changes to $|0_-\rangle = \frac{1}{\mathcal{N}}(|\alpha\rangle - |-\alpha\rangle)$ and $|1_-\rangle = \frac{1}{\mathcal{N}}(|i\alpha\rangle - |-i\alpha\rangle)$. Is this collapse a real physical evolution? Can one detect if a photon was lost without demolishing the qubit quantum information initialized in $|\psi\rangle = \alpha_0|0_+\rangle + \beta_0|1_+\rangle$? What would happen if another photon is then lost? Can be this quantum error correction protocol be maintained for ever?

E Quantum Particle

Let us consider a particle of mass M that can move in space. For simplicity we will assume that it can only move in a single direction, more precisely along the Ox axis. The ket $|x\rangle$ describes this particle when it is perfectly localized at point x (with $\langle x|x'\rangle = \delta(x-x')$). The general quantum state of such a particle is therefore the ket $|\Psi\rangle = \int dx\, \psi(x)|x\rangle$, where the complex function $\psi(x) = \langle x|\Psi\rangle$ is the wave function of the particle.

The position x of the particle is associated with the Hermitian operator $\hat{X}$, such that,

$$\hat{X}|x\rangle = x|x\rangle \quad \forall\, x \in R. \tag{E.1}$$

The momentum p of the 1D particle is associated with the Hermitian operator $\hat{P}$, and its eigenstates $|p\rangle$, such that

$$\hat{P}|p\rangle = p|p\rangle \quad \forall\, p \in R, \tag{E.2}$$

with

$$\left[\hat{X}, \hat{P}\right] = i\hbar\hat{1}. \tag{E.3}$$

E.1 Free Particle

When the particle is free, its Hamiltonian is $\hat{H}_{free} = \hat{P}^2/2m$. It has any of the $|p\rangle$ state vectors as energy eigenstates: Its energy spectrum is the continuum $[0, \infty[$. The wave function of the $|p\rangle$ state is

$$\langle x|p\rangle = \frac{1}{2\pi\hbar} e^{ipx/\hbar}. \tag{E.4}$$

In these states, the variance $Var(P)$ of the momentum is zero, and therefore, because of the Heisenberg inequality, the variance $Var(X)$ is infinite: The state is completely delocalized in position.

Let us also introduce the position and momentum displacement operators $\hat{D}_x(x_0)$ and $\hat{D}_p(p_0)$, which will be useful in the next sections:

$$\hat{D}_x(x_0)|x\rangle = |x + x_0\rangle, \quad \hat{D}_x(x_0) = e^{i\hat{P}x_0/\hbar}, \tag{E.5}$$

$$\hat{D}_p(p_0)|p\rangle = |p + p_0\rangle, \quad \hat{D}_p(p_0) = e^{-i\hat{X}p_0/\hbar}. \tag{E.6}$$

E.2 1D Harmonic Oscillator

When the particle is submitted to a binding external potential $V(\hat{x})$, it now has localized bound states and discrete values of energy. An important particular case is the harmonic oscillator when the binding Hamiltonian is quadratic:

$$\hat{H}_{harmonic} = \frac{\hat{P}^2}{2M} + \frac{M\omega_v^2 \hat{X}^2}{2}, \tag{E.7}$$

where ω_v is the vibration frequency. We present in the chapters and appendices of this book different implementations of harmonic oscillators: trapped electrons, ions, atoms, or molecules, vibrational states of elastically bound massive nano-objects (see Figure E.1)

It is convenient to introduce the non-Hermitian operator $\hat{a}_v$,

$$\hat{a}_v = \sqrt{\frac{1}{2M\hbar\omega_v}}\left(M\omega_v\hat{X} + i\hat{P}\right), \tag{E.8}$$

with

$$\left[\hat{a}_v, \hat{a}_v^\dagger\right] = \hat{1}, \tag{E.9}$$

which allows us to write the Hamiltonian as

$$\hat{H}_{harmonic} = \hbar\omega_0\left(\hat{a}_v^\dagger\hat{a}_v + 1/2\right). \tag{E.10}$$

One also introduces the number operator $\hat{N}_v = \hat{a}_v^\dagger\hat{a}_v$, which has the natural integers n_v as eigenvalues, and number states $|n_v\rangle$ as eigenstates: $\hat{N}_v|n_v\rangle = n_v|n_v\rangle$. The number states $|n_v\rangle$ are the energy eigenstates with eigenvalues $(n_v + 1/2)\hbar\omega_v$. They obey the following relations:

$$\hat{a}_v|n_v\rangle = \sqrt{n_v}|n_v - 1\rangle, \quad \hat{a}_v|n_v = 0\rangle = 0, \quad \hat{a}_v^\dagger|n_v\rangle = \sqrt{n_v + 1}|n_v + 1\rangle. \tag{E.11}$$

The operators $\hat{a}_v^\dagger$ and $\hat{a}_v$ are jump, or ladder, operators called raising and lowering operators, respectively.

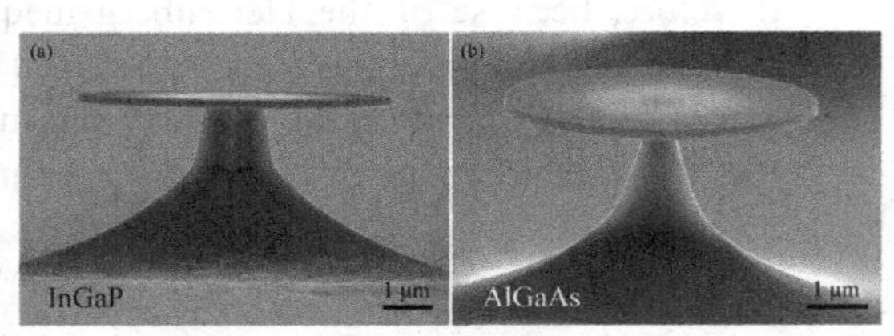

Fig. E.1 The mechanical vibrations of the thin disk, in the 100 MHz range, are coupled to the light trapped by internal reflection in a whispering gallery mode at the edge of the disks. From B. Guha et al. High frequency optomechanical disk resonators in the III-V ternary conductors, *Optics Express*, 2017 [84].

The ground state $|n_v = 0\rangle = |0\rangle$ is such that

$$\langle x|0\rangle = \left(\frac{M\omega_v}{\pi\hbar}\right)^{1/4} e^{-M\omega_v x^2/2\hbar}, \quad Var(X) = \frac{\hbar}{2M\omega_v} = \Delta x_0^2, \tag{E.12}$$

$$\langle p|0\rangle = \left(\frac{1}{\pi M\hbar\omega_v}\right)^{1/4} e^{-p^2/2M\hbar\omega_v}, \quad Var(P) = \frac{M\hbar\omega_v}{2} = \Delta p_0^2. \tag{E.13}$$

The ground state has a nonzero energy and a finite extension both in x and p that are called *zero point fluctuations*. They saturate the Heisenberg inequality $Var(X)\,Var(P) \geq \hbar^2/4$. This is also the case for the vacuum x-displaced states $\hat{D}_x(x_0)|0\rangle$, or the vacuum p-displaced states $\hat{D}_p(p_0)|0\rangle$, that are analogous to the coherent states $|\alpha\rangle$ (with $\alpha = (x_0 + ip_0)/\sqrt{2}$) of the electromagnetic field. Such states have been experimentally observed and characterized in trapped and cooled ions.

Let us give an order of magnitude of the value of position fluctuations in the ground state of different physical harmonic oscillators that we quoted at the beginning of this section:

- electron trapped in a Paul trap: $\nu_0 = 10^7$ Hz, $\Delta x_0 = 10^{-6}$ m
- trapped ion: $\nu_0 = 3$ MHz, $\Delta x_0 = 200$ nm
- trapped atom: $\nu_0 = 100$ Hz, $\Delta x_0 = 3$ nm
- molecular interatomic vibration of a diatomic molecule: λ=3 μm, $\Delta x_0 = 0.1$ nm
- vibrating nano-object: Δx_0 of the order of a nanometer.

The excited states $|n_v\rangle$ have x and p wavefunctions given by

$$\langle x|n_v\rangle = \left(\frac{M\omega_v}{\pi\hbar}\right)^{1/4} \frac{1}{\sqrt{2^n n!}} e^{-M\omega_v x^2/2\hbar} H_n\left(x\sqrt{M\omega_v/\hbar}\right), \tag{E.14}$$

$$\langle p|n_v\rangle = \left(\frac{1}{\pi M\hbar\omega_v}\right)^{1/4} \frac{1}{\sqrt{2^n n!}} e^{-p^2/2M\hbar\omega_v} H_n\left(p/\sqrt{M\hbar\omega_v}\right), \tag{E.15}$$

where H_n are the Hermite polynomials. They are characterized by the variances $Var(X) = (n+1)\Delta x_0^2$ and $Var(P) = (n+1)\Delta p_0^2$ that increase with the quantum number and are no longer minimal. Note that, as energy eigenstates, these states are stationary, and therefore not oscillatory, a striking difference with the classical, oscillating harmonic oscillator.

E.3 Exercises for Appendix E

Exercise E.1 Discuss the quantum fluctuation of a quantum particle in a harmonic oscillator. What is the fluctuating quantity? Take an ordinary macroscopic massive particle in a spring and estimate the fluctuations expected. Estimate the magnitude of the thermal fluctuations too.

Exercise E.2 Show that a quantum particle (a wave packet) evolving in free space under the Schrödinger equation undergoes diffusion. Compare this with the classical heat equation.

Exercise E.3 Solve the equation of motion of a classical harmonic oscillator of frequency ω with initial conditions given by the position x_0 and the momentum p_0. Solve the equation of motion of a quantum harmonic oscillator with initial conditions given by the coherent state $|\psi(t = 0)\rangle = |\alpha\rangle$, where $\alpha = (x_0 + ip_0)/\sqrt{(2)}$. Discuss: Is the quantum-classical correspondance a property of the harmonic oscillator or of the coherent state? How is this related to the Ehrenfest theorem?

F Quantum Electromagnetic Field

F.1 Classical Field

Let us start with classical electrodynamics and consider an orthonormal set of complex solutions of Maxwell equations in a vacuum, usually called modes of the electromagnetic field $\{\vec{u}_\ell(\mathbf{r},t)\}$ with

$$\frac{1}{V}\int d^3r\, \vec{u}_\ell(\mathbf{r},t)\cdot\vec{u}^*_{\ell'}(\mathbf{r},t) = \delta_{\ell\ell'}, \tag{F.1}$$

where V is the large volume into which the physical system is fully inserted; this is necessary to work with a countable number of modes. This set is a basis of Maxwell equations solutions. It allows us to write any classical electromagnetic field in vacuum, $\vec{E}(\mathbf{r},t)$, a real quantity,

$$\vec{E}(\mathbf{r},t) = \vec{E}^{(+)}(\mathbf{r},t) + \left(\vec{E}^{(+)}(\mathbf{r},t)\right)^*, \quad \vec{E}^{(+)}(\mathbf{r},t) = \sum_\ell E^{(+)}_\ell\, \vec{u}_\ell(\mathbf{r},t), \tag{F.2}$$

where $E^{(+)}_\ell$ is the complex amplitude of the mode. Its real and imaginary parts are called the *quadratures* of the field. Its modulus and argument are its amplitude and phase.

F.2 Quantum Field

We note $\hat{\vec{E}}(\mathbf{r},t)$ is the Hermitian operator in the Heisenberg representation that is associated with the real electromagnetic field $\vec{E}(\mathbf{r},t)$. One can show (see, for example, [82]) that it can be written as

$$\hat{\vec{E}}(\mathbf{r},t) = \hat{\vec{E}}^{(+)}(\mathbf{r},t) + \left(\hat{\vec{E}}^{(+)}(\mathbf{r},t)\right)^\dagger, \tag{F.3}$$

$$\hat{\vec{E}}^{(+)}(\mathbf{r},t) = \sum_\ell \hat{E}^{(+)}_\ell\, \vec{u}_\ell(\mathbf{r},t), \tag{F.4}$$

$$\hat{E}^{(+)}_\ell = \mathcal{E}_\ell \hat{a}_\ell, \tag{F.5}$$

where $\mathcal{E}_\ell = \sqrt{\hbar\omega_\ell/2\varepsilon_0 V}$ is the electric field amplitude of a single photon and $\hat{a}_\ell$ a photon annihilation operator that has the same mathematical properties as the lowering operator of a harmonic oscillator (relations (E.11)). This means that there is a mathematical analogy between a single mode quantum field and a material harmonic oscillator. One has, more precisely (relation (E.9)) that

$$\left[\hat{a}_\ell, \hat{a}^\dagger_{\ell'}\right] = \delta_{\ell\ell'}\hat{1}. \tag{F.6}$$

We can define, as for the harmonic oscillator, a number operator in mode ℓ, $\hat{N}_\ell = \hat{a}^\dagger_\ell \hat{a}_\ell$, and number states $|n_\ell\rangle$ that are eigenstates of the number operator with integer eigenvalues n_ℓ:

$$\hat{N}_\ell |n_\ell\rangle = n_\ell |n_\ell\rangle, \quad n_\ell \in N. \tag{F.7}$$

F.3 Single Mode Field

Let us now restrict ourselves to a single mode field of given propagation direction, polarization, frequency ω_ℓ, and transverse shape, like a TEM_{00} beam. The Hamiltonian in this case is $\hat{H}_\ell = \hbar\omega_\ell(\hat{N}_\ell + 1/2)$. The Hermitian operator $\hat{E}_\ell$ associated with the real electric field in this mode is

$$\hat{E}_\ell = \mathcal{E}_\ell \left(\hat{a}_\ell + \hat{a}^\dagger_\ell\right). \tag{F.8}$$

F.3.1 Particle Properties

The energy eigenstates are the number states $|n_\ell\rangle$. They are quantum states with exactly n_ℓ photons in the volume V.

The ground state of the system is the number state $|n_\ell = 0\rangle$, the vacuum, that describes darkness, i.e. the state of light where all light sources have been switched off, whereas the state $|n_\ell = 1\rangle$ describes a single photon, a state that is now routinely produced in experiments. One can also consider quantum states like $|\Psi_1\rangle = (|0\rangle + |1\rangle)/\sqrt{2}$, a superposition of a single photon state and vacuum. The state $|\Psi_1\rangle$ can be shown to have very different properties from the single photon state $|1\rangle$. For example, it has a nonzero mean value of the electromagnetic field, whereas the vacuum state and the single photon state have a zero mean field.

F.3.2 Polarization Properties

Light is a transverse field that is also characterized by its polarization. In each mode, the polarization state is described by a state vector of dimension 2 if one restricts oneself to single-photon states. These polarization states provide convenient qubit states. In particular, one can define single photon polarization states: $|V\rangle$ and $|H\rangle$ for linear vertical and horizontal polarizations; $|\pm D\rangle$ for linear diagonal polarizations in directions $\pm\pi/4$; $|\pm C\rangle$ for right and left circular polarizations. These different states obey the relations:

$$|\pm D\rangle = (|H\rangle \pm |V\rangle)/\sqrt{2}; \quad |\pm C\rangle = (|H\rangle \pm i|V\rangle)/\sqrt{2}. \tag{F.9}$$

These states are widely used in quantum cryptography [13].

F.3.3 Wave Properties

Let us now introduce the *quadrature operators* $\hat{X}_\ell$ and $\hat{P}_\ell$ as

$$\hat{a}_\ell = \left(\hat{X}_\ell + i\hat{P}_\ell\right)\big/\sqrt{2}, \tag{F.10}$$

$$\hat{\vec{E}}^{(+)}(\mathbf{r},t) = \mathcal{E}_\ell \frac{\hat{X}_\ell + i\hat{P}_\ell}{\sqrt{2}} \vec{u}_\ell(\mathbf{r},t), \tag{F.11}$$

$$\hat{X}_\ell = \frac{\hat{a}_\ell^\dagger + \hat{a}_\ell}{\sqrt{2}}, \qquad \hat{P}_\ell = i\frac{\hat{a}_\ell^\dagger - \hat{a}_\ell}{\sqrt{2}}. \tag{F.12}$$

In particular, relations (F.12) shows that $\hat{X}_\ell$ and $\hat{P}_\ell$ are, respectively, associated with the real and imaginary part of the complex amplitude of the field (its quadratures) in mode ℓ. They contain information about the phase and amplitude of the oscillating field in this mode.

In addition, we have

$$\left[\hat{X}_\ell, \hat{P}_{\ell'}\right] = i\delta_{\ell,\ell'}\hat{1}, \tag{F.13}$$

the same commutation relation as for a 1D particle (Section E.2), and therefore the same Heisenberg inequality for the variances

$$Var(X_\ell)\,Var(P_\ell) \geq \frac{1}{4}. \tag{F.14}$$

In the vacuum state, one has $Var(X_\ell) = Var(P_\ell) = 1/2$: These nonzero fluctuations of the field are called *vacuum fluctuations*. The existence of these fluctuations, which are present even when there is no photon, is the second important feature of the quantum electromagnetic field, in addition to the existence of the photon.

It is also useful to introduce the *phase rotated quadratures* $\hat{X}_\theta$ and $\hat{P}_\theta$ by

$$\hat{X}_\theta = \hat{X}\cos\theta + \hat{P}\sin\theta, \quad \hat{P}_\theta = \hat{P}\cos\theta - \hat{X}\sin\theta. \tag{F.15}$$

They correspond to a change of the origin of phases for the field oscillation. We still have $[\hat{X}_\theta, \hat{P}_\theta] = i\hat{1}$, and therefore in vacuum, as for $\hat{X}$ and $\hat{P}$, $Var(X_\theta) = Var(P_\theta) = 1/2$.

The displacement operator $\hat{D}(\alpha)$ $(\alpha = x + ip)$, defined by

$$\hat{D}(\alpha) = e^{\alpha\hat{a}^\dagger - \alpha^*\hat{a}} = e^{i\sqrt{2}(p\hat{X} - x\hat{P})} \tag{F.16}$$

is an extension to the whole complex plane of the position and momentum displacement operators that we introduced for the free or harmonically bound particle (relations (E.5) and (E.6)). There is an interesting composition relation for the displacement operator:

$$\hat{D}(\alpha)\hat{D}(\beta) = \hat{D}(\alpha + \beta)e^{(\alpha\beta^* - \alpha^*\beta)/2}. \tag{F.17}$$

F.4 Some Quantum States of the Field

F.4.1 Number State

Number states $|n\rangle$ describe states of light containing exactly n photons. As they form a basis of the single mode Hilbert space, the most general state of single mode light is the linear combination $|\Psi\rangle = \sum_{n=0}^{\infty} c_n|n\rangle$.

Number states are characterized by quantum field fluctuations around a zero mean field that are the same for all quadrature operators $\hat{X}_\ell$, $\hat{P}_\ell$ and phase-rotated quadratures $\hat{X}_\theta$ and $\hat{P}_\theta$:

$$\langle\hat{X}_\ell\rangle = \langle\hat{P}_\ell\rangle = \langle\hat{X}_\theta\rangle = \langle\hat{P}_\theta\rangle = 0, \tag{F.18}$$

$$Var(X_\ell) = Var(P_\ell) = Var(X_\theta) = Var(P_\theta) = \left(n + \frac{1}{2}\right)\mathcal{E}_\ell^2$$

F.4.2 Coherent State

One defines coherent states, denoted $|\alpha\rangle$, where α is any complex number, as eigenstates of the annihilation operator

$$\hat{a}|\alpha\rangle = \alpha|\alpha\rangle. \tag{F.19}$$

They are also called Glauber states or quasiclassical states. One can show that one also has $|\alpha\rangle = \hat{D}(\alpha)|0\rangle$: Coherent states can also be considered as displaced vacuum states. Their expression on the number state basis is

$$|\alpha\rangle = e^{-|\alpha|^2/2}\sum_{n=0}^{\infty}\frac{\alpha^n}{\sqrt{n!}}|n\rangle. \tag{F.20}$$

The mean photon number is equal to $\langle N\rangle = |\alpha|^2$. The photon number variance is $Var(N) = \langle N\rangle$ and the photon number distribution is $p(n) = \langle N\rangle^n/n!$, characteristic of a Poisson distribution: The coherent state describes an assembly of photons that are randomly distributed in the mode.

Coherent states have another interesting property. One can easily show that

$$\hat{E}_\ell|\alpha\rangle = \langle\alpha|\hat{E}_\ell|\alpha\rangle|\alpha\rangle + |\delta\phi\rangle, \tag{F.21}$$

where $|\langle\alpha|\hat{E}_\ell|\alpha\rangle|^2 = |\alpha|^2\mathcal{E}_\ell^2$ and $|\delta\phi\rangle|^2 = \mathcal{E}_\ell^2$. $\hat{E}_\ell$ is the Hermitian field operator in mode ℓ (Equation (F.5)). Equation (F.21) shows that when $|\alpha|^2$ is large compared to 1, a coherent state, in addition to being an exact eigenstate of the annihilation operator, is "almost" an eigenstate of the field operator. This implies that one makes a good approximation by replacing the field operator by its mean value $\langle\alpha|\hat{E}_\ell|\alpha\rangle$ when the state contains many photons.

Coherent states oscillate in time at frequency ω_ℓ, like a classical field. More precisely, if $|\psi(0)\rangle = |\alpha\rangle$ then $|\psi(t)\rangle$ at a later time is the coherent state

$|\alpha e^{-i\omega_\ell t}\rangle$. One can show that they are the states of light that are produced by classical electromagnetic sources, like antennas or single mode lasers operating well above their oscillation threshold. They still have a quantum property: They are affected by quantum fluctuations equal to the vacuum fluctuations, which enable them to be used in quantum cryptography [81]. Note that the vacuum state can be defined as a number state $|n = 0\rangle$ with zero photons as well as a coherent state $|\alpha = 0\rangle$ of zero mean amplitude.

F.4.3 Squeezed Vacuum State

We now introduce the squeezing operator $\hat{\sigma}(S)$, where S is real:

$$\hat{\sigma}(S) = e^{S(\hat{a}^{\dagger 2} - \hat{a}^2)/2}. \tag{F.22}$$

One now defines the squeezed vacuum state $|S_{0,S}\rangle$ by $|S_{0,S}\rangle = \hat{\sigma}(S)|0\rangle$. Its expression on the number state basis is

$$|S_{0,S}\rangle = \frac{1}{\sqrt{\cosh S}} \sum_{n=0}^{\infty} (\tanh S)^n \frac{\sqrt{2n!}}{2^n n!} |2n\rangle. \tag{F.23}$$

It expands only on the number state with an even number of photons, hence one of its names, "two-photon coherent state."

It is characterized by variances $Var(P) = e^{-S}/2$ and $Var(P) = e^{S}/2$. It is a minimum uncertainty state. When $S \to +\infty$ and $Var(P) \to 0$, the quantum state $|S_{0,S}\rangle$, with $S > 0$, gets closer and closer to the eigenstate of quadrature $\hat{X}$ of eigenvalue 0. Then, by applying the displacement operator $\hat{D}(\alpha = x_0/\sqrt{2})$, or $\hat{D}(\alpha = ip_0/\sqrt{2})$, one is able to generate all the eigenstates $|x = x_0\rangle$ of $\hat{X}$, or the eigenstates $|p = p_0\rangle$ of $\hat{P}$:

$$\hat{X}|x_0\rangle = x_0|x_0\rangle, \quad \hat{P}|p_0\rangle = p_0|p_0\rangle. \tag{F.24}$$

In practice, when S is positive and large, the squeezed vacuum state has a quadrature P that is much less noisy than the vacuum fluctuations at the expense of a simultaneous increase of the variance of the other quadrature operator X. As seen in Chapter 11, this property is exploited in gravitational wave detectors to reduce their background noise.

F.5 Detection of Photon Coincidences

There are different ways of acquiring information about quantum fields. The photodetectors described in Chapter 1 allow us to record a photocurrent I proportional to the number of photons in the case of low flux, or to the energy of the impinging beam during some measurement time T. Homodyne detection, detailed in Section 9.4, allows us to directly measure the quadratures of the quantum electromagnetic field. There is another method that gives information about the photon statistics, which was initiated in the 1950s by Hanbury Brown and Twiss [29], which consists of recording the number of counts that occur at the same time on two different photon detectors and is called a photon coincidence measurement. It can be shown that the coincidence rate at time t is proportional to

$$\langle \hat{\vec{E}}^{(-)}(\mathbf{r}, t)\hat{\vec{E}}^{(-)}(\mathbf{r}', t)\hat{\vec{E}}^{(+)}(\mathbf{r}, t)\hat{\vec{E}}^{(+)}(\mathbf{r}', t)\rangle, \tag{F.25}$$

where $\mathbf{r}$ and $\mathbf{r}'$ are the positions of the photodetectors, $\vec{E}^{(+)}$ is the complex electric field operator (containing annihilation operators), and $\vec{E}^{(-)}$ its Hermitian conjugate (containing creation operators). One notes that, as expected, the coincidence rate is zero in vacuum and single photon states because two annihilation operators appear on the right of Equation (F.25), and two creation operators appear on the left (*normal order*)

Coincidences are better evaluated by the normalized *second-order correlation function*, $g^{(2)}$, also called the intensity correlation function, defined by

$$g^{(2)}(\mathbf{r}, \mathbf{r}', t, t') = \frac{\langle \hat{\vec{E}}^{(-)}(\mathbf{r}, t)\hat{\vec{E}}^{(-)}(\mathbf{r}', t')\hat{\vec{E}}^{(+)}(\mathbf{r}, t)\hat{\vec{E}}^{(+)}(\mathbf{r}', t')\rangle}{\langle \hat{\vec{E}}^{(-)}(\mathbf{r}, t)\hat{\vec{E}}^{(+)}(\mathbf{r}, t)\rangle\langle \hat{\vec{E}}^{(-)}(\mathbf{r}', t')\hat{\vec{E}}^{(+)}(\mathbf{r}', t')\rangle}. \tag{F.26}$$

If one records equal time coincidences between photons in two single mode fields labeled 1 and 2, then the correlation function, named in this case $g^{(2)}(0)$, is equal to

$$g^{(2)}(0) = \frac{\langle \hat{a}_1^\dagger \hat{a}_2^\dagger \hat{a}_1 \hat{a}_2\rangle}{\langle \hat{a}_1^\dagger \hat{a}_1\rangle\langle \hat{a}_2^\dagger \hat{a}_2\rangle}. \tag{F.27}$$

For classical fields E having intensity and phase fluctuations, the normalized intensity correlation function has the following simple expression for a stationary field and for equal detection times:

$$g^{(2)}(0) = \frac{\overline{I(t)I(t)}}{\overline{I(t)}\ \overline{I(t)}}, \tag{F.28}$$

where the bar denotes a classical mean value and $I(t) = |E(t)|^2$. The Cauchy–Schwarz inequality [115] implies that $\overline{I(t)I(t)} - \overline{I(t)}^2 = \overline{(I(t) - \overline{I(t)})^2} \geq 0$, and therefore that, for classically fluctuating fields, $g^{(2)}(0) \geq 1$.

Let us now consider the configuration of Figure F.1. One uses a 50% beamsplitter to mix two input modes labeled 1 in and 2 in and measure the coincidence rate between the two output ports of the beamsplitter. Using the relation (H.28) between the output and input modes,

$$\hat{a}_1^{out} = \frac{1}{\sqrt{2}}\left(\hat{a}_1^{in} + \hat{a}_2^{in}\right) \quad \hat{a}_2^{out} = \frac{1}{\sqrt{2}}\left(-\hat{a}_1^{in} + \hat{a}_2^{in}\right), \tag{F.29}$$

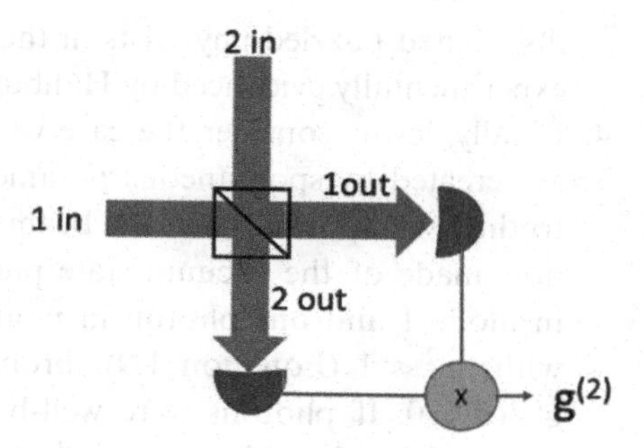

Fig. F.1 Hong–Ou–Mandel coincidence measurement set-up. The light to characterize is split by a beamsplitter and the two outputs sent to two photodetectors. The measured signal is the product of the two photocurrents in the CV regime and the simultaneous occurrence of single photon clicks on both outputs in the DV regime.

one obtains the following expression for the second-order correlation function expressed in terms of the input modes 1 and 2:

$$g^{(2)}(0) = \frac{\left\langle \left(\left(\hat{a}_1^{in\dagger}\right)^2 - \left(\hat{a}_2^{in\dagger}\right)^2 \right) \left(\left(\hat{a}_1^{in}\right)^2 - \left(\hat{a}_2^{in}\right)^2 \right) \right\rangle}{\left\langle \hat{a}_1^{in\dagger} \hat{a}_1^{in} \right\rangle^2 + \left\langle \hat{a}_2^{in\dagger} \hat{a}_2^{in} \right\rangle^2}. \tag{F.30}$$

If one sends nothing (i.e. vacuum) in port 2 of the beamsplitter, one has

$$g^{(2)}(0) = \frac{\langle (\hat{a}_1^{in\dagger})^2 (\hat{a}_1^{in})^2 \rangle}{\langle \hat{a}_1^{in\dagger} \hat{a}_1^{in} \rangle^2} \tag{F.31}$$

One can now envision different situations:

1. The input state is a number state. $g^{(2)}(0)$ is equal, in this case, to $1 - \frac{1}{n}$, and is zero for a single-photon state $|n = 1\rangle$. This value is in agreement with the simple explanation of the impossibility of "cutting in two" a photon, which is really an atom of light that goes either on output 1, or on output 2, but never on both at the same time. This phenomenon, called *photon antibunching*, is in strong disagreement with the classical result $g^{(2)}(0) \geq 1$.
2. The input state is a weak coherent state. $g^{(2)}(0)$ is equal, in this case, to the classical limit value 1: Among the randomly distributed photons with a Poisson distribution that constitute the coherent state, there exist "accidental" pairs of photons that lead to a simultaneous transmission of one photon of the pair and reflection of the other, and therefore contribute to the coincidence count.
3. The input state is a thermal state, $\rho \propto \sum_n e^{-n\hbar\omega/k_B T}|n\rangle\langle n|$. In this case $g^{(2)}(0)$=2: One has a situation, called "photon bunching," which is the opposite of the antibunching effect. Though this value can be accounted for in terms of a classically fluctuating field [115], the fact that in a thermal state the photons are bunched and not randomly

distributed puzzled physicists in the 1950s when this phenomenon was experimentally evidenced by Hanbury Brown and Twiss [29].

4. Finally, let us consider the case where one sends "twin photons" that are created by spontaneous parametric down conversion (Section I.2) to the two input ports of the beamsplitter. The input quantum state is now made of the vacuum state plus a tensor product of one photon in mode 1 and one photon in mode 2: $|\psi\rangle = |0\rangle + a|1\ :\ 1\rangle \otimes |1\ :\ 2\rangle$ with $|a|^2 \ll 1$ (Equation 1.9). From expression (F.30), one finds that $g^{(2)}(0) = 0$. If photons were well-behaved classical particles, a picture enforced by the existence of the antibunching effect for single photons, they would have a 50% probability of exiting the beamsplitter simultaneously on the different ports, the two photons being either both transmitted or both reflected by the beamsplitter. This would give rise to a nonzero coincidence count. The absence of such coincidences implies that the signal and idler photons always take the same path when they cross on the beamsplitter, an example of quantum interference. This striking non-classical behavior of quantum light was discovered by Hong, Ou, and Mandel [96]. Note that one obtains the same effect of zero coincidences when at the input, a state $|1 : 1\rangle \otimes |1 : 2\rangle$ made of two independent factorized single photon states in modes 1 and 2 is used, provided they are perfectly indistinguishable [17].

F.6 Exercises for Appendix F

Exercise F.1 Show that the quadrature operator mean field and variances in a coherent state $|\alpha\rangle$ are

$$\langle \hat{X} \rangle = x, \quad \langle \hat{P} \rangle = p, \quad Var(X) = Var(P) = 1/2. \tag{F.32}$$

What is the variance of the photon number and the time evolution of the electric field in number states?

Exercise F.2 Compute the photon flux of a 1 mW light beam of wavelength $\lambda = 1\ \mu\text{m}$, assumed to be described by a coherent state $|\alpha\rangle$. Determine the amplitude of the photon number fluctuations. What is the signal-to-noise ratio after averaging for 1 ms?

Exercise F.3 Derive the following relations:

$$\hat{\sigma}\hat{a}\hat{\sigma}^\dagger = \hat{a}\cosh S - \hat{a}^\dagger \sinh S, \tag{F.33}$$

$$\hat{\sigma}\hat{a}^\dagger\hat{\sigma}^\dagger = \hat{a}^\dagger \cosh S - \hat{a}\sinh S. \tag{F.34}$$

Hint: One may use the relation, valid for any linear operators A and B,

$$e^A B e^{-A} = B + [A, B] + \frac{1}{2!}[A, [A, B]] + \frac{1}{3!}[A, [A, [A, B]]] + \cdots \tag{F.35}$$

Exercise F.4 Show that the quadrature operator mean field and variances in a squeezed vacuum state $|S\rangle$ are:

$$\langle \hat{X} \rangle = \langle \hat{P} \rangle = 0, \quad Var(X) = \frac{1}{2}e^{S}, \quad Var(P) = \frac{1}{2}e^{-S}, \tag{F.36}$$

so that the system remains in a minimum state, but now with an imbalance between the two quadratures.

Exercise F.5 Prove that $\lim_{S\to\infty} \hat{\sigma}(S)|\alpha\rangle = |x\rangle$ and that $\lim_{S\to-\infty} \hat{\sigma}(S)|\alpha\rangle = |p\rangle$, where $|x\rangle = |\sqrt{2}Re(\alpha)\rangle$ and $|p\rangle = |\sqrt{2}Im(\alpha)\rangle$ are eigenstates of the quadrature operators $\hat{X}$ and $\hat{P}$.

Exercise F.6 Determine the expression of the quantum state in the two modes a and b at the output of a 50% beamsplitter when the input state is the two-photon state $|1 : a, 1 : b\rangle$.

Exercise F.7 One considers the Hong–Ou–Mandel coincidence count configuration using a beam splitter with amplitude reflection and transmission coefficients r and t with $r^2 + t^2 = 1$. Determine the expression of $g^{(2)}$ in this case. What is the physical meaning of these terms. Show that the annulation of $g^{(2)}$ when $r = t$ can be interpreted as originating from a destructive interference between these terms.

Exercise F.8 One considers a linear cavity with decay rate κ driven by a field of amplitude ϵ_d. Take the cavity to be at frequency ω and the drive to be at ω_d. Take the displaced annihilation operator by the drive $\hat{a}(t) = \alpha e^{-i\omega_d t} + \delta\hat{a}(t)$ to express the number operator $\hat{n} = \hat{a}^\dagger\hat{a}$ and show that the autocorrelation function for the photon number in an initially coherent state is given by $\langle \delta n(t)\delta\hat{n}(0)\rangle = \bar{n}e^{-\frac{\kappa}{2}|t|}$. This describes the photon number fluctuations around the mean value $\bar{n} = |\alpha|^2$.

G Interaction between Light and Atoms

G.1 Interaction Hamiltonian

In the dipole approximation, the quantum light–atom interaction term in the Hamiltonian has the form

$$\hat{H}_{int} = -\hat{\vec{D}} \cdot \hat{\vec{E}}(\mathbf{r}_0), \tag{G.1}$$

where $\hat{\vec{D}} = q(\hat{\vec{R}} - \mathbf{r}_0)$ is the electric dipole operator of the atom, $\hat{\vec{R}}$ the position operator of the electron, $\mathbf{r}_0$ the position of the nucleus, taken as a classical quantity in a first approximation, and $\hat{\vec{E}}(\mathbf{r}_0)$ the quantized electric field operator taken at the position of the nucleus. If the system is an atom, it is invariant under any geometrical rotation and therefore has no preferred direction. This implies that the mean value of any vectorial operator, and, in particular, the average value of the dipole moment in energy eigenstates, is zero. In the case of the qubit approximation for the atomic levels, we have

$$\langle +|\hat{\vec{D}}|+\rangle = \langle -|\hat{\vec{D}}|-\rangle = 0, \quad \langle +|\hat{\vec{D}}|-\rangle = \vec{d}, \tag{G.2}$$

where $\vec{d}$ is the mean atomic dipole.

G.2 Semiclassical Approach

G.2.1 Evolution of the Bloch Vector

As a first step, we neglect spontaneous emission. If the field applied to the atom is macroscopic and contains many photons it can be treated, according to the discussion of Equation (F.21), as a first approximation, as a classical oscillating field $\vec{E}(t) = E\vec{\varepsilon}\cos(\omega t + \phi)$. As the atom is treated as a quantum object and the field as a classical object, this approach is called "semiclassical." The semiclassical Hamiltonian for a two-level atom is then the time-dependent operator

$$\hat{H}_{SC} = \hbar\omega_0|+\rangle\langle+| + \hbar\Omega_R\cos(\omega t + \phi)(\hat{\sigma}_+ + \hat{\sigma}_-), \tag{G.3}$$

with $\Omega_R = -\vec{d} \cdot \vec{\varepsilon}\, E/\hbar$. We assume that the applied field is quasi-resonant with the atom: $\omega \simeq \omega_0$.

In order to remove the time dependence and derive the expression of the evolution operator, we apply the unitary transformation $\hat{U} = e^{i\hat{\sigma}_z\omega t/2}$ to the state vector $|\psi(t)\rangle$,

$$|\tilde{\psi}(t)\rangle = \hat{U}|\psi(t)\rangle, \tag{G.4}$$

which is a rotation around Oz at the angular velocity ω. The rotated state now obeys a Schrödinger equation with the effective Hamiltonian $\hat{H}_{eff}$,

$$\begin{aligned}\hat{H}_{eff} &= \hat{U}\hat{H}_{SC}\hat{U}^{\dagger} + i\hbar\frac{d\hat{U}}{dt}\hat{U}^{\dagger}\\ &= -\hbar\delta|+\rangle\langle+| + \hbar\Omega_R\cos(\omega t + \phi)(\hat{\sigma}_+ e^{i\omega t} + \hat{\sigma}_- e^{-i\omega t})\end{aligned} \tag{G.5}$$

with $\delta = \omega - \omega_0$. We now keep the time-independent terms in the Hamiltonian $\tilde{\hat{H}}_{eff}$ and neglect the fast oscillating ones that have a very weak effect on the long-term evolution: This approximation is made very often and has many different names, e.g. rotating-wave approximation, quasi-resonant approximation. Here, we will use the term lent to the astronomers: "Secular approximation," meaning an evolution averaged over very long times. Finally, one obtains the time-independent Hamiltonian

$$\begin{aligned}\hat{H}'_{SC} &\simeq -\hbar\delta|+\rangle\langle+| + \frac{\hbar\Omega_R}{2}(\hat{\sigma}_+ e^{-i\phi} + \hat{\sigma}_- e^{i\phi})\\ &= -\hbar\delta\left(\hat{s}_z + \frac{1}{2}\right) + \frac{\hbar\Omega_R}{2}(\hat{\sigma}_x\cos\phi + \hat{\sigma}_y\sin\phi).\end{aligned} \tag{G.6}$$

The evolution of the state vector in the rotating frame is now very simple:

$$|\tilde{\psi}(t)\rangle = \exp\left(-i\Omega_R t(\hat{s}_x\cos\phi + \hat{s}_y\sin\phi) + i\delta t\hat{s}_z\right)|\tilde{\psi}(t=0)\rangle, \tag{G.7}$$

where $\hat{\vec{s}} = \hat{\vec{\sigma}}/2$ is the spin 1/2 angular momentum operator. One recognizes in the exponential factor a *rotation operator* around the direction of 3D coordinates $(\cos\phi, \sin\phi, -\delta)$. In the resonant case $\delta = 0$, this vector is contained in the xOy plane with a direction depending on the phase of the applied field. This is what is called the "Rabi nutation": The Bloch vector extremity periodically reaches the "poles" A and B of the Bloch sphere, i.e. the atomic levels $|+\rangle$ and $|-\rangle$, and periodically crosses the equator where the state vector acquires a maximum transverse component. Note that the length of the Bloch vector is a constant of motion, so that its extremity is always on the surface of the Bloch sphere if the initial state is one of the two qubit states.

So, *by applying a pulse of resonant electromagnetic field of well-controlled amplitude, phase, and duration and starting from the ground state $|-\rangle$, one can reach any point on the surface of the sphere*, and therefore any pure qubit state. One can, in particular, use a π pulse to completely transfer the qubit to the excited state $|+\rangle$, or a $\pi/2$ pulse to transfer it to an equatorial state of maximum coherence. This technique is widely used in quantum physics. For a 2π pulse, one can show, by developing in series the exponential, that

$e^{-i\Omega_R t\hat{s}_x} = e^{-i\pi\hat{\sigma}_x} = -\hat{1}$: The 2π rotation changes the sign of the state vector. This phase change is used, for example, for the QND measurement of a photon (Chapter 3).

A combination of pulses can also be used. This is the case, for example, of the Ramsey technique. It comprises two $\pi/2$ pulses of electromagnetic field at frequency ω separated by a time T that are applied on an atom of frequency ω_0 in state $|-\rangle$. The first pulse brings the Bloch vector in the equatorial plane. At exact resonance, this vector is at the same angular velocity as the rotating frame, on which is appears fixed. The second $\pi/2$ pulse completes the π pulse and brings the Bloch vector to the state $|+\rangle$. When $\omega \neq \omega_0$, the Bloch vector is no longer fixed in the rotating frame in which it rotates during time T by an angle $(\omega - \omega_0)T$. When $\omega - \omega_0 = \pi/T$, this angle is equal to π, and the Bloch vector induced by the second $\pi/2$ pulse brings it back to the initial state $|-\rangle$. One can show more precisely that the probability to be in $|+\rangle$ state is

$$Proba(|+\rangle) = \cos^2(\omega - \omega_0)T/2. \tag{G.8}$$

This oscillation of the probability is called Ramsey fringes. This set-up is widely used in frequency metrology to get very narrow resonance signals.

G.2.2 Spontaneous Emission

An atom alone cannot escape interacting with the vacuum state of the quantized electromagnetic field, characterized by annihilation operators in different modes $\hat{a}_\ell$ (see Appendix F). If the atom is initially in its excited state $|+\rangle$, ($|-\rangle$ being the ground state) and the quantum field in the vacuum state $|0\rangle$, the initial state of the total system, written as $|+;0\rangle$, is coupled by the interaction term $\widehat{H}_{int}$ to the states $|-;1:\ell\rangle$ corresponding to the atom in its ground state and one photon in the mode labeled ℓ of the electromagnetic field. We assume that the system (two-level atom + quantized field) is isolated. It is therefore described at any time t by the state vector

$$|\Psi(t)\rangle = \gamma_0(t)\,|+;0\rangle + \sum_\ell \gamma_\ell(t)\,|-;1:\ell\rangle\,. \tag{G.9}$$

One deduces from the Schrödinger equation that the evolution of the different coefficients is given by the coupled equations

$$i\hbar\frac{d\gamma_0}{dt} = \hbar\omega_0\gamma_0 + \sum_\ell V_\ell\gamma_\ell, \tag{G.10}$$

$$i\hbar\frac{d\gamma_\ell}{dt} = \hbar\omega_\ell\gamma_\ell + \gamma_0 V_\ell^*, \tag{G.11}$$

with $V_\ell = \langle +;0|\,\hat{H}_{int}\,|-;1:\ell\rangle = -\vec{d}\cdot\vec{\varepsilon}_\ell\mathcal{E}_\ell$, $\hbar\omega_0$ the atomic energy, and $\hbar\omega_\ell$ the photon energy, the atom being at position $\mathbf{r}_0 = 0$.

Using the initial conditions $\gamma_0(0) = 1$ and $\gamma_\ell(0) = 0$, one can formally integrate the Equation (G.11) and get

$$\gamma_\ell(t) = \frac{V_\ell^*}{i\hbar}\int_0^t \gamma_0\,(t')\,e^{i\omega_\ell(t'-t)}dt'. \tag{G.12}$$

By inserting this expression into Equation (G.10), one obtains the following exact integro-differential equation for γ_0:

$$i\hbar\frac{d\gamma_0}{dt} = -i\omega_0\gamma_0 - \sum_\ell \frac{|V_\ell|^2}{\hbar^2}\int_0^t e^{i\omega_\ell(t'-t)}\gamma_0(t')dt'. \tag{G.13}$$

Using $\gamma_0(t) = \alpha\,(t)\,e^{-i\omega_0 t}$ and

$$\mathcal{N}(\tau) = \frac{1}{\hbar^2}\sum_\ell |V_\ell|^2\,e^{i(\omega_0-\omega_\ell)\tau}, \tag{G.14}$$

Equation (G.13) becomes

$$\frac{d}{dt}\alpha\,(t) = -\int_0^t \mathcal{N}(\tau)\,\alpha\,(t-\tau)\,d\tau, \quad (\tau = t - t'). \tag{G.15}$$

The derivative of α at instant t depends on previous values of α belonging to a time interval equal to the width of the "memory function" $\mathcal{N}(\tau)$. When $\tau = 0$ the exponential are all equal to 1, whereas when $\tau \neq 0$ the oscillations of different frequencies average rapidly to zero if $\omega_0 \neq \omega_l$. $\mathcal{N}(\tau)$ therefore tends to zero when τ is larger than a memory time, which turns out to be very short, of the order of a few optical periods. So if t is large compared to this correlation time, and assuming that α does not vary much during this correlation time, one can write

$$\frac{d}{dt}\alpha(t) \approx -\left[\int_0^t \mathcal{N}(\tau)d\tau\right]\alpha\,(t) \approx -\left[\int_0^{+\infty} \mathcal{N}(\tau)d\tau\right]\alpha(t), \tag{G.16}$$

because $\mathcal{N}(\tau)$ is close to zero between t and infinity. Writing

$$\int_0^{+\infty} \mathcal{N}(\tau)d\tau = \frac{\Gamma}{2} + i\delta\omega, \tag{G.17}$$

one finally gets an expression for $\gamma_0(t)$ and $\gamma_\ell(t)$ valid for any time t within the short memory approximation (also known as the Markov approximation):

$$\gamma_0(t) = e^{-\frac{\Gamma}{2}t}e^{-i(\omega_0+\delta\omega)t}, \tag{G.18}$$

$$\gamma_\ell(t) = \frac{V_\ell^*}{i\hbar}\frac{1 - e^{-\frac{\Gamma}{2}t}e^{-i(\omega_\ell-\omega_0+\delta\omega)t}}{\frac{\Gamma}{2} - i(\omega_\ell - \omega_0 - \delta\omega)}. \tag{G.19}$$

Equation (G.18) implies that the population of the excited level decreases in time as $e^{-\Gamma t}$: This is the phenomenon of spontaneous emission.

G.2.3 Bloch Equations

Let us now determine the evolution of the two-level atom in the presence of spontaneous emission of rate Γ when the state $|-\rangle$ is the ground state and $|+\rangle$ the first excited state, so that spontaneous emission cannot populate other atomic levels, and $\rho_{++} + \rho_{--}$ remains equal to 1 during the evolution. The two-level system is not isolated as it interacts with the quantized field and must be described by a reduced density matrix ρ obeying a master equation in the Lindblad form, which we write for the density matrix in the frame rotating at velocity ω, $\tilde{\rho} = \hat{U}\rho\hat{U}^\dagger$ (Equation (G.5)):

$$\frac{d\tilde{\rho}}{dt} = \frac{1}{i\hbar}\left[\hat{H}'_{SC}, \tilde{\rho}\right] - \frac{\Gamma}{2}(\tilde{\rho}|+\rangle\langle+| + |+\rangle\langle+|\tilde{\rho}) + \Gamma|-\rangle\langle+|\tilde{\rho}|+\rangle\langle-|, \quad \text{(G.20)}$$

where $\hat{H}'_{SC}$ is the atom+field Hamiltonian $-\hbar\delta|+\rangle\langle+| + \frac{1}{2}\hbar\Omega_R(\hat{\sigma}_+e^{-i\phi} + \hat{\sigma}_-e^{i\phi})$ introduced in Equation (G.7).

More precisely, the components of the density matrix in the rotating frame evolve as

$$\frac{d\tilde{\rho}_{++}}{dt} = -\Gamma\tilde{\rho}_{++} - \Omega_R\Im(\tilde{\rho}_{+-}e^{-i\phi}), \quad \text{(G.21)}$$

$$\frac{d\tilde{\rho}_{+-}}{dt} = \left(i\delta - \frac{\Gamma}{2}\right)\tilde{\rho}_{+-} + i\frac{\Omega_R}{2}e^{i\phi}(\tilde{\rho}_{--} - \tilde{\rho}_{++}), \quad \text{(G.22)}$$

where $\Im$ denotes the imaginary part. Γ is the spontaneous emission decay rate for the population. These equations are known as the *optical Bloch equations* and couple the evolution of populations to the coherence and the evolution of coherence to population difference. One notes that the spontaneous emission decay rate for the population is twice the spontaneous emission decay rate for the coherence.

In contrast with the previous situation (Section G.1) of negligible spontaneous emission in which the density matrix components endlessly oscillate, in the presence of spontaneous emission the density matrix, after a damped oscillatory regime, reaches the steady state ρ^{ss}:

$$\left(\tilde{\rho}^{ss}_{--} - \tilde{\rho}^{ss}_{++}\right) = \frac{(\Gamma/2)^2 + \delta^2}{(\Gamma/2)^2 + \delta^2 + \Omega_R^2/2}, \quad \text{(G.23)}$$

$$\tilde{\rho}^{ss}_{+-} = \frac{\Gamma/2 - i\delta}{(\Gamma/2)^2 + \delta^2 + \Omega_R^2/2}\Omega_R. \quad \text{(G.24)}$$

The Bloch equations can be generalized to more than two levels. They have a wide range of validity and are able to account for the majority of physical situations, such as stimulated emission and laser theory, optical pumping, saturated absorption, coherent population trapping, light shifts, electromagnetically induced transparency, and "slow light" [46].

G.3 Cavity Quantum Electrodynamics

We now consider the full quantum approach of light–matter interaction. In order to simplify the problem we restrict ourselves to the situation where the two-level system with Bohr frequency ω_0 interacts with a single mode optical cavity of resonance frequency ω. The expression of the quantized field in the cavity is given by Equations (F.3)–(F.5), where the volume V

is now a physically well-defined quantity as it is the number of photons confined in the cavity. The quantized field interacts with a two-level atom placed at the origin of coordinates. This situation is usually known as *cavity quantum electrodynamics (CQED)*.

G.3.1 Jaynes–Cummings Hamiltonian

For the time being, we assume that the two levels $|+\rangle$ and $|-\rangle$ have an infinite lifetime (which is almost the case for the hyperfine levels in the ground state of hydrogen and alkali atoms, a configuration widely used in experiments), and that the frequency ω of the cavity mode is close to the frequency ω_0 of the atom. The interaction Hamiltonian (G.1) simplifies to

$$\hat{H}'_{int} = -i\vec{d}\cdot\vec{\varepsilon}_\ell \mathcal{E}_\ell \left(\hat{a}_\ell\hat{\sigma}_+ - \hat{a}_\ell^\dagger\hat{\sigma}_-\right) = i\frac{\hbar\Omega_V}{2}\left(\hat{a}_\ell\hat{\sigma}_+ - \hat{a}_\ell^\dagger\hat{\sigma}_-\right), \tag{G.25}$$

with $\Omega_V = -2\vec{d}\cdot\vec{\varepsilon}_\ell\mathcal{E}_\ell/\hbar$. The two terms describe the two resonant processes consisting of: one photon absorbed + atom raised, and one photon created + atom de-excited. The total Hamiltonian of the cavity+atom system, called the Jaynes–Cummings Hamiltonian, is time independent and can be written as

$$\hat{H}_{JC} = \hbar\omega_0|+\rangle\langle+| + \hbar\omega\hat{a}_\ell^\dagger\hat{a}_\ell + i\hbar\frac{\Omega_V}{2}\left(\hat{a}_\ell\hat{\sigma}_+ - \hat{a}_\ell^\dagger\hat{\sigma}_-\right). \tag{G.26}$$

It acts in the following way on the basis states of the atom+field system:

$$\hat{H}_{JC}|-;n\rangle = n\hbar\omega|-;n\rangle - i\sqrt{n}\frac{\Omega_V}{2}|+;n-1\rangle. \tag{G.27}$$

$$\hat{H}_{JC}|+;n-1\rangle = [\hbar\omega_0 + (n-1)\hbar\omega]\,|+;n-1\rangle + i\sqrt{n}\frac{\Omega_V}{2}|-;n\rangle, \tag{G.28}$$

The subspace $\mathcal{M}_n$ spanned by the vectors $|+;n\rangle$ and $|-;n+1\rangle$ is therefore stable with respect to the action of the Jaynes–Cummings Hamiltonian. It can be independently diagonalized inside each of these spaces. One gets the following expressions for the eigenvectors and eigenvalues:

$$E_{+n} = \hbar\left(n\omega - \frac{\delta}{2} + \frac{1}{2}\sqrt{n\Omega_V^2 + \delta^2}\right), \tag{G.29}$$

$$E_{-n} = \hbar\left(n\omega - \frac{\delta}{2} - \frac{1}{2}\sqrt{n\Omega_V^2 + \delta^2}\right), \tag{G.30}$$

$$|\psi_+\rangle = \cos\theta_n|-;n\rangle + i\sin\theta_n|+;n-1\rangle, \tag{G.31}$$

$$|\psi_-\rangle = i\sin\theta_n|-;n\rangle + i\cos\theta_n|+;n-1\rangle, \tag{G.32}$$

with $\delta = \omega - \omega_0$, $\tan 2\theta_n = -\Omega_V\sqrt{n}/\delta$ $(0 \le \theta_n \le \pi/2)$ (Figure G.1).

We note that, compared to the case where there is no interaction between the field and the atom, the degeneracy between the energies is lifted at resonance, where $\theta_n = \pi/4$. A gap, equal to $\hbar\Omega_V$ at resonance, appears between the two energy levels. This is what is called an "avoided crossing," or an anti-crossing. The eigenstates $|\psi_\pm\rangle$ are entangled states, and it is not possible to separate an atomic part and a field part: They are called dressed

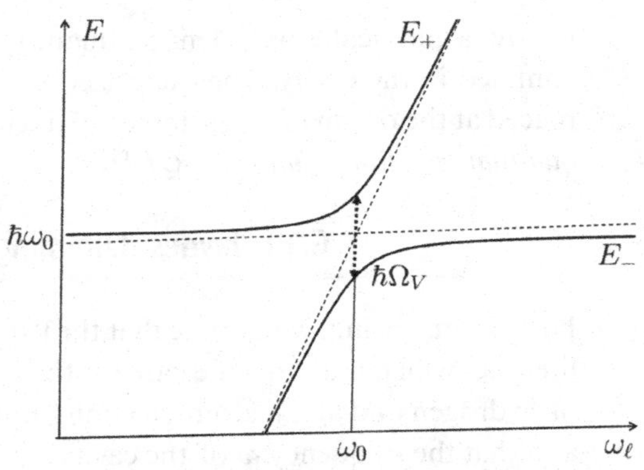

Fig. G.1 Energy levels of a two-level system of Bohr frequency ω_0 interacting with a quantized cavity mode of frequency $\omega = \omega_\ell$. Note the onset of an energy gap at exact resonance.

atomic states, meaning dressed by cavity photons. One could equally have named them as dressed photonic states (dressed by the atomic matter inserted into the optical cavity). Also, note that Ω_V is inversely proportional to the cavity volume V, so that the "dressing" effects will be more apparent in micro- or nanocavities.

G.3.2 Evolution of the System: Resonant Case

The knowledge of the eigenvectors of the system allows us to determine different kinds of evolution according to different possible initial conditions:

1. If the atom is initially in the excited state with no photons present, a situation corresponding to spontaneous emission, but now inside an optical cavity, not in free space. One easily finds that

$$|\psi(t)\rangle = e^{-i\omega_0 t}\left(\cos\frac{\Omega_V t}{2}|+;0\rangle - \sin\frac{\Omega_V t}{2}|-;1\rangle\right). \tag{G.33}$$

The probability of finding the atom in the excited state is therefore

$$P_+(t) = \cos^2\frac{\Omega_V t}{2}. \tag{G.34}$$

One finds an oscillating behavior, not an exponential decay. The cavity walls reflect the photon emitted by the excited atom, which can now reabsorb it, hence the temporal oscillations, which are called "vacuum Rabi oscillations," or single photon Rabi oscillations. Note that the atom periodically goes back completely to its excited state. An important physical conclusion is that spontaneous emission is not an intrinsic property of the atom, but depends also on its physical environment. At time t such that $\Omega_V t/2 = \pi/4$, the system is the entangled state $e^{-i\omega_0 t}(|+;0\rangle - |-;1\rangle)/\sqrt{2}$, which exhibits perfect quantum atom-field correlations: If the atom is detected to be in the ground state, then the cavity is certainly in a single photon Fock state.

2. If the atom is initially in its ground state $|-\rangle$ with n photons present in the cavity, the state vector of the system is found to be

$$|\psi(t)\rangle = e^{-in\omega_0 t}\left(\cos\frac{\sqrt{n}\Omega_V t}{2}|+;0\rangle - \sin\frac{\sqrt{n}\Omega_V t}{2}|-;1\rangle\right). \tag{G.35}$$

It is again a Rabi nutation, but with larger angular velocity.

3. If the atom is initially in the ground state with the cavity optical mode being on a coherent state α, one finds, using the decomposition of the coherent state on the Fock state basis $|\alpha\rangle = e^{-|\alpha|^2/2}\sum_n(\alpha^n/\sqrt{n!}|n\rangle$, that

$$|\psi(t)\rangle = e^{-|\alpha|^2/2}\sum_n \frac{\alpha^n}{\sqrt{n!}}e^{-in\omega_0 t}\left(\cos\frac{\sqrt{n}\Omega_V t}{2}|+;0\rangle - \sin\frac{\sqrt{n}\Omega_V t}{2}|-;1\rangle\right). \tag{G.36}$$

This situation will be examined in more detail in the Exercises: The evolution is a damped Rabi oscillation followed by a quantum "revival" (Figure G.3), a typical quantum behavior.

G.3.3 Evolution of the System: Dispersive Limit

When δ is large, the energy levels and the system eigenstates are, in the lowest multiplicity $n = 1$,

$$E_{+1} = \hbar\left(\omega_0 + \frac{\Omega_V^2}{4\delta}\right), \tag{G.37}$$

$$E_{-1} = \hbar\left(\omega - \frac{\Omega_V^2}{4\delta}\right), \tag{G.38}$$

$$|\psi_+\rangle = |-;1\rangle + i\frac{\Omega_V}{\delta}|+;0\rangle, \tag{G.39}$$

$$|\psi_-\rangle = -i\frac{\Omega_V}{\delta}|-;1\rangle + |+;0\rangle. \tag{G.40}$$

It is now easy to determine the evolution of the system during an interaction time T when it starts at time 0 in the states $|+;0\rangle$ or $|-;1\rangle$:

$$|+;0\rangle \to |+;0\rangle e^{-i(\omega_0 T+\Omega_V^2 T/4\delta)}, \quad |-;1\rangle \to |-;1\rangle e^{-i(\omega T-\Omega_V^2 T/4\delta)}. \tag{G.41}$$

We see that the interaction of the atom with the cavity has not led to an energy transfer because of the large energy mismatch between the cavity and the atom frequencies, and that the system has remained in the same state, *except for a phase shift* $\pm\Omega_V^2 T/4\delta$, that depends on the presence or not of a photon in the cavity. As we will see more precisely later, this system is the archetype of a quantum nondemolition measurement (see Chapters 3 and 5).

G.4 Effect of Damping in Cavity QED

G.4.1 Master Equation

Excited atoms inserted into a cavity decay by spontaneous emission and emit photons not only in the cavity mode, but also in other modes. In addition, the cavity photons have a finite lifetime because of cavity transmission and losses. These two independent relaxation processes are, respectively, characterized by two decay constants Γ and κ, in the master equation. For example, in a typical experiment using the cesium atom and its transition in the near infrared, coupled to a high finesse Fabry–Perot cavity, $\Gamma \simeq 6 \times 10^7 \text{s}^{-1}$ and $\kappa \simeq 2 \times 10^7 \text{s}^{-1}$.

The master equation of the total system atom+cavity electromagnetic mode is therefore

$$\frac{d\rho}{dt} = -\frac{i}{\hbar}\left[\hat{H}_{JC}, \rho\right] - \kappa\left(\hat{a}^\dagger \hat{a}\rho + \rho\hat{a}^\dagger\hat{a}\right) - \Gamma\left(\hat{\sigma}_z\rho + \rho\hat{\sigma}_z\right) + 2\kappa\hat{a}\rho\hat{a}^\dagger + 2\Gamma\hat{\sigma}_-\rho\hat{\sigma}_+. \tag{G.42}$$

This equation can be solved numerically, but not analytically. Often, the problem simplifies if one is interested only in the evolution of a small number of observables. One can, as a first step, deduce from Equation (G.42) the coupled evolution equations of $\hat{a}_\ell$ and $\hat{\sigma}_+$ in the Heisenberg representation, where $\hat{a}_\ell$ is proportional to the complex field $\hat{E}^{(+)}$ and $\hat{\sigma}_+$ is proportional to the complex atomic dipole. One has, at exact resonance,

$$\frac{d}{dt}\hat{\sigma}_+ = \left(-i\omega_0 - \frac{\Gamma}{2}\right)\hat{\sigma}_+ - \frac{\Omega_V}{2}\hat{\sigma}_z\hat{a}_\ell,$$
$$\frac{d}{dt}\hat{a}_\ell = \left(-i\omega_0 - \frac{\kappa}{2}\right)\hat{a}_\ell + \frac{\Omega_V}{2}\hat{\sigma}_+. \tag{G.43}$$

This is not a closed system of equations because the evolution of $\hat{\sigma}_+$ is linked to the evolution of the operator $\hat{\sigma}_z\hat{a}_\ell$. We should therefore include in the system of equations the evolution equation for this operator, which will be linked to that of another operator, and so on and so forth. Unfortunately, the set of this whole hierarchy of equations is not solvable.

We can reduce the set of coupled equations by replacing $\hat{\sigma}_z\hat{a}_\ell$ by $-\hat{a}_\ell$ in Equation (G.43). One can show that this simplification is correct in the regime of cavity-assisted spontaneous emission, i.e. when there is one or zero photons in the cavity. In this regime, the evolution only involves the quantum states $|+;0\rangle$, $|-;1\rangle$, and $|-;0\rangle$. We now have a closed system of linear equations:

$$\frac{d}{dt}\hat{\sigma}_+ = \left(-i\omega_0 - \frac{\Gamma}{2}\right)\hat{\sigma}_+ \frac{\Omega_V}{2}\hat{a}_\ell,$$
$$\frac{d}{dt}\hat{a}_\ell = \left(-i\omega_0 - \frac{\kappa}{2}\right)\hat{a}_\ell + \frac{\Omega_V}{2}\hat{\sigma}_+. \tag{G.44}$$

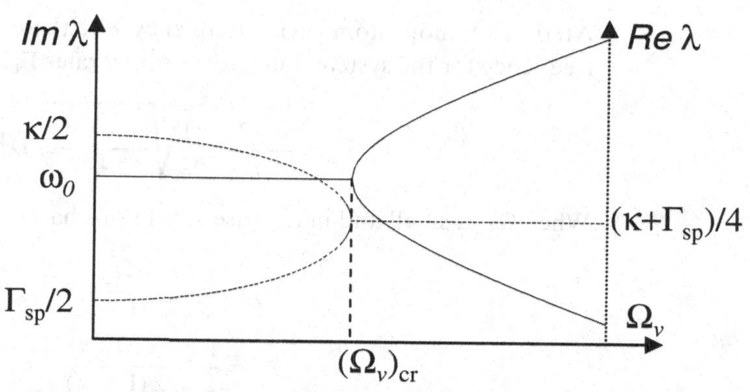

Fig. G.2 Real (dotted line: damping rate) and imaginary (solid line: oscillation frequency) parts of the eigenvalues as a function of coupling Ω_V in the case $\kappa > \Gamma$. One observes a transition between two different regimes that occurs when Ω_V crosses the value $(\Omega_V)_{cr} = |\kappa - \Gamma|/2$.

These equations are simply the evolution equations of two coupled classical oscillators, each one subject to damping. Their solution in the form of exponential functions is now simple to derive. Figure G.2 gives the variation of the real (associated with damping) and imaginary (associated with oscillations) parts of the eigenvalues λ of the 2×2 matrix. Two regimes appear, which we discuss in the next section.

G.4.2 Strong Coupling and Weak Coupling Regimes

Strong Coupling: $\Omega_V > \frac{|\kappa-\Gamma|}{2}$

This is a regime of weak (or identical) dampings. The eigenvalues λ are given in this case by

$$\lambda_{1,2} = -i\omega_0 \pm \frac{i}{2}\sqrt{\Omega_V^2 - \frac{|\kappa - \Gamma|^2}{4}} - \frac{\kappa + \Gamma}{4}. \tag{G.45}$$

One notes that the system is damped at the constant rate $(\kappa + \Gamma)/4$, and that two oscillation frequencies, separated by $\sqrt{\Omega_V^2 - (\kappa - \Gamma)^2/4}$, appear. When there is no relaxation, these two frequencies are separated by Ω_V, as expected from the discussion of Section G.3. When the relaxation is not zero, the difference between these frequencies is reduced, and is zero when $\Omega_V = \frac{|\kappa-\Gamma|}{2}$. The population of the excited state does not decay in a monotonic way, but undergoes damped oscillations. In this regime there is an interchange between atomic and field coherence, which is harnessed for quantum computing using atomic (or ionic) qubits.

Weak Coupling: $\Omega_V < \frac{|\kappa-\Gamma|}{2}$,

This regime corresponds to strongly damped atom and cavity,

$$\lambda = -i\omega_0 \pm \frac{1}{2}\sqrt{\frac{|\kappa - \Gamma|^2}{4} - \Omega_v^2} - \frac{\kappa + \Gamma}{4}. \tag{G.46}$$

At the common atom-cavity frequency ω_0, there is a single oscillation frequency for the system, but two damping rates $\Gamma_\pm$, equal to

$$\Gamma_\pm = \frac{\kappa + \Gamma}{4} \pm \frac{1}{2}\sqrt{\frac{|\kappa - \Gamma|^2}{4} - \Omega_V^2}. \tag{G.47}$$

When Ω_V is small and in the case $\kappa > \Gamma$, one has

$$\Gamma_+ = \frac{\kappa}{2} - \frac{\Omega_V^2}{2\,(\Gamma - \kappa)}, \tag{G.48}$$

$$\Gamma_- = \frac{\Gamma}{2} + \frac{\Omega_V^2}{2\,(\Gamma - \kappa)}. \tag{G.49}$$

It is easy to see that for small values of the coupling constant, Γ_- gives the relaxation rate of the atomic dipole. We see that when one inserts an atom into a resonant cavity, even when strongly damped, *the atom has a reduced lifetime*. This is the so-called Purcell effect. In the bad cavity limit (that is, when $\kappa \gg \Gamma$ with $\kappa \gg i\Omega_V$) and using the expressions of Γ and Ω_V, one finds that

$$\Gamma_- \simeq \frac{\Gamma}{2}\frac{3c\lambda^3}{8\pi\kappa V^2}. \tag{G.50}$$

The Purcell effect is stronger in a cavity of dimensions close to the value of the wavelength. An increase of the damping rate of the order of 100 to 1000 has been observed experimentally.

G.5 Exercises for Appendix G

Exercise G.1 Show that the Bloch vector of section D.1 $\vec{U}(t)$ obeys the evolution equation $d\vec{U}/dt = \vec{\Omega}_R \times \vec{U}$. Give a classical analog of such an evolution.

Exercise G.2 Derive the time evolution of the atomic energy and dipole at exact resonance and for $\phi = 0$. Discuss.

Exercise G.3 Write the semiclassical Hamiltonian for an assembly of N indistinguishable atoms. Show that it can be described in terms of the collective Bloch vector introduced in Section D.2. Show that the first collective excited state $|\Phi_N\rangle$ evolves N times faster than a single atom. Show that spontaneous emission from the singly excited $|\Phi_N\rangle$ collective state is also N times faster than that of a single atom (single excited atom superradiance). Consider the case $N = 2$: Show that one can define a "subradiant" state that has no spontaneous decay.

Consider the case where all the atoms of the sample are initially excited. What is the collective state corresponding situation? Show that the system will remain in the states of maximum angular momentum $|S = N/2, M_S\rangle$ while decaying by spontaneous emission. Calculate the matrix element between two successive states.

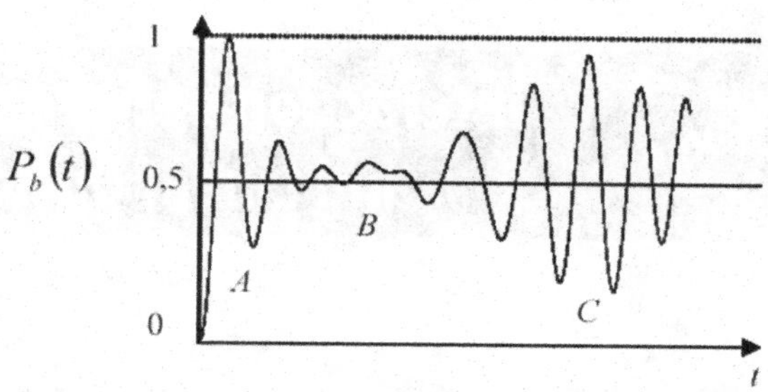

Fig. G.3 Evolution of the excited state population in the case of an illumination by a coherent state. The system undergoes damped Rabi oscillations (phase A) around value 0.5, on which it stabilizes for a while (phase B). The system is then subject to a temporary "quantum revival" (phase C).

Describe qualitatively the evolution of the system (N-excited atoms superradiance).

Exercise G.4 The initial quantum state of the cavity mode is now a coherent state $|\alpha\rangle = \sum_n c_n|n\rangle$ where $c_n = \alpha^n/n!\, e^{-|\alpha|^2/2}$. Give the expression of the time evolution of the excited state population P_+. Show qualitatively that $P_+(t)$ undergoes a damped Rabi nutation tending to the value 0.5 and followed by a revival of the oscillation. This phenomenon is called "quantum revival" (see Figure G.3).

H Interaction between Light Beams and Linear Optical Media

We now consider a physical situation that is often encountered in quantum optics. It consists of one or several light beams that interact with matter and that are measured at the output. In contrast with the case in Section G.3, light is continuously produced at one end and absorbed at the other: It is an open system, which we will assume to be stationary. Consequently, we will not be concerned by the time evolution of the system, but rather by the *spatial propagation* of the different light beams, in the form of input–output relations, that constitute another implementation of the quantum maps studied in Chapter 4. In such situations, the electromagnetic energy and the photon number are no longer physically relevant parameters as they grow unbounded with time. One instead deals with the electromagnetic Poynting vector or the photon flux.

We consider a linearly polarized single mode TEM_{00} beam of waist w_0, frequency ω, and wave vector k along the Oz axis. The complex electric field at position z on the beam axis, and the corresponding quantum operator in the Heisenberg representation, are given by

$$E^{(+)}(z,t) = Ee^{i(kz-\omega_0 t)} \quad \text{and} \quad \hat{E}^{(+)}(z,t) = \mathcal{E}_\ell \hat{a} e^{i(kz-\omega_0 t)}, \tag{H.1}$$

respectively, where $\mathcal{E}_\ell = \sqrt{\hbar\omega_0/2\varepsilon_0 cST}$, $S = \pi w_0^2$ is the beam cross section, and T the measurement time [66]. The photon number operator $\hat{a}^\dagger \hat{a}$ measures the photon number in the cylindrical volume of length cT and transverse section S, and $\hat{a}^\dagger \hat{a}/T$ is the photon number flux.

H.1 Propagation of a Light Beam in a Sample of Two-Level Systems: Semiclassical Approach

Let's consider a light beam, which we will treat as classical, propagating along the $0z$ axis through a sample of length L containing N two-level atoms. The evolution of the atomic density matrix of each atom, written in the frame rotating at frequency ω, is ruled by the optical Bloch equations (G.21) and (G.22). When $t \gg \Gamma^{-1}$ the atom reaches a stationary regime given by Equations (G.23) and (G.24) in the quasi-resonant case ($\omega \simeq \omega_0$). For the complex atomic dipole, one finds that

$$\mathbf{D}^{(+)} = \frac{d^2}{2\hbar} \frac{i\Gamma/2 - \delta}{\delta^2 + \Gamma^2/4 + \Omega_R^2/2} E. \tag{H.2}$$

The macroscopic complex polarisation density $\mathbf{P}^{(+)}$ induced by the light beam in the sample, a source term in the Maxwell equations of dielectrics, derives from the dipole moment of a single atom, assuming that the atoms are identical but not correlated, and the sample diluted enough so that there are no local field effects:

$$\mathbf{P}^{(+)} = \frac{N}{V}\frac{d}{\hbar}\frac{i\Gamma/2-\delta}{\delta^2+\Gamma^2/4+\Omega_R^2/2}\Omega_R, \tag{H.3}$$

from which one defines the complex susceptibility $\chi = \chi' + i\chi''$ of the medium by

$$\mathbf{P}^{(+)} = \varepsilon_0 \chi E. \tag{H.4}$$

We first consider the low intensity regime, without a pumping term. The $\Omega_R^2/2$ term can be neglected, so that the complex susceptibility χ does not depend on E: The atomic sample is a linear dielectric medium. We know that in this regime the field propagates as $E(z) = E(0)e^{i(kz-\omega t)}$, with the wavevector $k = n\omega/c$, where n is the refractive index of the medium, related to the susceptibility by the relation

$$n^2 = 1+\chi; n \simeq 1 + \frac{\chi'}{2} + i\frac{\chi''}{2} \tag{H.5}$$

for a dilute sample of small susceptibility. If k' and k'' are the real and imaginary parts of the wavevector, the input–output relation for this medium is

$$E(L) = E(0)e^{ik'L}e^{-k''L}, \tag{H.6}$$

$$k' = \frac{\omega}{c}\left(1 - \frac{Nd^2}{2\varepsilon_0\hbar V}\frac{\delta}{\delta^2+\Gamma^2/4}\right), \tag{H.7}$$

$$k'' = \frac{\omega}{c}\frac{Nd^2}{2\varepsilon_0\hbar V}\frac{\Gamma/2}{\delta^2+\Gamma^2/4}. \tag{H.8}$$

The real part k' describes the dispersive effect due to the presence of the quasi-resonant atomic medium: The phase velocity of the wave is modified, mainly in the wings of the resonance. The imaginary part k'' describes changes of the beam power. As k'' is positive, the beam is attenuated by its resonant interaction with the medium. The intensity attenuation factor for the beam crossing the whole sample is

$$A = \exp -\left(\frac{N\omega L d^2}{\varepsilon_0\hbar V}\frac{\Gamma/2}{\delta^2+\Gamma^2/4}\right). \tag{H.9}$$

It is maximum at exact resonance, with a resonance width equal to Γ. The energy is actually scattered away from the light beam by spontaneously emitted photons: This is the phenomenon of resonance fluorescence.

Let us now add an incoherent pump mechanism, like collisions or illumination by incoherent light, that continuously brings the atoms from $|-\rangle$ to $|+\rangle$ at a rate Λ. We must then add the term $\Lambda\tilde{\rho}_{--}$ to Equation (G.21). The Bloch equations become

$$\frac{d\tilde{\rho}_{++}}{dt} = \Lambda\tilde{\rho}_{--} - \Gamma\tilde{\rho}_{++} - \Omega_R \Im(\tilde{\rho}_{+-}e^{-i\phi}), \tag{H.10}$$

$$\frac{d\tilde{\rho}_{+-}}{dt} = \left(i\delta - \frac{\Gamma}{2}\right)\tilde{\rho}_{+-} + i\frac{\Omega_R}{2}e^{i\phi}(\tilde{\rho}_{--} - \tilde{\rho}_{++}). \tag{H.11}$$

When the light beam is switched off, the population of the excited state is $\tilde{\rho}_{++} = \Lambda/\Gamma = N_+$. The macroscopic steady-state complex polarization of the sample can be shown to be

$$\mathbf{P}^{(+)} = -\frac{N_+}{V}\frac{d}{\hbar}\frac{i\Gamma/2 - \delta}{\delta^2 + \Gamma^2/4 + \Omega_R^2/2}\Omega_R. \tag{H.12}$$

It is the opposite (except for the number of atoms N_+) of the polarization given by Equation (H.3) in the absence of pumping. As a result, k'' is now negative: *The intensity of the light beam grows as it propagates through the sample*. The medium is a *coherent light amplifier*, as the phase coherence is preserved in the amplification process. Its gain in power, G, is equal at exact resonance to

$$G = \exp\frac{2N\omega L d^2}{\varepsilon_0\hbar\Gamma V}. \tag{H.13}$$

From the optical Bloch equations, one can calculate the steady-state population difference $\tilde{\rho}_{--} - \tilde{\rho}_{++}$, which is negative, corresponding to *population inversion* compared to the case of no pumping. This phenomenon of optical amplification is a manifestation of *stimulated emission* that occurs in the medium and is at the basis of laser operation when the amplifying medium is inserted in a resonant cavity.

At high intensities, one cannot neglect the intensity-dependent term $\Omega_R^2/2$ in the denominator. Nonlinear effects arise, such as the existence of an intensity dependence of the index of refraction (self-Kerr effect), and the possibility of beam coupling if two light beams are launched simultaneously in the medium (cross-Kerr effect).

H.2 Quantum Input–Output Relations for Optical Amplifiers and Attenuators

We now switch to a quantum description of the propagation in a linear medium, which we treat in the most general way. At the classical level, linear optical amplifiers and attenuators obey the following rule for the complex output field $E^{(+)}_{out}$:

$$E^{(+)}_{out} = \sqrt{G}\,E^{(+)}_{in}, \tag{H.14}$$

where G is real, greater than one for an amplifier and smaller than one for an attenuator. To avoid confusion, we will replace G by A in the attenuator case. At the quantum level, the corresponding relation,

$$\hat{a}^{out} = \sqrt{G}\,\hat{a}_{in}, \tag{H.15}$$

cannot be valid, except when $G = 1$, because it is not a commutator-preserving, canonical relation. One therefore needs to introduce an ancilla mode, characterized by a noise operator $\hat{B}$, which is coupled to the amplifier mode. The actual input–output, canonical transformation (symplectic in the present case), is then

$$\hat{a}^{out} = \sqrt{G}\hat{a}^{in} + \hat{B}^{in}, \tag{H.16}$$

where $\hat{B}^{in}$ is an operator acting on the ancilla mode. The canonical commutation relation $[\hat{a}^{out}, \hat{a}^{\dagger}_{out}] = \mathbf{1}$ is ensured when

$$\left[\hat{B}^{in}, \hat{B}^{\dagger}_{in}\right] = 1 - G. \tag{H.17}$$

Let us first consider the amplifier case $G > 1$. Equation (H.17) is satisfied when

$$\hat{B}^{in} = \sqrt{G-1}\,\hat{b}^{\dagger}, \tag{H.18}$$

where $\hat{b}^{\dagger}$ is the bosonic creation operator in the ancilla mode. This mode must be in the vacuum state in order to retrieve the classical expression for the mean value of the field.

In the attenuator case, $A = G < 1$, one has similarly that

$$\hat{B}^{in} = \sqrt{1-A}\,\hat{b}, \tag{H.19}$$

where $\hat{b}$ is the annihilation operator of the ancilla mode, the exact expression of which depends on the physical system. For example, the beamsplitter, described in Section H.3, is the simplest possible attenuator. In this case, the ancilla mode is the "mirror mode" obtained by reflection of the input mode on the beamsplitter surface.

Relations (H.18) and (H.19), giving the expression of excess noise terms that are necessary to enforce the laws of quantum mechanics, have interesting very general consequences:

- In the attenuator case, the attenuation factor A being real, we have for the variances of the quadrature operators X and P,

$$Var(X^{out}) = \frac{1}{2}(1-A) + A\,Var(X^{in}), \quad Var(P^{out}) = \frac{1}{2}(1-A) + A\,Var(P^{in}). \tag{H.20}$$

  If we inject a coherent state, $Var(X^{in}) = 1/2$, then $Var(X^{out}) = 1/2$ whatever the attenuation. If instead we inject a squeezed vacuum state ($Var(X^{in}) < 1/2$) on the X quadrature, the variance of the attenuated field $Var(X^{out}) \simeq (1-A)/2$ is larger than $Var(X^{in})$: squeezing is destroyed by losses.
- In the amplifier case, similarly one has

$$Var(X^{out}) = \frac{1}{2}(G-1) + G\,Var(X^{in}), \quad Var(P^{out}) = \frac{1}{2}(G-1) + G\,Var(P^{in}) \tag{H.21}$$

If we inject a coherent state, $Var(X^{in}) = 1/2$, then $Var(X^{out}) = (2G-1)/2 > 1/2$: amplification adds excess noise. If we instead inject a perfectly squeezed vacuum state $Var(X^{in}) \simeq 0$ on the X quadrature, the variance of the amplified field $Var(X^{out}) \simeq (G-1)/2$ is larger than vacuum noise when $G \geq 2$: A gain of 2 is enough to completely degrade a perfectly squeezed input beam.

This degradation due to the amplification process is characterized by the noise figure NF, defined by

$$NF = \frac{(S/N)^{in}}{(S/N)^{out}} = \frac{N^{out}}{G\,N^{in}}, \tag{H.22}$$

where S/N is the signal-to-noise ratio, characteristic of each specific amplifier, and N the noise, i.e. the variance of the signal. For an input signal in a coherent state ($Var(X) = Var(P) = \frac{1}{2}$) and using Equation (H.21), the noise figure of the amplifier is equal to

$$NF = \frac{2G-1}{G}. \tag{H.23}$$

At very high gain the noise figure tends to 2 or 3dB. This is what is called the "3dB penalty," which sets a lower limit for the degradation of the signal-to-noise ratio valid for any phase insensitive amplifier, independently of the physical process of amplification.

- The previous expressions are the same for the two quadratures, as we dealt with a phase-insensitive amplifier. We now consider the case of a *phase-sensitive amplifier*, which has different gains on the two quadratures:

$$X^{out} = \sqrt{G_X}\,X^{in}, \quad P^{out} = \sqrt{G_P}\,P^{in}. \tag{H.24}$$

In order to obey the commutation rule $[\hat{X}^{out}, \hat{P}^{out}] = i$, we must have

$$G_X\,G_P = 1. \tag{H.25}$$

In this case one does not need to introduce an ancilla mode, and the noise figure is 1 for the two quadratures, but if there is amplification on X, for example, ($G_X > 1$) there is necessarily attenuation on P ($G_P < 1$).

There are different implementations of optical amplifiers. Stimulated emission, considered in Section H.1, is the most useful one, as it is at the heart of any laser. The added noise in this case is related to spontaneous emission. Nondegenerate parametric down conversion, considered in the Appendix I, is another possible source of gain for the signal mode (Section I.2). The added noise comes from spontaneous down conversion, and the associated ancilla mode is the idler mode. Degenerate parametric down conversion is an example of a phase sensitive amplifier that operates without added noise. It is used in circuit QED (Appendix K).

H.3 Input–Output Relations for the Beamsplitter

Let us now consider the simplest linear two-mode optical device, which is ubiquitous in quantum optics, namely the *beamsplitter*. It consists of a coated dielectric surface and couples two modes of identical frequencies with transmission and reflection coefficients in amplitude r and t (with $r^2+t^2=1$), labeled 1 and 2. The classical input–output relation for the field complex amplitudes at the position $z=0$ where the two beams intersect on the beamsplitter surface is

$$E_1^{out} = tE_1^{in} + rE_2^{in}, \tag{H.26}$$

$$E_2^{out} = -rE_1^{in} + tE_2^{in}, \tag{H.27}$$

the minus sign being required by the conservation of the sum of powers of the two beams power. The quantum input–output relation, assuming no transfer of energy to the dielectric medium, has a very similar expression when expressed in terms of the photon annihilation operator:

$$\hat{a}_1^{out} = t\hat{a}_1^{in} + r\hat{a}_2^{in}, \tag{H.28}$$

$$\hat{a}_2^{out} = -r\hat{a}_1^{in} + t\hat{a}_2^{in}.$$

One checks that this input–output relation conserves, in addition to the total power or photon number, the canonical commutation relations $[\hat{a}_i^{out}, \hat{a}_j^{\dagger out}] = \delta_{i,j}$, as required for free propagating beams. In addition, the transformation is linear, so it is a symplectic transformation, a concept that is discussed in Section 9.2.

The input–output relation (H.28) for the annihilation operators in the Heisenberg representation can also be written for the quantum state. Since the two-mode system is isolated, one can write it in terms of a unitary transformation for the state vector,

$$|\Psi^{out}\rangle = \exp\phi\left(\hat{a}_1\hat{a}_2^\dagger - \hat{a}_1^\dagger\hat{a}_2\right)|\Psi^{in}\rangle, \tag{H.29}$$

with $\cos\phi = t$ and $\sin\phi = r$.

In the case where a single photon is sent on the beamsplitter, $|\Psi^{in}\rangle = |1,0\rangle$. Expanding the exponential in series and separating the odd and even terms, one easily finds the simple relation

$$|\Psi^{out}\rangle = \cos\phi\,|1,0\rangle + \sin\phi\,|0,1\rangle = t\,|1,0\rangle + r\,|0,1\rangle. \tag{H.30}$$

One can also express the relation in terms of a map for the Wigner function (Equation (9.69)):

$$W^{out}(x_1,p_1,x_2,p_2) = W^{in}(tx_1+rx_2, tp_1+rp_2, -rx_1+tx_2, -rp_1+tp_2). \tag{H.31}$$

By cascading several input–output relations like Equation (H.3) and phase shift propagation factors $\hat{a}^{out} = e^{ikL}\hat{a}^{in}$ over a distance L, one can easily find the input–output relation of any network of beamsplitters and

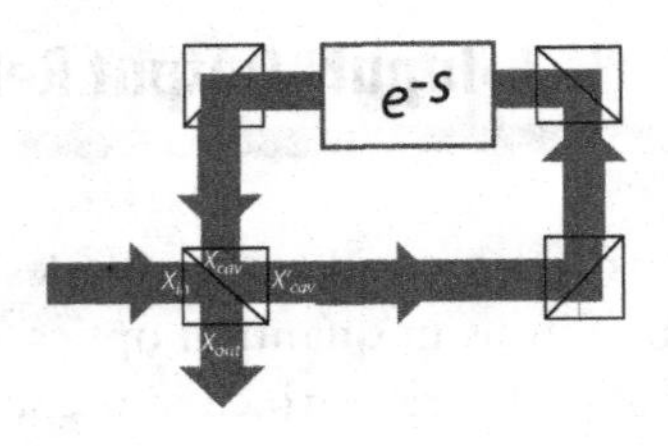

Fig. H.1 Intracavity squeezing device.

propagation, i.e. of any N-port interferometer. Actually, it can be shown [151] that any $N \times N$ multimode unitary operator can be implemented using such networks.

H.4 Exercises for Appendix H

Exercise H.1 In order to enhance the interaction between the nonlinear medium and the field, one inserts the medium inside a resonant cavity (see Figure H.1). Its round-trip length L is such that $e^{ikL} = 1$. The cavity mirrors are perfectly reflecting, except for one, which has amplitude reflection and transmission coefficients r and t. This couples the input and output fields $\hat{a}_{in}$ and $\hat{a}_{out}$ to the intracavity fields $\hat{a}_{cav}$ and $\hat{a}'_{cav}$ before and after the mirror. The nonlinear medium is a degenerate parametric down conversion medium with squeezing factor e^{-S} for quadrature X. Write the cavity round trip equation for the intracavity field quadrature X, including the effect of propagation and squeezing by the medium. Deduce from these equations the output quadrature operator $\hat{X}_{out}$ as a function of the input $\hat{X}_{in}$. When no field enters the cavity, what is the squeezing of the output field. When is it maximum?

I Interaction between Light Beams and Nonlinear Optical Media

As we saw in the Appendix H, matter modifies the properties of the light going through it. Dielectric media get polarized by the impinging electric field $E = (E^+ e^{i(kr-\omega t)} + \textit{complex conjugate})$ and acquire a polarization per unit volume $P(\mathbf{r}, t)$ that is a source term in the Maxwell equations. At high intensities, the medium reacts in a nonlinear way [27], and the real polarization has the following expression, in a rather schematic writing:

$$P = \varepsilon_0 \left(\chi^{(1)} E + \chi^{(2)} E^2 + \chi^{(3)} E^3 + \cdots \right). \tag{I.1}$$

Let us consider as an example of nonlinear response the effects linked to the second term, called parametric or second-order, nonlinear interaction, and leave aside higher order nonlinear effects. More precisely, we consider the case where E is the sum of two fields propagating in the same direction:

$$E = E_1 \exp i(k_1 z - \omega_1 t) + E_2 \exp i(k_2 z - \omega_2 t) + \text{complex conjugate}, \tag{I.2}$$

where E_1 and E_2 are complex amplitudes. The square term E^2 in Equation (I.1) introduces, among other terms, a cross term $E_1 E_2$ oscillating at the sum frequency

$$\omega_3 = \omega_1 + \omega_2 \tag{I.3}$$

in the polarization, which in turn induces the generation of a new field oscillating at this frequency, hence the term *three-wave mixing* given to this situation.

I.1 Three-Wave Mixing: Classical Approach

I.1.1 Propagation Equations

One can show that the Maxwell equations applied to this case involve three fields, the amplitude of which vary with z because of their mutual coupling. More precisely, one finds that [82]

$$\begin{aligned}
\frac{dE_1}{dz} &= i\frac{\omega_1}{n_1 c}\chi^{(2)} E_2^* E_3 e^{-i\Delta k z}, \\
\frac{dE_2}{dz} &= i\frac{\omega_2}{n_2 c}\chi^{(2)} E_1^* E_3 e^{-i\Delta k z}, \\
\frac{dE_3}{dz} &= i\frac{\omega_3}{n_3 c}\chi^{(2)} E_1 E_2 e^{i\Delta k z},
\end{aligned} \tag{I.4}$$

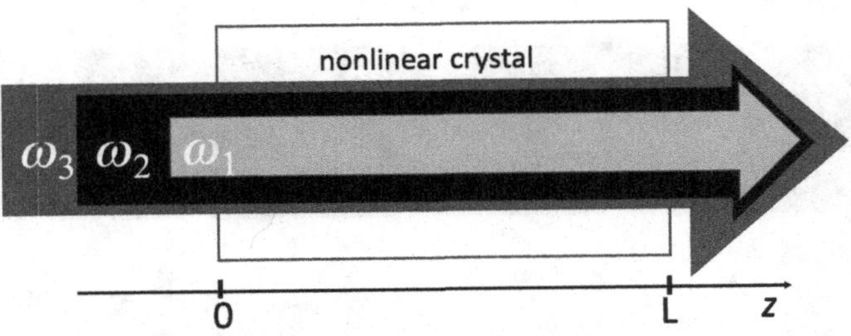

Fig. I.1 Three-wave coupling in a nonlinear crystal. The nonlinear interaction induces a transfer of energy from the pump beam to the other modes, or to the generation of waves at different frequencies (frequency doubling or parametric down conversion).

where $\Delta k = k_1 + k_2 - k_3$ and n_i is the index of refraction of the medium at frequency ω_i $(i = 1, 2, 3)$. The nonlinear interaction has the highest efficiency when the perfect phase-matching condition $\Delta k = 0$ is ensured. This we assume for the following.

One can show that these coupled equations have two constants of propagation: $\Pi_1 + \Pi_2 + \Pi_3$ and $\Pi_1/\omega_1 - \Pi_2/\omega_2$, where Π_i is the Poynting vector of beam i: $\Pi_i = 2n_i\varepsilon_0 c E_i E_i^*$. The first constant is simply the conservation of electromagnetic energy: The nonlinear medium "assists" the energy transfer between the beams, but does not change the total energy. The second constant is known as the Manley–Rowe constant of propagation. Though completely classical, its physical interpretation is simple in terms of photons, as we will see below.

I.1.2 Generation of Beams at New Frequencies

Many different configurations lead to interesting phenomena:

1. Consider the situation where a nonlinear medium of length L is illuminated by three waves. The first one, called the signal, is a weak wave of frequency ω_1 and amplitude $E_1(0)$ at $z = 0$. The second wave, called the idler, has a frequency ω_2 and a null amplitude $E_2(0) = 0$ at $z = 0$. The third wave is an intense pump of frequency ω_3. The three frequencies are linked by the relation $\omega_3 = \omega_1 + \omega_2$.

   In the case where the energy transfer from the pump to the other beams is small, the pump beam amplitude E_3 can be considered constant. The solution of Equation (I.4) in this case is

$$E_2(L) = i\sqrt{\frac{n_1\omega_2}{n_2\omega_1}}\frac{E_3}{|E_3|}E_1(z=0)\sinh(mL), \tag{I.5}$$

$$E_1(L) = E_1(z=0)\cosh(mL), \tag{I.6}$$

   with $m = \chi^{(2)}|E_3|\sqrt{\frac{\omega_1\omega_2}{n_1 n_2 c^2}}$. We observe that the idler beam grows from zero and that the signal beam amplitude is linearly amplified as it propagates through the nonlinear medium. This system is called a parametric amplifier.

2. The nonlinear medium is illuminated by an intense pump at frequency ω_3, the two other waves having zero amplitude at the origin $z = 0$. The solution of Equation (I.4) in this case is $E_1(z) = E_2(z) = 0$ for all z; no signal and idler light is generated in this situation.
3. The nonlinear medium is illuminated by intense beams E_1 and E_2, assumed constant, while $E_3(z) = 0$. One has that

$$E_3(L) = i\frac{\omega_3}{n_3 c}\chi^{(2)} E_1 E_2 L. \tag{I.7}$$

A beam at frequency $\omega_3 = \omega_1 + \omega_2$ appears from vacuum. This is the phenomenon of sum frequency generation, a widely used configuration to generate coherent beams at high frequencies. If one shines in the medium a single intense beam of frequency ω_1, and with a new phase matching configuration $k_3 = 2k_1$, one generates a beam at frequency $\omega_3 = 2\omega_1$: This is the frequency doubling configuration.

I.2 Three-Wave Mixing: Quantum Approach

I.2.1 Parametric Fluorescence

When one shines an intense laser beam of frequency ω_3 through a crystal having a $\chi^{(2)}$ nonlinearity [89], one experimentally observes that light is generated in a cone around the pump beam (see Figure 6.1). This light, called parametric fluorescence, has a broad spectrum of frequencies. All are lower than ω_3, hence the name *spontaneous parametric down conversion* also given to this phenomenon. As we saw in Section I.1.2 (item 2), the generation of such light is not predicted by the classical approach. As we will see in this section, it is very simply explained in the full quantum description of light–matter interaction.

This phenomenon bears analogy with the atomic fluorescence by spontaneous emission. As we saw in Section G.2, spontaneous emission cannot be explained by classical electrodynamics, but it can be simply accounted for by quantum electrodynamics.

Starting from the expression of the electromagnetic energy in a nonlinear dielectric medium, one can show that parametric down conversion, in the case of perfect phase-matching and classical pump, is described at the quantum level by the multimode Hamiltonian

$$\hat{H}_{int} = \hbar \sum_{\ell,\ell'} G_{\ell,\ell'} \alpha_{pump}(\omega_\ell + \omega'_\ell)\, \hat{a}^\dagger_\ell \hat{a}^\dagger_{\ell'} + \text{Hermitian conjugate}, \tag{I.8}$$

where $\alpha_{pump}(\omega)$ is the pump amplitude at frequency ω, $\hat{a}^\dagger_\ell$ and $\hat{a}^\dagger_{\ell'}$ are the creation operators of photons in the signal and idler pairs of modes, and $G_{\ell,\ell'}$ is a coefficient that depends on the nonlinear medium and the mode spatial overlap.

Note that the operator $\hat{a}_\ell^\dagger \hat{a}_\ell - \hat{a}_{\ell'}^\dagger \hat{a}_{\ell'}$ commutes with the Hamiltonian and is therefore a constant of propagation. There is a simple interpretation to this conservation law in terms of photons: Each time a signal photon is emitted (annihilated), an idler photon is also emitted (annihilated), hence the name "twin photons" given to these simultaneously generated pairs of photons. As the quantities Π_i/ω_i are proportional to the photon flux $\Pi_i/\hbar\omega_i$ in the corresponding mode, we now understand the Manley–Rowe relation introduced in Section I1.1, written as $\Pi_\ell/\hbar\omega_\ell - \Pi_{\ell'}/\hbar\omega_{\ell'} = constant$, as anticipating the phenomenon of simultaneous creation or destruction of photons in the signal and idler modes.

Assuming that the coupling constant $G_{\ell,\ell'}$ is small and starting from vacuum, the quantum state of the system at the output of the crystal of length L is, at first order of perturbation theory,

$$|\psi(t = L/c)\rangle = |0\rangle - i \sum_{\ell,\ell'} G_{\ell,\ell'} \alpha_{pump} L/c |1 : \ell\rangle \otimes |1 : \ell'\rangle. \tag{I.9}$$

We see that light is indeed emitted by the system (in the form of photon pairs), in agreement with the experimental observations, whereas the classical approach of Section I.1.2 predicts in this case that no light would be emitted by the nonlinear crystal.

The parametric fluorescent light consists of an assembly of photon pairs. It is a highly multimode entangled state, with strong signal-idler correlations: If one idler photon is detected, the state vector conditionally collapses into a single photon state in the associated signal mode $|1 : s\rangle$ with probability $|G_{\ell,\ell'} \alpha_{pump} L/c|^2$. This configuration is widely used by experimentalists to conditionally generate single photon states in a well-defined mode. This mode usually has a small temporal extension, around the photon detection time, equal to the photon detector response time. However, this state is conditional with a low success rate: An experiment requiring the simultaneous conditional generation of two or more single photons using this technique will take longer and longer observation times.

I.2.2 Nondegenerate Parametric Down Conversion

Let us now focus on a single signal-idler pair of modes, of frequencies ω_s and ω_i, which can be selected by appropriate frequency filters or irises of small diameter. Three modes only (signal, idler, pump), labeled by indices s, i, p, are implied in the interaction. In the case of exact phase matching, and bearing in mind that in a medium of index n the complex field is related to the annihilation operator by $\hat{E} = \hat{a}\sqrt{\hbar\omega/2n\varepsilon_0 c}$, the propagation equations for the annihilation operators of the three coupled modes, derived from the Hamiltonian

$$\hat{H}_{int} = \hbar G_{si} \hat{a}_p \hat{a}_s^\dagger \hat{a}_i^\dagger + \text{Hermitian conjugate}, \tag{I.10}$$

are simply

$$c\frac{d\hat{a}_s}{dz} = iG_{si}\hat{a}_i^\dagger \hat{a}_p,$$
$$c\frac{d\hat{a}_i}{dz} = iG_{si}\hat{a}_s^\dagger \hat{a}_p,$$
$$c\frac{d\hat{a}_p}{dz} = iG_{si}\hat{a}_s\hat{a}_i, \tag{I.11}$$

setting $z = ct$ to relate the time evolution map to a spatial propagation map.

If one uses an intense pump, treated as a classical wave of amplitude α_{pump}, a first-order development is not sufficient. One must use the full unitary transformation derived from the Hamiltonian (I.10), replacing $\hat{a}_p$ by α_{pump} and developing in series the exponential:

$$|\Psi_{out}\rangle = e^{-iS(\hat{a}_s^{in\dagger}\hat{a}_i^{in\dagger} + \hat{a}_s^{in}\hat{a}_i)}|0 : s; 0 : i\rangle$$
$$= \frac{1}{\cosh S}\sum_{n=0}^{\infty}(\tanh S)^n|n : s; n : i\rangle, \tag{I.12}$$

with $S = G_{si}\alpha_{pump}L/c$. This is also an entangled state involving the creation of higher order twin photons ($|n : s\rangle \otimes |n : i\rangle$) in the signal and idler modes. This state is an eigenstate of $\hat{N}_- = \hat{N}_s - \hat{N}_i$ with eigenvalue 0 that can be harnessed for making measurements below the standard quantum limit (see Chapters 10 and 11).

The input–output equations for the annihilation operators can be shown to be

$$\hat{a}_s^{out} = \cosh S\,\hat{a}_s^{in} + \sinh S\,\hat{a}_i^{in\dagger},$$
$$\hat{a}_i^{out} = \sinh S\hat{a}_s^{in} + \cosh S\hat{a}_i^{in\dagger}. \tag{I.13}$$

This transformation is the same as the classical parametric down conversion transformation [82], except for the hats on top of the quantum expressions. As for the beamsplitter case, the input–output relations for the annihilation operators have simpler form than the input–output equations for the whole two-mode quantum state. It is easy to check that the transformation is symplectic, as it conserves the commutators and is linear, but not unitary, as it does not conserve the signal+idler mode total energy. Its Wigner function is therefore Gaussian.

If one is interested only in the signal mode, one must make a partial trace over the idler space. One obtains the following density matrix:

$$\rho_{out} = \frac{1}{\cosh^2 S}\sum_{n=0}^{\infty}(\tanh S)^{2n}|n : s\rangle\langle n : s|, \tag{I.14}$$

which has the same form as a thermal state (expression (2.34)). When $S \gg 1$, the different coefficients are all close to 1. The quantum state of the signal beam is close to the minimal information state characteristic of very high temperatures.

An interesting situation occurs when the signal and idler modes have the same frequency $\omega_p/2$ and are identical, except for their polarization: horizontal for the signal, and vertical for the idler, with respective annihilation operators $\hat{a}_H$ and $\hat{a}_V$. The Hamiltonian can be written as

$$\hat{H} = i\hbar G\left(\hat{a}_V^\dagger \hat{a}_H^\dagger - \hat{a}_V \hat{a}_H\right) = i\hbar G\left(\left(\hat{a}_{+D}^{2\dagger} - \hat{a}_{+D}^2\right) - \left(\hat{a}_{-D}^{2\dagger} - \hat{a}_{-D}^2\right)\right)/2, \quad \text{(I.15)}$$

where $\hat{a}_{\pm D} = (\hat{a}_H \pm \hat{a}_V)/\sqrt{2}$ are the annihilation operators for the linear polarization modes in the diagonal directions $\pm\pi/4$. The two forms of the Hamiltonian show that one can generate at its output either two squeezed states in orthogonal polarizations, or a highly entangled two-mode state, by a simple change of the experimental configuration.

Note that when $G \to \infty$, and because of the minus sign in the second term of the Hamiltonian, one has a complete and simultaneous noise reduction on the quadratures $\hat{X}_{+45}$ and $\hat{P}_{-45}$. This may seem to contradict the usual X, P Heisenberg inequality. This is not the case because the corresponding operators do commute:

$$\left[\hat{X}_{+45}, \hat{P}_{-45}\right] = \frac{1}{2}\left[\hat{X}_H + \hat{X}_V, \hat{P}_H - \hat{P}_V\right] = i\hbar - i\hbar = 0. \quad \text{(I.16)}$$

A similar commutation relation was used by Einstein in 1935 as the basis of his arguments [62]. This is the subject of Section 9.5.

I.2.3 Degenerate Parametric Down Conversion

By appropriately tuning the experimental parameters, one can reach the degenerate situation where the idler mode is identical to the signal mode, with $\omega_s = \omega_i = \omega_p/2$, the so-called subharmonic generation case. Neglecting the pump depletion, and in the case of spontaneous parametric down conversion for which the input state is vacuum, one now gets

$$|\Psi_{out}\rangle = e^{S(\hat{a}_s^{2in\dagger} - \hat{a}_s^{2in})}|0 : s; 0 : i\rangle \quad \text{(I.17)}$$

$$= \frac{1}{\cosh S}\sum_{n=0}^{\infty} \tanh^n S|2n : s\rangle. \quad \text{(I.18)}$$

It is a state containing even numbers of photons, more precisely a *squeezed vacuum state* $|S_{0,S}\rangle$, as defined in Section F.3. Subharmonic generation is actually the simplest and most efficient way to generate a squeezed vacuum state, a basic resource for quantum measurements.

The input–output operator relations can be derived from Equations (I.13) by using $\hat{a}_s \equiv \hat{a}_i$:

$$\hat{X}_s^{out} = e^S \hat{X}_s^{in}, \quad \hat{P}_s^{out} = e^{-S}\hat{P}_s^{in}, \quad \text{(I.19)}$$

which implies for the quadrature mean values that

$$\langle \hat{X}_s^{out}\rangle = e^S\langle \hat{X}_s^{in}\rangle, \quad \langle \hat{P}_s^{out}\rangle = e^{-S}\langle \hat{P}_s^{in}\rangle. \quad \text{(I.20)}$$

In the case where $S > 0$, the device amplifies the mean value of quadrature X with a power gain $G = e^{2S}$, and deamplifies quadrature P: It is a phase sensitive amplifier, which can be used as a regular amplifier if one is able to lock the phase of the input field on the right quadrature. As far as the variances are concerned,

$$Var\left(X_s^{out}\right) = \frac{1}{2}e^{2S}\, Var\left(X_s^{in}\right), \quad Var\left(P_s^{out}\right) = \frac{1}{2}e^{-2S}\, Var\left(P_s^{in}\right). \tag{I.21}$$

The device amplifies the mean field and the noise by the same amount, so that the signal to noise ratio is conserved. The noise factor NF, defined in Equation (H.2), is 1; it is not constrained by the 3 dB penalty valid for phase insensitive amplifiers considered in Appendix H.

I.3 Use of an Optical Cavity

Expression (I.21) shows that in order to get a high noise reduction by squeezing, one must have a large value of S, and therefore use an intense pump beam. Large squeezing is only achieved for very large pump power.

There is another way, which is to use a cavity that is resonant for the signal and idler modes. The nonlinear medium-cavity system is called an *optical parametric oscillator* (OPO). It is analogous to a laser in which the stimulated emission gain is replaced by parametric amplification. Like in the laser, there is an oscillation threshold value $|\alpha_{pump}^{threshold}|$ for the pump amplitude.

In the nondegenerate configuration, the OPO emits above threshold coherent, laser-like signal and idler beams that are at the same time fully macroscopic, like lasers, and have perfectly intensity-correlated quantum fluctuations, whatever the pump amplitude value fluctuations,

$$\forall\, \alpha_{pump}, \quad Var\left(\hat{N}_s - \hat{N}_i\right) = 0. \tag{I.22}$$

This very efficient quantum noise manipulation can be used to improve the sensitivity of spectroscopic measurements or to conditionally generate a sub-Poissonian beam having an intensity noise well below the shot noise level (Section 8.2.2).

The degenerate configuration is considered in Exercise H.1. There is also an oscillation threshold, above which the system emits a coherent laser-like optical beam. Below the threshold, the signal field has a zero mean value, and its variance is squeezed. One shows in Exercise H.1 that the squeezing is enhanced by the resonant cavity intensity build-up. It can be very large *for a finite value of the pump*, namely when the pump approaches from below the threshold value $|\alpha_{pump}^{threshold}|$.

I.4 Higher Order Nonlinear Effects

Parametric interaction arises from the introduction of nonlinear optical effects at the lowest order. Higher order terms exist also. They are responsible for new interesting physical effects, especially if the parametric interaction is zero and does not hide the weaker higher order terms.

These effects are detailed at the classical and quantum level in different textbooks. The most important nonlinear effects are the four-wave mixing effects, involving two quantum modes and two classical pump modes. We have also encountered in saturated atomic media and Josephson junctions self-Kerr and cross-Kerr media having intensity dependent indices of refraction, based on terms like $\hat{a}_1^{\dagger 2}\hat{a}_1^2$ or $\hat{a}_1^\dagger\hat{a}_1\hat{a}_2^\dagger\hat{a}_2$ in the Hamiltonian.

Nonlinear effects often involve Hamiltonians containing terms that are the product of more than two creation and/or annihilation operators. This implies that the derived input–output relation for the annihilation operator is not linear and that the corresponding transformation is not symplectic. This has important consequences. For example, such higher order effects can be used to generate non-Gaussian states directly from vacuum.

I.5 Exercises for Appendix I

Exercise I.1 Find the static effective Hamiltonian of a nonlinear driven system as follows. Take a nonlinear crystal that induces a field description given by the Hamiltonian $\hat{H}/\hbar = \omega_o\hat{a}^\dagger\hat{a} + g_4(a^\dagger + \hat{a})^4$ under the pump drive $\hat{V} = if \cos\omega_d t(a^\dagger - \hat{a})$. For this, solve the Schrödinger equation for $\alpha(t)$, where $\hat{a}|\alpha\rangle = \alpha|\alpha\rangle$ setting $g_4 = 0$ to get the linear response of the field. Transform the Hamiltonian to the frame displaced by $\alpha(t)$ to get rid of the trivial harmonic behavior. The final step in this standard procedure is to go to the rotating frame at the observation frequency ω_r (r for read out) and take the time average of the Hamiltonian (the rotating wave approximation). What remains are the dominant processes at ω_r. See also [170].

Exercise I.2 We are now interested in the parametric processes arising from $g_3(a^\dagger + \hat{a})^3$. Can it be used to create squeezing?

Exercise I.3 Consider a linear oscillator described by $\hat{H}^{(0)} = \frac{\hat{p}^2}{2m} + \frac{m\omega_0^2}{2}\hat{X}^2$ with a small perturbation given by $\hat{H}^{(1)} = \alpha\hat{X}^3 + \beta\hat{X}^4$. Using perturbation theory, compute the correction to the transitions frequency inbetween state with large quantum numbers ($n \gg 1$) as a function of its amplitude [114].

Exercise I.4 Write down the classical propagation equations (I.5) and (I.6) in the frequency doubling configuration ($E_1(0) = E_2(0)$, $E_3(0) = 0$, $\omega_1 = \omega_2$), replacing annihilation operators $\hat{a}$ by c-numbers α in propagation equations like equation (I.4). Give the expression of the generated second harmonic field assuming an undepleted pump, and calculate the doubling efficiency for pump power 1 W of wavelength $\lambda = 1\,\mu\text{m}$ focused on a surface $(100\,\mu\text{m})^2$, where $L = 1\,\text{cm}$, $n \simeq 1$, and $\chi^{(2)}$=5 pm/V. What could be changed to reach higher doubling efficiencies?

One now drops the hypothesis that the pump field propagates undepleted. Show that there is a complete energy transfer from the field at frequency ω_1 to its second harmonic $\omega_3 = 2\omega_1$ for long interaction lengths.

Exercise I.5 One wants to derive the characteristics of a classical, nondegenerate, OPO, i.e. of intracavity nondegenerate parametric generation with resonant signal and idler fields, so that the propagation equations (I.11) can be treated to first order in the coupling. The conversion efficiency being weak, one assumes that the pump is constant throughout its propagation in the nonlinear crystal. The pump, signal, and idler fields are treated as classical fields α_p, α_s, α_i. The cavity has a single coupling mirror of small intensity transmission $T \ll 1$. Determine the intracavity signal and idler fields round trip equations and show that in a steady state regime they have a trivial solution $\alpha_s = \alpha_i = 0$. Find the value of the pump threshold that provides a nonzero solution for the signal and idler fields. What is the relation in this case between the signal and idler output classical intensities?

Exercise I.6 One now considers the quantum theory of the above threshold OPO. Write the cavity input–output relations for the operators $\hat{a}_s$ and $\hat{a}_i$ in the above threshold conditions. Show, to first order in the interaction, that

$$\left(\hat{N}_s - \hat{N}_i\right)_{out} = \left(\hat{N}_s - \hat{N}_i\right)_{in}.$$

One assumes that the input state is vacuum. Calculate $Var(\hat{N}_s - \hat{N}_i)_{out}$ and show that the signal and idler output beams of the OPO are perfectly intensity correlated, hence the name "twin beams." Suggest a possible physical interpretation of this effect.

Compare with the case of the two output beams 1,2 of a 50% beamsplitter, illuminated by a coherent state of real amplitude α and vacuum in the other port, by calculating $Var(N_1 - N_2)$. Suggest a possible physical interpretation of this effect.

J Optomechanics

Light carries not only energy, but also momentum, which can induce mechanical effects on the motion of matter with which it interacts. This domain of physics is often called *optomechanics*

Light can move microscopic particles, such as atoms or ions, and also macroscopic objects containing many atoms, like mirrors or membranes. Light-induced forces can be used at the quantum level to generate interesting quantum matter–field states.

J.1 Forces Exerted by Light on an Atom

J.1.1 Radiation Pressure and Dipole Force

We consider a two-level atom (Appendix D) submitted to a beam of light of frequency $\omega \simeq \omega_0$ and wavevector $\mathbf{k}$ along $0z$. The atom is free to move. We call M its mass, $\hat{\mathbf{R}}$ the center of mass position operator, and $\hat{\mathbf{P}}$ the atomic momentum operator.

Simple arguments based on momentum and energy conservation permit a first approach of these phenomena: When a photon is absorbed by the atom initially in internal ground state $|-\rangle$ and momentum eigenstate $|\mathbf{p}\rangle$, and promotes it to the excited state $|+\rangle$, the momentum of the atom is then $\mathbf{p}' = \mathbf{p} + \hbar\mathbf{k}$. It undergoes a velocity change $\hbar k/M$, typically of a few cm/s. When the atom in its excited state $|+\rangle$ emits a photon of wavevector k by spontaneous emission, so that its velocity changes by the quantity $\hbar k/M$ in a random direction. If this cycle absorption+ spontaneous emission is able to repeat itself many times, the mechanical effect due to the spontaneous photons averages to zero, whereas the momentum exchange due to the light beam photons is always in the same direction. The net effect is therefore what is called the radiation pressure force, consisting of a mean force in the direction of the incident beam, plus a fluctuating force in the transverse plane.

Let us write the conservation of energy in the single photon absorption process as

$$E_- + \mathbf{p}^2/2M + \hbar\omega = E_+ + (\mathbf{p} + \hbar\mathbf{k})^2/2M, \tag{J.1}$$

which implies the following expression for the resonant frequency of the field able to induce such a transition:

$$\omega = \omega_0 + \mathbf{k} \cdot \frac{\mathbf{p}}{M} + \frac{\hbar k^2}{2M}, \tag{J.2}$$

where $\hbar\omega_0 = E_+ - E_-$. The second term, $\mathbf{k}\cdot\mathbf{v}$, where $\mathbf{v}$ is the velocity of the atom, is simply the well-known and classical Doppler shift, while the last term, called the *recoil shift*, exists even when the atom is initially at rest. It is related to the velocity $\hbar k/M$ acquired because of the recoil effect. For an atom moving at the average thermal velocity at room temperature and for a transition in the visible range, the Doppler shift is of the order of a few hundred kHz, while the recoil term is smaller, of the order of a few tens of kHz.

Here, we will use the semiclassical approach for the atom–field interaction, where the intense field is treated classically. We write its complex amplitude as $E^{(+)}(\mathbf{r}) = E_0(\mathbf{r})e^{-i(\omega t-\mathbf{k}\cdot\mathbf{r})}$ and its polarization $\vec{\varepsilon}$. The Hamiltonian of the system is

$$\hat{H} = \hat{H}_{at} + \frac{\hat{\mathbf{P}}^2}{2M} - \hat{\mathbf{D}}\cdot\vec{\varepsilon}\,E(\mathbf{r}). \tag{J.3}$$

The external state of the atom is now supposed to be localized in a wavepacket of small dimensions, so that one can treat the position operator of the atom $\hat{\vec{R}}$ as a classical object $\vec{R}_{at}$ [82]. We can now calculate the mean force

$$\langle\mathbf{F}\rangle = \frac{d}{dt}\langle\hat{\mathbf{P}}\rangle = \frac{1}{i\hbar}\left\langle\left[\hat{\mathbf{P}},\hat{H}\right]\right\rangle, \tag{J.4}$$

which gives, after a few calculations and factorization between the internal and external degrees of freedom,

$$\langle\mathbf{F}\rangle = \langle\hat{\mathbf{D}}\cdot\vec{\varepsilon}\rangle\vec{\nabla}(E(\mathbf{r}))_{\mathbf{r}=\mathbf{r_{at}}}. \tag{J.5}$$

The mean dipole induced by the external field can be calculated using the Bloch equations (G.21) and (G.22) for the internal variables. One finds (Equation (H.2)) that

$$D(\mathbf{r}) = \varepsilon_0\alpha(\omega)E(\mathbf{r}) + \text{complex conjugate}, \tag{J.6}$$

where $\alpha(\omega) = \alpha'(\omega) + i\alpha''(\omega)$ is the complex single atom polarizability, the two terms of which are given by

$$\alpha(\omega) = \frac{d^2}{\varepsilon_0\hbar}\frac{-\delta + i\Gamma/2}{\delta^2 + \Omega_R^2/2 + \Gamma^2/4}, \tag{J.7}$$

with $\delta = \omega - \omega_0$. The mean force therefore comprises two terms, $\mathbf{F}_{dip}$ and $\mathbf{F}_{RR}$:

$$\mathbf{F}_{dip} = \varepsilon_0\alpha' E_0\vec{\nabla}(E_0(\mathbf{r}))_{\mathbf{r}=\mathbf{r_{at}}}, \tag{J.8}$$

$$\mathbf{F}_{RR} = -\varepsilon_0\alpha'' E_0^2\,\mathbf{k}. \tag{J.9}$$

The first force, $\mathbf{F}_{dip}$, is related to the reactive, or dispersive, part of the atomic response and is maximum in the wings of the resonance $\omega_0 \pm \Gamma/2$. It depends on the gradient of the light intensity and is called the *dipole force*. For negative detuning ($\omega < \omega_0$) the atoms are attracted towards regions of high intensity. A widely used experimental technique to create a trap for the atoms is to shine onto them a highly focussed and intense light beam detuned from the atomic resonance. The evanescent wave at the surface of a dielectric in the regime of total reflection varies on the scale of the wavelength. It therefore yields a strong dipole force that is used by experimentalists to realize a mirror for atoms.

The second force, $\mathbf{F}_{RR}$, is related to the dissipative part of the atomic response and is maximum at exact resonance. It is parallel to the field wavevector $\mathbf{k}$, i.e. to the propagation direction of the beam. It is called the *radiation pressure force*, and its interpretation in terms of momentum exchange for absorbed and re-emitted photons was given at the beginning of this section. In the case of atoms with velocity $\mathbf{v}$, and neglecting the recoil force, it can also be written as

$$\mathbf{F}_{RR} = \hbar\mathbf{k}\frac{\Gamma}{2}\frac{\Omega_R^2/2}{(\delta - kv_z)^2 + \Omega_R^2/2 + \Gamma^2/4}, \tag{J.10}$$

and saturates at high intensities to the value $\mathbf{F}_{RR} = \hbar\mathbf{k}\frac{\Gamma}{2}$, roughly equal to one recoil per lifetime.

J.1.2 Doppler Cooling

The radiation pressure force that we have just derived always pushes the atoms in the same direction. It is efficient to deflect atoms, but not to trap them at some position in the empty space. Let us now consider the case of two beams of identical powers propagating in opposite directions on the Oz axis and in the case $\omega < \omega_0$. We assume that we can add the radiation pressure forces $\mathbf{F}_{RR}$ induced by the two beams, as if they were acting independently on the atoms. For the total force, in the low field regime, one finds that

$$\mathbf{F}_{tot} = \hbar\mathbf{k}\frac{\Gamma}{2}\left(\frac{\Omega_R^2/2}{(\delta - kv_z)^2 + \Gamma^2/4} - \frac{\Omega_R^2/2}{(\delta + kv_z)^2 + \Gamma^2/4}\right). \tag{J.11}$$

This expression is odd in v_z, which implies that $\mathbf{F}_{tot}$ is proportional to v_z for small v_z values:

$$\mathbf{F}_{tot} \approx Akv_z. \tag{J.12}$$

The detailed calculation of $\mathbf{F}_{tot}$ shows that the coefficient A is negative: The net force is therefore a *friction force*, which slows down the atoms wherever they are situated in the region of intersection of the two light beams. However, this force cannot trap them. In order to create an atomic trap, one must use the so-called MOT configuration (MOT stands for magneto optical trap), which consists of adding a magnetic field to the previous set-up. The configuration is chosen in such a way that the Zeeman shifts $\hbar\omega_B$ are opposite for the atomic transitions that interact with the two counterpropagating beams. The total MOT force is:

$$\mathbf{F}_{MOT} = \hbar\mathbf{k}\frac{\Gamma}{2}\left(\frac{\Omega_R^2/2}{(\delta - kv_z + \omega_B)^2 + \Gamma^2/4} - \frac{\Omega_R^2/2}{(\delta + kv_z - \omega_B)^2 + \Gamma^2/4}\right). \tag{J.13}$$

The applied magnetic field is inhomogeneous, and produced in such a way that, close to $z = 0$, ω_B is proportional to z, $\omega_B = (d\omega_B/dz)z$, so that

$$\mathbf{F}_{MOT} \approx A(kv_z - \omega_B) = Akv_z - A\frac{d\omega_B}{dz}z. \tag{J.14}$$

In addition to the force due to the Doppler shift, the MOT force now comprises a term that is proportional to $-z$. If the magnetic field gradient is properly chosen, the MOT force, along the Oz axis, is directed to the $B = 0$ point. In order to have a 3D trap at this point, one must add two more pairs of counter-propagating beams, along the Ox and Oy axes. The configuration turns out to be convenient and efficient, and leads to trapped and cold clouds of atoms, with temperatures of the order of 100 μK.

Because of the stochastic aspect of spontaneous emission, the atom undergoes a Brownian motion in momentum space that is responsible for the existence of a lower limit for the temperature reached in the MOT configuration, called the Doppler temperature, and equal to $\hbar\Gamma/2k_BT$. In order to reach lower temperatures and higher densities, and ultimately reach the regime of Bose condensation, one must use more sophisticated techniques, the description of which is out of the scope of this book but can be found in [44].

J.2 Trapping and Cooling Ions

J.2.1 Paul and Penning Traps

The Earnshaw theorem establishes that a charged particle cannot be subjected to a static restoring force around some empty point of space in all three dimensions because of the Gauss equation $\nabla \cdot \mathbf{E} = 0$. Two main kinds of ion traps have been developed, called the Penning and Paul traps.

1. The Penning trap is the superposition of a static magnetic field B_0 along Oz and of a quadrupole electrostatic potential V:

$$V(\mathbf{r}) = \frac{M\omega^2}{2}\left(z^2 - (x^2 + y^2)/2\right). \tag{J.15}$$

   The potential shape along Oz is that of a harmonic oscillator, whereas the cyclotron circular motion in the xOy plane due to the magnetic field compensates the effect of the expelling electric field and prevents the charged particle from escaping in this plane. In this configuration the ion motion mixes a vibrational slow motion along Oz motion due the Oz trapping potential, a magnetron motion and a fast cyclotron motion. The oscillation of the charged particle induces a current in the electrical circuit connected to the electrodes that can be measured and used for detection. This current also dissipates the energy of the trapped particle (which can be an ion, but also a single electron) by single quanta of the oscillator (see Chapter 1).

2. In a Paul trap [140] the particle is submitted to a quadrupolar electrostatic potential of the form

$$V(x,y,z) = A\left(x^2 + y^2 - 2z^2\right).$$

This potential is not confining in the z direction, a necessary consequence of Earnshaw's theorem. One can bypass the theorem by violating the static hypothesis and add a field $A = A_0 \cos\omega_d t$ that varyies rapidly with time, quickly enough that, because of the inertia of the particle, its effect will be only that of its average. Neglecting the micromotion induced by the rapid drive, one gets a static, trapping, effective potential that reads [50]

$$\bar{V} = \frac{A_0^2}{m\omega_d^2}\left(\bar{x}^2 + \bar{y}^2 + 4\bar{z}^2\right),$$

where the overbars denote the time average.
3. Other configurations are used to simultaneously trap several ions and line them up along the 0z axis, a convenient way to create a linear array of qubits used as a *quantum register*. These arrays can be harnessed for quantum computing applications [20].

J.2.2 Sideband Cooling

The ion trapped in a harmonic oscillator in the direction $0z$ has energy eigenstates denoted $|\pm; n_v\rangle$, where $|+\rangle$ is the excited atomic state of energy $\hbar\omega_0$ and $|-\rangle$ its ground state. $|n_v\rangle$ is a vibrational, or phonon, eigenstate of energy $(n_v + 1/2)\hbar\omega_v$, where ω_v is the vibration frequency. In order to manipulate the external motion of the ion and, in particular, to cool it down to its ground motional state $|n_v = 0\rangle$, one submits the trapped particle to a laser irradiation of frequency ω and wavenumber k along the Oz axis, which will be treated as a classical wave of complex amplitude $Ee^{-i(\omega t - kz)}$. The Hamiltonian governing the coupling between the ion and the field in the frame rotating at the laser frequency ω and at the secular approximation (see Section G.2) is

$$\hat{H}_{ion} = \frac{\hat{P}^2}{2M} + \frac{M\omega_v^2\hat{Z}^2}{2} + \hbar(\omega_0 - \omega)\hat{\sigma}_z + \frac{\hbar\Omega_R}{2}\left(\hat{\sigma}_+ e^{ik\hat{Z}} + \hat{\sigma}_- e^{-ik\hat{Z}}\right), \quad (J.16)$$

where $\Omega_R = -\vec{d}\cdot\vec{\varepsilon}\, E/\hbar$ is the Rabi frequency and $\hat{Z} = z_0(\hat{a}_v + \hat{a}_v^\dagger)$ the position operator along the laser beam. $z_0 = \sqrt{Var(Z)} = \sqrt{\hbar/2M\omega_v}$ is the zero point motion extension in the trap ground state. We observe that this coupling depends on both the internal and external degrees of freedom. Spontaneous emission introduces dissipative terms in the Bloch equations that are the same as in Section G.2. They imply a finite width Γ for the resonances.

Since $e^{ik\hat{Z}} \simeq 1 + kz_0(\hat{a}_v + \hat{a}_v^\dagger) + (kz_0)^2(\hat{a}_v + \hat{a}_v^\dagger)^2/2 + \cdots$, the last term of the Hamiltonian (J.16) describes processes where the ion is promoted from its ground internal state to its excited internal state $|+\rangle$ while one or several phonons are either emitted or absorbed. If one neglects the very small recoil energy, the resonances occur at frequencies ω of the field equal to $\omega_0 + (n_v - n'_v)\omega_v$. The new resonances, separated by multiples of the oscillation frequency ω_v, are called sidebands. In the perturbative regime for the applied laser field their intensities are proportional to $|\langle n_v|e^{ikz_0(\hat{a}_v+\hat{a}_v^\dagger)}|n'_v\rangle|^2$ with $n'_v = n_v, n_v \pm 1$. Assume now that $kz_0 \ll 1$, a situation called the Lamb–Dicke regime, which occurs when the oscillator wavepacket is small compared to the wavelength of the applied field. We are then able to approximate the exponential by $1 + kz_0(\hat{a}_v + \hat{a}_v^\dagger)$. In this situation, one has only two sidebands of frequencies $\omega_0 \pm \omega_v$ around the main transition at frequency ω_0. Two regimes can actually occur:

- $\omega_v \ll \Gamma$ (weak confinement regime)
  The separation between the resonances is much smaller than their width. One can ignore this separation and neglect the effect of quantization of the harmonic oscillator. One then returns to the Doppler cooling situation described above and concerning a free particle interacting with the laser beam.
- $\omega_v \gg \Gamma$ (strong confinement regime)
  The separation between the resonances is much larger than their width, so that the resonances involving the sidebands are well separated from the "zero-phonon" line. Now, assume that the internal initial state of the ion, pure or mixed, projects on many vibrational number states $|n_v\rangle$, like for example in a thermal state. One tunes the laser frequency to the low frequency sideband $\omega_0 - \omega_v$, thus inducing transitions $|-; n_v\rangle \to |+; n_v - 1\rangle$ where the atom is excited and the ion vibration is reduced. The ion then spontaneously decays to the ground state through transitions $|+; n_v - 1\rangle \to |-; n_v - 1\rangle$. If the two level transition is closed, the cycle (ion excitation by laser – spontaneous emission) repeats itself many times. The net effect is then a global decrease of the vibrational excitation of the ion until the ion reaches the ground state, from where it cannot be re-excited to higher vibrational states. This technique, called sideband cooling, is very efficient and allows experimentalists to reach the ground state of the system $|-; n_v = 0\rangle$, and to generate exotic vibrational states like squeezed states or Schrödinger cat states, as described in Chapter 3. Note that this method can also be applied to trapped atoms having closed transition cycles.

J.3 Optomechanics of Macroscopic Objects

J.3.1 Radiation Pressure in an Optical Cavity

Mechanical effects due to photons are not restricted to atoms and ions. Light also exerts forces on macroscopic massive objects: The first effect is the radiation pressure force acting along the direction of propagation of the beam due to an exchange of momentum between the field and the

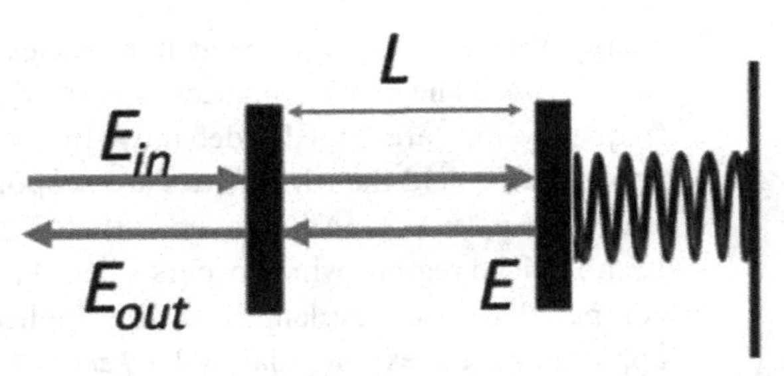

Fig. J.1 Sketch of an opto-mechanical set-up, consisting of a linear Fabry–Perot cavity with a fixed coupling mirror and a moving, harmonically bound end mirror, submitted to the radiation pressure force proportional to the intracavity field intensity. CC BY 4.0.

macroscopic mass. It is analog to the one acting on atoms that we described at the beginning of this appendix. The second possible mechanical effect is a torque around the beam direction, due to an exchange of angular momentum when the light beam is in a Laguerre–Gauss mode which carries orbital angular momentum. We will not consider this second effect here.

We consider a two-mirror linear Fabry–Perot cavity of length L (see Figure J.1) made of a coupling mirror of small energy transmission T and of a perfect mirror that can move along the cavity axis $0z$. For a mirror of small transmission, the amplitude reflection coefficient is $\sqrt{R} = \sqrt{1-T} \simeq 1 - T/2$. The equation for the intracavity field complex amplitude E in a Fabry–Perot cavity is [82]:

$$E = \frac{\sqrt{T}E_{in}}{T/2 + i(\omega - \omega_{cav})\tau_{cav}} = \frac{2}{\sqrt{T}} \frac{\kappa/2}{\kappa/2 - i(\omega - \omega_{cav})} E_{in}, \tag{J.17}$$

where $\tau_{cav} = 2L/c$ is the cavity round-trip time, ω_{cav} a resonance frequency of the cavity, and $\kappa = T/\tau_{cav}$ the cavity energy decay constant. At exact resonance the energy build-up is $4/T$. This field exerts a radiation pressure force F_R on the moving mirror because of the momentum transfer of $2\hbar k$ per round trip time, equal to

$$F_R = 2\Phi/c = 2\hbar k \frac{|\alpha|^2}{\tau_{cav}} = \frac{\hbar\omega}{L}|\alpha|^2, \tag{J.18}$$

where Φ is the intracavity power and $|\alpha|^2$ the intracavity photon number.

The mirror is harmonically bound. We call M its mass, Ω_M its oscillation frequency, and Γ_M its damping rate. The intracavity field pushes the mirror, which changes the cavity length L and therefore the intracavity resonance build-up. Such an intensity-dependent cavity length is reminiscent of the Kerr effect, where the index of refraction, and therefore the optical path, depends on the light intensity.

More precisely, the cavity resonant frequency $\omega_{cav} = p\pi c/L$, where p is an integer. Therefore,

$$\omega_{cav}(L) = \omega_{cav}(L_0) + (L - L_0)\frac{\partial \omega_{cav}}{\partial L} = \omega_{cav}(L_0) + X\frac{\omega_{cav}}{L_0}, \tag{J.19}$$

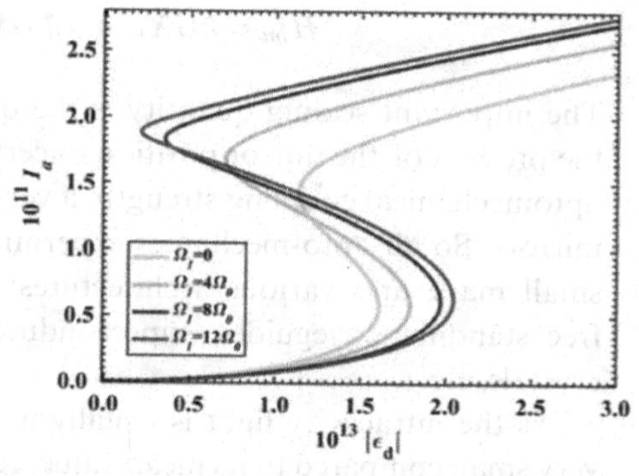

Fig. J.2 Mean intracavity photon number as a function of pump field. The negative slope parts turn out to be unstable, leading to a regime of bistability. From Yan Gao et al. Optical bistability in an opto-mechanical system with *N*-type atoms under non-resonant conditions, *Photonics*, 2020 [72].

where X is the displacement $L - L_0$, so that the intracavity field amplitude is given by

$$E = \frac{2E_{in}}{\sqrt{T}} \frac{\kappa/2}{\kappa_0/2 - i(\omega - \omega_{cav}(L_0)) - iGX}, \tag{J.20}$$

where $G = \omega_{cav}/L_0$ is the strength of the opto-mechanical coupling. The displacement X obeys the evolution equation

$$\frac{d^2X}{dt^2} + \Gamma_M \frac{dX}{dt} + \Omega_M^2 X = \frac{F_R}{M}. \tag{J.21}$$

From the previous equations it is easy to show that the steady-state intracavity intensity $I = |E|^2$ is the solution of the following equation:

$$\left(\kappa_0^2/4 + (\omega - \omega_{cav}(L_0) + mI)^2\right) I = TI_{in}, \tag{J.22}$$

where $m = 4\hbar k^2/M\Omega_M^2$. Figure J.2 plots the intracavity power I as a function of cavity-field detuning: One observes that the usual resonance curve is distorted when the intensity increases. Several possible values of I for a given detuning are possible. The curve exhibits the phenomenon of *bistability*, one of the important phenomena of nonlinear physics. Close to the bistability point the derivative of the light intensity versus mirror displacement is very large. This optical set-up is an incredibly sensitive position sensor.

J.3.2 Interaction Hamiltonian

We can now treat the system (field + mirror motion) as two coupled quantum harmonic oscillators in a fully quantum way. The phonon annihilation operator is denoted by $\hat{b}$. In the absence of interaction, the Hamiltonian of the field–cavity system (removing all zero-point energies) is $\hat{H}_0 = \hbar\omega_{cav}\hat{a}^\dagger\hat{a} + \hbar\Omega_M\hat{b}^\dagger\hat{b}$. As ω_{cav} varies with X, so does the single photon energy ω_{cav}. One can therefore write the photon–phonon coupling Hamiltonian as

$$\hat{H}_{int} = \hbar G \hat{X} \hat{a}^\dagger \hat{a} = \hbar G x_0 \hat{a}^\dagger \hat{a} \left(\hat{b} + \hat{b}^\dagger \right). \quad \text{(J.23)}$$

The important scaling quantity is the quantity $g_0 = Gx_0 = G\hbar/2M\Omega_M$, the product of the mirror position uncertainty in the ground state with the optomechanical coupling strength, a very small frequency for macroscopic mirrors. So all opto-mechanics experiments involve nano-objects of very small mass, and various architectures: microdisks, membranes, toroids, free standing waveguides, superconducting circuits, defects in photonic crystals, etc.

As the intracavity light is usually macroscopic, its fluctuations $\delta\hat{a}$ are very small compared to its mean value, equal to $\alpha = \sqrt{N}$. One can therefore linearize the interaction Hamiltonian and keep the dominant quantum terms:

$$\hat{H}_{int} \simeq \hbar\omega_{cav} \frac{x_0}{L_0} \alpha (\delta\hat{a}^\dagger + \delta\hat{a}) \left(\hat{b} + \hat{b}^\dagger \right). \quad \text{(J.24)}$$

This Hamiltonian couples well-defined quadratures of the two oscillators. It is quadratic with respect to the creation and annihilation operators and therefore induces a symplectic quantum map (see Section 9.2). Consequently, any input Gaussian state (like a coherent state, or vacuum) will be transformed into another Gaussian state: squeezed mirror states and EPR cavity–mirror entanglement can be produced in the present configuration, but not non-Gaussian states, which need to be generated a third-order Hamiltonian. This configuration can be realized in circuit QED devices.

J.3.3 Applications

We assume that we are in the experimental configuration where κ_0, which gives the width of the cavity resonances, is smaller that their spacing Ω_M, so that one can precisely tune the input laser frequency to one of the three resonances. In particular, if the laser frequency is tuned to the lower sideband $\omega = \omega_{cav} - \Omega_M$, one retrieves the situation of sideband cooling, already encountered for cooling ions. The mirror motion can be very efficiently cooled using this technique, and several experiments have been able to populate the ground state of the microscopic oscillators with a fairly high probability [43]. This regime also allows one to transfer quantum states from the photons to the phonons. If the laser frequency is tuned to the exact cavity resonance sideband $\omega = \omega_{cav}$, one can perform a Quantum Non Demolition measurement of the mirror position, as explained in Chapter 11. One can use this nonlinear device to generate a squeezed state in the beam reflected from the cavity. If the laser is blue detuned and in the strong coupling regime, it is possible to manipulate the mirror motion at the quantum level via parametric amplification. One can squeeze its quantum fluctuations and develop quantum correlations between the mirror and field variables.

Many other quantum phenomena have been predicted and experimentally studied in the domain of opto-mechanics, and in particular the generation of nonclassical and entangled states of the mirror motion [10, 43].

J.4 Exercises for Appendix J

Exercise J.1 How many optical photons would it take to stop a black body with the mass of a small molecule approaching at a speed of a few meters per second? How many microwave photons would it take? What would be the frequency of a single photon capable of stopping it?

Exercise J.2 Consider a mass m in a harmonic potential defined by the small oscillation frequency ω. What is the condition on the temperature to observe the zero point motion? Evaluate the condition for realistic experimental parameters.

Exercise J.3 Standard quantum limit for a position measurement of an harmonically bound mass m [99].

Consider an intense light beam characterized by a real mean field, a photon number N in measurement time T, and a wavenumber k_0, that is reflected by a moving mirror of mass m and position x harmonically bound with an oscillation frequency Ω_M. Using an interferometer, one measures the phase variations $\delta\phi^{out}$ of the reflected beam. Relate the intensity and phase of the beam to its two field quadratures X and P. Show that any measured phase variation $\delta\phi^{out}$ can be attributed to a mirror position variation $\delta x = \delta\phi/2k_0 = (1/2k_0\sqrt{2N})\delta P_{in}$, where P_{in} is the phase quadrature of the beam. The mirror is also submitted to the radiation pressure force. Calculate the displacement x_{RP} due to this force and relate it to N and to the amplitude quadrature X_{in}.

Show that the position noise $\delta x_{measured}$ inferred from the phase measurement has two contributions: the actual phase fluctuations of the input beam and the mirror displacement noise due to the fluctuations of the radiation pressure force, more precisely,

$$\delta x_{measured} = \frac{A}{\sqrt{2N}}\delta P_{in} + B\sqrt{2N}\delta X_{in},$$

with values of A and B to be determined. What are the dominant sources of quantum noise at low intensity and at high intensities? Assuming that the phase and amplitude noises of the input beam are not correlated, show that there a minimum position noise for a photon number N_{min} that one will determine. Give the value of this minimum, $\Delta x_{standard} = \sqrt{\hbar/m\Omega_M^2 T}$, called the standard quantum limit of position in the literature.

Is the squeezing of the P or X quadrature useful to improve the accuracy of the position measurement? Show that squeezing of a definite rotated quadrature X_θ permits us to go below the standard quantum limit of position $\Delta x_{standard}$. This procedure is under development on the gravitational wave antennas.

Exercise J.4 Consider a Fabry–Perot cavity, of resonance frequency ω_c, with a fixed mirror and a mirror hooked to a spring, of resonance frequency ω_s, and of equilibrium position x_0. Derive the optomechanical Hamiltonian $\hat{H}/\hbar = \omega_c\hat{a}^\dagger\hat{a} + \omega_s\hat{b}^\dagger\hat{b} - g_0\hat{a}^\dagger\hat{a}(\hat{b}^\dagger + \hat{b})$ coupling the optical mode $\hat{a}$ and the mechanical mode $\hat{b}$. Do it by linearizing the cavity frequency dependence with respect to the mirror position. What is the force per photon made by the radiation pressure in the mirrors? How much is the mirror displaced by the intracavity field? What is the condition between g_0 and ω_s to observe single photons effects?

Exercise J.5 Linearized regime of optomechanics [42]. Show that under a strong light drive ($\hat{V} = i\hbar\Omega(\hat{a}e^{i\omega_L t} - \hat{a}^\dagger e^{-i\omega_L t})$) the Hamiltonian can be written in the rotating frame at the light frequency ($\hat{U}(t) = \exp(-i\omega_L\hat{a}^\dagger\hat{a}t)$) as $\hat{H}/\hbar = -\Delta_0\hat{a}^\dagger\hat{a} + \omega_s\hat{b}^\dagger\hat{b} - g_0 a^\dagger\hat{a}(\hat{b}^\dagger + \hat{b}) + i\Omega(\hat{a} - \hat{a}^\dagger)$, where $\Delta_0 = \omega_L - \omega_c$. Derive the equations of motion for the mean values of $\hat{a}$ and $\hat{b}$ and linearize them. Find the effective Hamiltonian that describes the system from the linearized equations. Show that the displacement of the $\hat{a}$ mode (α_{lin}) amplifies the coupling constant g_0 between $\hat{a}$ and $\hat{b}$ to $g_0|\alpha_{lin}|$. In experiments this enhancement can be of several orders of magnitude [10].

Exercise J.6 Consider a two-level atom in presence of a coherent drive and dissipation. Show that the optical Bloch equations yield for the population of the excited state and the coherences of the atom

$$\begin{aligned} \partial_t\rho_{ee} &= -\Gamma\rho_{ee} + \frac{\mathrm{i}}{2}\left(\Omega_R^*\rho_{eg} - \Omega_R\rho_{ge}\right), \\ \partial_t\rho_{eg} &= \left(\mathrm{i}\delta - \frac{\Gamma}{2}\right)\rho_{eg} - \frac{\mathrm{i}\Omega_R}{2}\left(\rho_{gg} - \rho_{ee}\right), \end{aligned} \tag{J.25}$$

where Ω_R is the Rabi frequency of the atom in the coherent drive, Γ is the spontaneous emission rate of the atom, and δ is the detuning between the atomic frequency and the Rabi drive. Show that, in a steady state, the force over the atom is $F = \hbar k\Gamma P_e$, where $\hbar k$ is the momentum of the photon. Show that the steady-state population can be written as $P_e = s/2(1+s)$, where s is the saturation parameter and reads $s = 2|\Omega_R|^2/(\Gamma^2 + 4\delta^2)$. Show that by taking into account the Doppler shift ($-kv$) of atoms moving at a velocity v ($\delta \to \delta - kv$), a velocity-dependent force appears. Show that two counter-propagating beams create viscosity $F \sim -\eta v$ that will cool down the atoms (for $\delta < 0$).

K Basics of Circuit Quantum Electrodynamics

Quantum mechanics rules the behavior of atoms, molecules, photons, and other elementary particles, but its applicability to "macroscopic objects" is still unclear and is the topic of much experimental and theoretical work. It has become rather evident in recent decades [187] that it is not the "size" that matters but how "isolated" a system may be: Large objects like cats (among other multi-atom systems) have naturally many coupling channels to their environment and are de facto not isolated at all. This is the reason why most (at least in part) large objects do not behave quantum mechanically. Nonetheless, under extremely controlled conditions, macroscopic quantum effects exist and are routinely explored in the laboratory (see Appendix J). Two kinds of macroscopic quantum effects may be distinguished [54]. The first kind corresponds to a macroscopic amount of individual quantum systems acting together. We may take as examples of this type superfluidity, superconductivity, and any other Bose–Einstein condensation of particles. These are microscopic effects that "survive" over the ensemble. The second kind of macroscopic quantum effects are those exhibited by macroscopic variables themselves. They correspond to systems where the microscopic structure can be neglected and the full system admits a description in terms of a few variables that are ruled by quantum equations of motion. This appendix treats in some detail an example of this second type, namely the voltages and currents in macroscopic circuits. The purpose of this appendix is to give a short overview of the domain, called *circuit quantum electrodynamics*, (cQED, not to be confused with CQED, reserved for as cavity quantum electro dynamics). This part of quantum physics is currently enjoying an impressive development as one of the most promising quantum platforms to implement quantum computing devices and processes, and hopefully, in some future, to build universal quantum computers. We present here a simple theoretical approach to the domain that will help the reader understand the basic aspects of the domain, as well as the experiments described in the experimental chapters of this textbook. More detailed descriptions of the domain can be found in [19].

K.1 Quantum Mechanics of Electrical Circuits

K.1.1 The LC Resonant Circuit

Let us consider a harmonic electrical circuit comprising a capacitor (capacitance C) and an inductor (inductance L). It is described classically by its energy H given by

$$H = \frac{Q^2}{2C} + \frac{\varphi^2}{2L}, \tag{K.1}$$

where $Q(t) = \int I(t)dt$ and $\varphi(t) = \int V(t)dt$ are the charge of the capacitor and the flux over the inductor. I, Q, V, and φ are macroscopic variables describing the collective behavior of the elementary charges carriers. Their evolution obeys Hamilton equations:

$$\frac{d\varphi}{dt} = \frac{\partial H}{\partial Q}, \quad \frac{dQ}{dt} = -\frac{\partial H}{\partial \varphi}, \tag{K.2}$$

which allows us to identify φ and Q as *canonically conjugate variables*. The procedure for deriving the quantum description of this circuit is straightforward: One replaces the classical time-dependent conjugate variables by operators $\hat{\varphi}$ and $\hat{Q}$ and the Poisson bracket ($\{\varphi, Q\} = 1$) by the commutator

$$\frac{1}{i\hbar}\left[\hat{\varphi}, \hat{Q}\right] = \hat{1}. \tag{K.3}$$

Here, $\hat{\varphi}$ and $\hat{Q}$ appear as the quantum analogs of the position and momentum of a harmonic oscillator, or of the quadrature components of the electromagnetic field.

As for these two latter cases, one defines an annihilation operator for this linear resonator:

$$\hat{r} = \sqrt{\frac{1}{2\hbar Z_0}}\left(\hat{\varphi} + iZ_0\hat{Q}\right), \tag{K.4}$$

where $Z_0 = \sqrt{L/C}$ is the circuit impedance. As for a mechanical harmonic oscillator, the Hamiltonian of the circuit reads $\hat{H} = \hbar\omega_0(\hat{r}^\dagger\hat{r} + 1/2)$ and its energy is quantized: $E_n = \hbar\omega_0(n + 1/2)$, where we have defined the resonator Fock states to be $\hat{r}^\dagger\hat{r}|n\rangle = n|n\rangle$. Here, ω_0 is the oscillation frequency $\omega_0 = 1/\sqrt{LC}$ of the LC circuit.

K.1.2 The Josephson Junction and the Transmon

A quantum harmonic oscillator, like a LC circuit, has properties that are very close to the classical one. In order to have measurable quantum effects, some kind of nonlinearity is required, like the parametric down conversion and light–atom coupling for the electromagnetic field. For the electrical circuit it is provided by a *Josephson junction*. Here, we give its main properties.

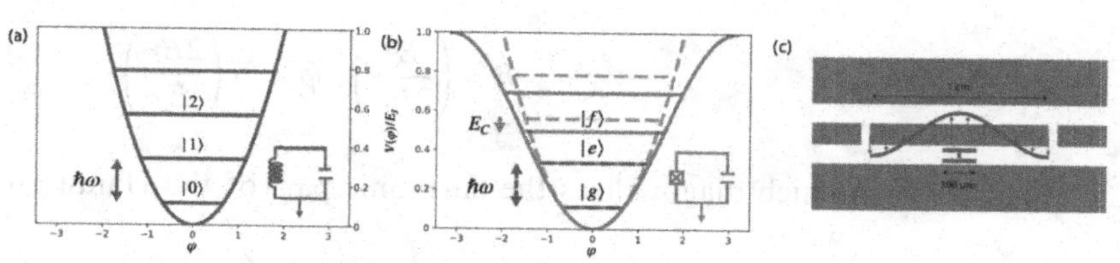

Fig. K.1 (a) Harmonic LC-oscillator with equally spaced energies. (b) Anharmonic oscillator: The Josephson Junction is designated by a boxed cross. The oscillator becomes anharmonic. (c) A strip line microwave resonator with a transmon: The qubit-cavity system in circuit QED. The arrows represent the electric field in the linear resonator at a given time.

A Josephson junction consists of two superconductors separated by a thin non-conducting layer (a few nanometers typically), which acts as a tunnel barrier for Cooper pairs. It is modeled as a nonlinear lossless inductance with a Hamiltonian contribution given by the Josephson potential

$$H_J = -E_J \cos\left(2\pi\frac{\Phi}{\Phi_0}\right). \tag{K.5}$$

Here, the Josephson energy E_J is a parameter that depends on the fabrication of the device and $\Phi_0 = h/2e$ is known as the flux quantum. Thus, replacing the regular coil (L) in an LC circuit by a Josephson junction (see Figure K.1) one then has a system described by a Hamiltonian

$$\hat{H} = \frac{\hat{Q}^2}{2C} - E_J \cos\left(2\pi\frac{\hat{\Phi}}{\Phi_0}\right) = \frac{\hat{Q}^2}{2C} + \frac{E_J}{2}\frac{\hat{\Phi}^2}{(\Phi_0/2\pi)^2} - \frac{E_J}{4!}\frac{\hat{\Phi}^4}{(\Phi_0/2\pi)^4} + \cdots. \tag{K.6}$$

The anharmonic inductive terms in the Hamiltonian break the degeneracy of the oscillator. This is shown in the first two panels of Figure K.1.

It must be noted that this Hamiltonian is identical to that of a mechanical pendulum $\left(\hat{H} = \frac{p_\theta^2}{2J} - J\omega_o^2\cos(\theta)\right)$, where J is the moment of inertia, θ its angle with the vertical, p_θ its conjugate momentum, and ω_o its small oscillation frequency, and it is thus its exact electrical analog.

At this point it is convenient to introduce dimensionless variables to simplify the notation ($\hat{n} = \hat{Q}/2e$ and $\hat{\varphi} = 2\pi\hat{\Phi}/\Phi_0$) and write[1]

$$\hat{H} \simeq 4E_C\hat{n}^2 + \frac{1}{2}E_J\hat{\varphi}^2 - \frac{1}{4!}E_J\hat{\varphi}^4, \tag{K.7}$$

where $E_C = e^2/2C$. The first two terms correspond to a harmonic contribution to the energy while the last term corresponds to a nonlinearity.[2] To simplify the expression further we define the annihilation operator

[1] The factor 2 in the definition of $\hat{n}$ has its motivation in that the current carriers in superconducting materials are Cooper pairs. The operator $\hat{n}$ counts the Cooper pairs that tunnel across the junction.

[2] The reader may gain intuition by thinking of the variable φ in the *electronic* oscillator as an analog to the position x of a *mechanical* oscillator (or the angle θ of a pendulum). The nonlinearity in the Hamiltonian plays the role of confining the quartic potential. One can define wavefunctions $\psi(\varphi)$ or $\psi'(x)$, the absolute square of which yields the probability density for the values of the canonical variables.

$$\hat{b} = \left(\frac{E_J}{2E_C}\right)^{\frac{1}{4}} \hat{\varphi} + 2i\left(\frac{2E_C}{E_J}\right)^{\frac{1}{4}} \hat{n}, \tag{K.8}$$

which diagonalizes the harmonic part of the Hamiltonian. One can then write

$$\hat{H}_T = \hbar\omega'\left(\hat{b}^\dagger\hat{b} + \frac{1}{2}\right) - \frac{E_C}{2}\hat{b}^\dagger\hat{b}^\dagger\hat{b}\hat{b} + \cdots, \tag{K.9}$$

where $\hbar\omega' = \sqrt{8E_C E_J} - E_C$.

This is the Hamiltonian of a Kerr electronic oscillator and in the regime $E_C/E_J \gg 1$ it is called a "transmon" [105]. To this day, it is the technology of choice for superconducting qubits [7, 37, 55]. The physical implementation of the transmon is ensured by two millimiter-size aluminium pads (Figure K.1(c)) deposited over a sapphire substrate and connected by a few-nanometers wide layer of aluminium oxide. This simple nonlinear quantum dipole is regarded as an "artificial atom" since it can be described by a single degree of freedom φ following the Schrödinger equation.

To see clearly the use of the transmon as a qubit, we rewrite the Hamiltonian, omitting a constant term, as $\hat{H}_T \sim \hbar\hat{\omega}_b\hat{b}^\dagger\hat{b}$ where $\hat{\omega}_b = \omega' - E_C(\hat{b}^\dagger\hat{b} - 1)/2\hbar$. We see that the transition frequency $\hat{\omega}_b$ is a function of the number of excitations in the oscillator. The technological implications are far reaching: Since the transition frequency from $|g\rangle = |-\rangle$ to $|e\rangle = |+\rangle$ is different from all others, one has isolated a two-level quantum system in a macroscopic device that can be easily manufactured and reproduced. Moreover, having resonances in the microwave domain, these artificial atoms are easily manipulated and measured.

In ordinary nonlinear circuits, the quantum effects are inconspicuous and to exhibit them two conditions have to be met. First, the circuit should have low dissipation: Dissipation broadens the discrete energies of the circuit and if their linewidth is of the order of their splitting, then the spectrum will effectively be continuous. The quantum condition (the existence of discrete energies) is naturally broken. Note that the splitting of the lines is directly given by the nonlinearity of the system. The condition is met by building circuits around nonlinear superconducting Josephson junctions operating at cryogenic temperatures. The second condition is to be able to prepare and maintain the system in a pure state. Uncontrolled thermal excitations are then to be avoided. Typical values of L, C for superconducting circuits are of the order of 5 nH, 5 pF, respectively, which yields $\omega_0 \simeq 6\,\text{GHz}$ in the microwave domain. To avoid thermal excitation one should satisfy $\hbar\omega_0 \gg k_B T$, which sets the working temperature to be $T \ll 100\,\text{mK}$. Under these conditions, the zero point spread of the oscillator can be directly observed, making manifest the quantum fluctuations of mesoscopic voltages (~μV) and currents (~μA).

K.2 Quantum Effects in Superconducting Circuits

A typical circuit QED (cQED) system comprises an artificial atom coupled to a linear resonator. The linear resonator can be either a linear LC circuit (Figure K.1(c)) or a 3D microwave cavity. This configuration is completely analogous to the cavity QED (CQED) architectures and many terms and notions are borrowed from it. We now turn to discuss the measurement and control of quantum circuit in the cQED architecture.

K.2.1 "Atom–Cavity" Coupling

The Hamiltonian in Equation (K.9) is familiar in quantum optics: It describes a nonlinear medium where the index of refraction depends on the power of the applied pump field. This is called the Kerr effect (Appendix I). Whereas in optics, the Kerr nonlinearity is small and requires intense laser illumination to be observable, in circuit QED, the nonlinear effects are very strong and observable even at the single photon level.

The transmon is coupled to a resonator of frequency ω_r (and of annihilation operator $\hat{r}$). In the case of a capacitive coupling, the transmon cavity (TC) system is described by:

$$\hat{H}_{TC} = \hat{H}_L + \hat{H}_{NL}, \tag{K.10}$$

$$\hat{H}_L = \hbar\omega'\hat{b}^\dagger\hat{b} + \hbar\omega_r\hat{r}^\dagger\hat{r} + \hbar g\left(\hat{b}^\dagger\hat{r} + \hat{r}^\dagger\hat{b}\right), \tag{K.11}$$

$$\hat{H}_{NL} = -\frac{E_C}{2}\hat{b}^\dagger\hat{b}^\dagger\hat{b}\hat{b}. \tag{K.12}$$

The coupling can be seen as a single excitation exchange between the cavity and the transmon and the linear Hamiltonian $\hat{H}_L$ can be diagonalized using a symplectic transformation:

$$\hat{b}' = \hat{b}\cos A + \hat{r}\sin A, \quad \hat{r}' = -\hat{b}\sin A + \hat{r}\cos A. \tag{K.13}$$

A particular choice of A allows us to remove the coupling term, so that $\hat{H}_{TC}$ is transformed into $\hat{H}'_{TC} = \hat{H}'_L + \hat{H}'_{NL}$ with

$$\hat{H}'_L = \hbar\tilde{\omega}'\hat{b}'^\dagger\hat{b}' + \hbar\omega'_r\hat{r}'^\dagger\hat{r}', \tag{K.14}$$

$$\hat{H}'_{NL} = \chi_b\hat{b}'^\dagger\hat{b}'^\dagger\hat{b}\hat{b} + \chi_c\hat{r}'^\dagger\hat{r}'^\dagger\hat{r}'\hat{r}' + \chi_{br}\hat{b}'^\dagger\hat{b}\hat{r}'^\dagger\hat{r}. \tag{K.15}$$

In analogy with nonlinear optics (Sections 3.5, 11.5, and Appendix I), the first two terms of $\hat{H}'_{NL}$ are called *self-Kerr nonlinearities* for the two new modes of the system. One of them is "transmon-like" and it has the greater nonlinearity. The other one is "cavity-like" and is the most linear of the two: The cavity-mode now has an inherited nonlinearity. The last term in the Hamiltonian is called *cross-Kerr nonlinearity*.

One replaces in the following the transmon number operator $\hat{b}'^{\dagger}\hat{b}'$ by $\hat{b}'^{\dagger}\hat{b}' = (\hat{\sigma}_z + \hat{1})/2$. This approximation of the anharmonic oscillator by a two-level system holds as long as the system is weakly excited, so that its population is restricted to the first two levels of the ladder: Microwave pulses should not contain frequencies addressing the higher excited transitions. In particular, the pulses addressing the qubit need to be much longer in time than $1/\hbar E_C$ (~10 ns). This sets a speed limit to the operations that can be made. Under the rotating wave approximation the Hamiltonian takes the form

$$\hat{H}_{TC} = \frac{1}{2}\hbar\tilde{\omega}'\hat{\sigma}_z + \hbar\left(\omega_r' + \chi_{br}\,\hat{\sigma}_z/2\right)\hat{r}'^{\dagger}\hat{r}'. \qquad \text{(K.16)}$$

The last term shows that the interaction with the transmon has shifted the frequency of the resonant circuit by a quantity that depends on the state of the transmon. This is know as the Jaynes–Cummings dispersive Hamiltonian and was originally developed to study real atom–field interactions (Appendix G.3.3)

K.2.2 Dispersive Readout and Control

The Hamiltonian of the joint system is given by Equation (K.16). Its interpretation is that the cavity has a resonance frequency $\omega_r' \pm \chi_{br}/2$ that depends on the state of the atom. Figure K.2 shows the experimental consequence of this shift: When a microwave classical continuous signal is sent to the cavity at its bare frequency the signal will be reflected with a phase encoding the state of the qubit.

One thus measures the state of the qubit by measuring the phase of the reflected signal. This is normally done by heterodyne detection

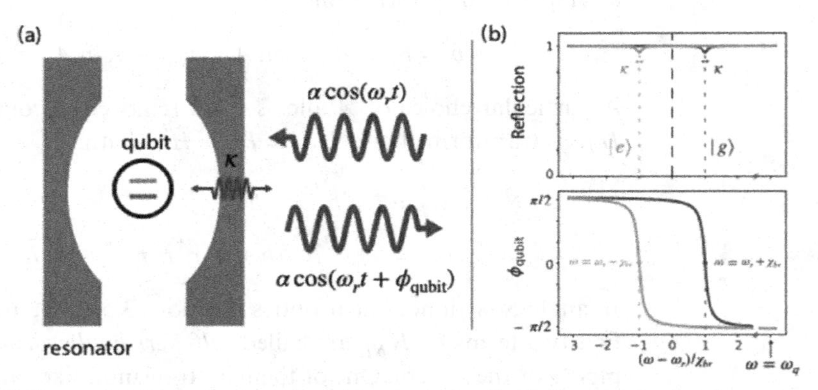

Fig. K.2 A field α cos $(\omega_r t)$ is sent to the cavity and its response in reflection is collected and reads α cos $(\omega_r t + \phi_{qubit})$. When the incident field is at frequency $\omega = \omega_q$, the qubit undergoes a Rabi oscillation and the phase of the reflected field is independent of the qubit state. If the incident is at frequency $\omega = \omega_r$, the phase of the reflected field is $\phi_{qubit} = \pm\pi/2$, depending on the state of the qubit e or g.

(Section 9.4). A readout field $\alpha \cos(\omega_r t)$ is sent to the cavity ($|\alpha|^2 = \langle \hat{n} \rangle \sim 1$). The response in reflection is collected and reads ideally as $\alpha \cos(\omega_r t + \phi_{qubit})$. The phase takes the values $\phi_{qubit} = \pm \pi/2$ depending on the state of the qubit e or g. This simple detection mechanism, reminiscent of the one used in CQED (see Section G.3 and [88]), allows for quantum nondemolition measurement with extremely high efficiency.

Note that sending a coherent field at the qubit frequency $\omega_q = \tilde{\omega}'$ does not allow us to gain information about the state, as the reflected wave has a phase of $-\pi/2$ regardless of the qubit state. It can be used to drive the qubit by a well-adjusted Rabi oscillation. After the readout pulse, the quantum state of the system qubit+signal mode $|\phi\rangle$ is given by

$$|\phi\rangle \propto |g\rangle \otimes \left| e^{i\phi^{(g)}_{qubit}} \alpha \right\rangle + |e\rangle \otimes \left| e^{i\phi^{(e)}_{qubit}} \alpha \right\rangle, \tag{K.17}$$

where α represents a coherent state in the readout line. If the driving field is applied at ω_r, then $|e^{i\phi^{(g)}_{qubit}}\alpha\rangle$ and $|e^{i\phi^{(e)}_{qubit}}\alpha\rangle$ are orthogonal and the state is entangled. This is the Schrödinger cat situation. If, instead, the driving field is applied at ω_q then $\phi^{(g)}_{qubit} = \phi^{(e)}_{qubit}$ and the state is factorizable ($\propto [\hat{U}(|g\rangle + |e\rangle)] \otimes |e^{-i\pi/2}\alpha\rangle$). A Rabi evolution can be implemented.

K.2.3 Quantum Limited Amplification

Having energies in the microwave domain (a few GHz), a typical quantum signal has an energy of 10^5 lower than in the optical domain (hundreds of THz): Single microwave photons are difficult to detect. The solution is to amplify the continuous variable electric signal in a way that does not spoil its quantum fluctuations.

The quantum input–output theory of an amplifier is detailed in Section H.2: A simple and general quantum mechanical argument shows that, whereas a phase insensitive amplifier cannot preserve the signal-to-noise ratio, and must pay at least a "3 dB penalty," a phase sensitive amplifier, which amplifies only oscillating fields having a given phase, is not subjected to this burden and may perfectly preserve the signal-to-noise ratio. In Section I.2 we showed that degenerate parametric down conversion in a nonlinear $\chi^{(2)}$ medium provides a noiseless phase-sensitive amplifier, which amplifies a given quadrature of the field with a gain G, and de-amplifies the other quadrature with a reduction factor $1/G$, while preserving the signal-to-noise ratio in the amplification.

To achieve such a quantum limited amplification in cQED, one relies on the intrinsic nonlinearity of the Josephson junction. A field of amplitude E_0 is used to drive a nonlinear circuit, transforming the operator $\hat{b}(t)$ as $\hat{b}(t) \to \hat{U}^\dagger \hat{b}(t) \hat{U} = E_0 e^{i(\theta - \omega_p t)} + \hat{b}(t)$.

The pumped Josephson Hamiltonian (pJ), after the pertinent canonical transformations [155] and taking the pump frequency to be twice the

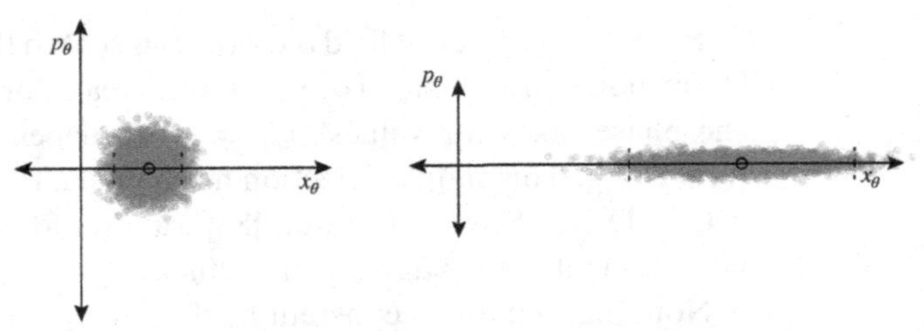

Fig. K.3 Noiseless parametric amplification. The input state is a coherent state the mean value of which is aligned on the x_θ quadrature. The dashed lines represent the standard deviation of its spread. In this configuration, the mean value and the standard deviation of the quantum fluctuations are amplified in the same way, thus preserving the signal-to-noise ratio in the amplification. The price to pay is that noiseless amplification is restricted to input coherent states aligned on the x_θ axis. For input states aligned on the other quadrature axis p_θ, the process actually "deamplifies," i.e. reduces, the mean field and the standard deviation.

oscillators frequency ($\omega_p = 2\omega'$), can be written as

$$\hat{H}_{pJ} \sim \hbar\omega'\hat{b}^\dagger\hat{b} + g\left(e^{i(2\omega' t+\theta)}\hat{b}^2 + e^{-i(2\omega' t+\theta)}(\hat{b}^\dagger)^2\right). \tag{K.18}$$

This a single-mode squeezing Hamiltonian, exactly what is required to get quantum limited amplification on one of the quadratures of the intra-cavity field.

The power gain G is typically +20 dB in the first amplification step for the x_θ quadrature defined as $x_\theta = x\cos\theta + p\sin\theta$. In Figure K.3, we show a theoretical sketch of the Wigner function of the state, before and after amplification, in terms of density of points. The initial state is a coherent state with phase independent quantum noise. The amplified state is represented by a phase-dependent noise characteristic of a squeezed state. Note that the mean value of the amplified quadrature is much bigger than the initial state, allowing for detection of the signal.

K.3 Comparison of cQED with CQED

cQED adopted many historical ideas from CQED, in particular from their implementation with Rydberg atoms. For the sake of comparison, we will name a few advantages and disadvantages of these competing platforms [86].

Circuit QED platforms have several advantages that make them good candidates for implementing quantum computing tasks: They have large dipole antenas (mm in size) and small mode volumes (well below the mode wavelength in microwave guides) that provide an enhanced speed. Operations are typically done in tens of nanoseconds. Circuit printing relies on well-established fabrication technologies, which are easily made and reproduced, and is very stable in time. The atomic coherence is limited typically below the millisecond timescale for a relatively large size (mm). The system operates in dilution refrigerators below 20 mK required to have a low thermal background at a few GHz frequencies.

Cavity QED platforms in the microwave domain, for example using circular Rydberg atoms, are characterized by very high Q cavities (lifetimes over 100ms), very long atomic lifetimes (>30 ms), and large dimensions (~125 nm), producing a large dipole coupling to the cavity and allowing for a speed of operation at the tens of kilohertz range. The typical frequency of operation is 50 GHz, demanding a temperature of operation barely below 1 K. The state of the art so far is achieved with a thermal beam of traveling qubits that cross the cavity horizontally or vertically during a limited amount of time (typically microseconds). Laser trapping of atoms allows us to overcome this limitation of CQED.

Cavity QED platforms in the optical domain, for example using trapped ions, quantum dots, or NV centers in diamond, do not need cooling, use true nanobjects, for which the fabrication technique is well mastered, thanks to the progress of microelectronics. They can be integrated in large numbers and directly coupled to telecom optical fibers. Their coherence time can reach the order of seconds or even minutes.

K.4 Exercises for Appendix K

Exercise K.1 From the two Josephson relation i) $I_J = I_0 \sin\varphi$ and ii) $V_J = \Phi_0/2\pi\dot{\varphi}$, derive the Josephson potential energy $U_J = -E_J \cos\varphi$. What is the meaning of I_0 and Φ_0? See [121].

Exercise K.2 Find the expression for the Josephson inductance L_J defined by $\dot{I} = L_J^{-1} V_J$. Is it an ordinary "linear" inductance?

Exercise K.3 Find expressions for L_J and I_J as a function of $U_J(\varphi)$. Generalize this expression to find the contribution to the circuit Hamiltonian of ordinary components such as a linear inductor and an external current source, for example.

Exercise K.4 Discuss the circuit quantization of these Hamiltonian contributions and why one can unambiguously state that the quantum version of this classical systems is obtained "by simply putting hats." Propose a Hamiltonian contribution type for which this will not be the case.

Exercise K.5 Use the above relations to derive the Lagrangian, the equation of motion, and then the Hamiltonian of a linear transmission line. Consider a 1D electron system with inductance per unit length l and capacitance per unit length c. How is the argument adapted to a single mode lumped element LC oscillator? See [77].

Exercise K.6 Consider the transmon as a Josephson junction connected to the ground via a large capacitance. Use the conservation of current (Kirchhoff's current law) to derive the transmon Hamiltonian.

Exercise K.7 Show that since $[\hat{\varphi}, \hat{n}] = i$, where $\hat{n}$ is the Cooper pair number operator, the Josephson potential is simply a charge tunneling operator: $\cos\hat{\varphi} = \sum_n \frac{1}{2}(|n\rangle\langle n+1| + |n+1\rangle\langle n|)$ [76].

References

[1] Scott Aaronson and Alex Arkhipov. "The computational complexity of linear optics." In: *Proceedings of the Forty-Third Annual ACM Symposium on Theory of Computing.* 2011, pp. 333–342.

[2] Junaid Aasi, J. Abadie, B. P. Abbott, et al. "Enhanced sensitivity of the LIGO gravitational wave detector by using squeezed states of light." In: *Nature Photonics* 7.8 (2013), pp. 613–619.

[3] Gerardo Adesso, Sammy Ragy, and Antony R. Lee. "Continuous variable quantum information: Gaussian states and beyond." In: *Open Systems & Information Dynamics* 21.01n02 (2014), p. 1440001.

[4] Yakir Aharonov, David Z. Albert, and Lev Vaidman. "How the result of a measurement of a component of the spin of a spin-1/2 particle can turn out to be 100." In: *Physical Review Letters* 60 (1988), pp. 1351–1354.

[5] Francesco Albarelli, Marco Genoni, Matteo Paris, and Alessandro Ferraro. "Resource theory of quantum non-Gaussianity and Wigner negativity." In: *Physical Review A* 98.5 (2018), p. 052350.

[6] Frank Arute. "Quantum supremacy using a programmable superconducting processor." In: *Nature* 574.7779 (2019), pp. 505–510.

[7] Frank Arute and John M. Martinis. "Quantum supremacy using a programmable superconducting processor." In: *Nature* 574.7779 (2019), pp. 505–510.

[8] Alain Aspect. "Bell's theorem: The naive view of an experimentalist." In: *Quantum [Un]Speakables.* Springer, 2002, pp. 119–153.

[9] Alain Aspect, Philippe grangier, and Gérard Roger. "Experimental Realization of Einstein–Podolsky–Rosen–Bohm Gedankenexperiment: A New Violation of Bell's Inequalities." In: *Physical Review Letters* 49, (1982), pp. 91–94.

[10] Markus Aspelmeyer, Tobias J. Kippenberg, and Florian Marquardt. "Cavity optomechanics." In: *Reviews of Modern Physics* 86.4 (2014), p. 1391.

[11] Gennaro Auletta, Mauro Fortunato, and Giorgio Parisi. *Quantum Mechanics.* Cambridge University Press, 2009.

[12] Charles Bamber and Jeff S. Lundeen. "Observing Dirac's classical phase space analog to the quantum state." In: *Physical Review Letters* 112.7 (2014), p. 070405.

[13] Stephen Barnett. *Quantum Information.* Oxford University Press, 2009.

[14] Stephen M. Barnett, John Jeffers, and David T. Pegg. "Quantum retrodiction: Foundations and controversies." In: *Symmetry* 13.4 (2021), p. 586.

[15] John Bell. "On the Einstein Podolsky Rosen paradox." In: *Physics* 1.3 (1964), pp. 195–200.

[16] John Bell. "On the problem of hidden variables in quantum mechanics." In: *Reviews of Modern Physics* 38 (1966), pp. 447–452.

[17] Jérôme Beugnon, M. P. A. Jones, J. Dingjan, et al. "Quantum interference between two single photons emitted by independently trapped atoms." In: *Nature* 440.7085 (2006), pp. 779–782.

[18] Gunnar Björk, Jonas Söderholm, Luis L. Sánchez-Soto, et al. "Quantum degrees of polarization." In: *Optics Communications* 283.22 (2010), pp. 4440–4447.

[19] Alexandre Blais Arne L. Grimsmo, S. M. Girvin, and Andreas Wallraff. "Circuit quantum electrodynamics." In: *Reviews of Modern Physics* 93 (2021), p. 025005.

[20] Rainer Blatt and David Wineland. "Entangled states of trapped atomic ions." In: *Nature* 453.7198 (2008), pp. 1008–1015.

[21] David Bohm. *Quantum Theory.* Prentice Hall, 1951.

[22] Niels Bohr. "On the constitution of atoms and molecules. Part I - Binding of electrons by positive nuclei." In: *Philosophical Magazine* 26 (1913), pp. 1–24.

[23] Max Born. "Statistical interpretation of quantum mechanics." In: *Science* 122.3172 (1955), pp. 675–679.

[24] V. Bouchiat, D. Vion, P. Joyez, D. Esteve, and M. H. Devoret. "Quantum coherence of charge states in the single electron box." In: *Journal of Superconductivity* 12.6 (1999), pp. 789–797.

[25] D. Bouwmeester, J.-W. Pan, K. Mattle, M. Eibl, H. Weinfurter, and A. Zeilinger "Experimental quantum teleportation." In *Nature* 390.6660 (1997), pp. 575–579 [Nature Publishing Group].

[26] Warwick P. Bowen et al. "Polarization squeezing of continuous variable Stokes parameters." In: *Physical Review Letters* 88.9 (2002), p. 093601.

[27] Robert W. Boyd. *Nonlinear Optics*, 4th ed. Academic Press, 2020.

[28] Daniel J. Brod et al. "Photonic implementation of boson sampling: a review." In: *Advanced Photonics* 1.3 (2019), p. 034001.

[29] R. Hanbury Brown and Richard Q. Twiss. "Correlation between photons in two coherent beams of light." In: *Nature* 177.4497 (1956), pp. 27–29.

[30] Vladimir Bužek and M. Hillery. "Quantum copying: Beyond the no-cloning theorem." In: *Physical Review A* 54.3 (1996), p. 1844.

[31] Y. Cai, J. Roslund, V. Thiel, C. Fabre, and N. Treps. "Quantum enhanced measurement of an optical frequency comb." In: *npj Quantum Information* 7.1 (2021), pp. 1–8.

[32] Tigrane Cantat-Moltrecht, Rodrigo Cortiñas, Brice Ravon, et al. "Long-lived circular Rydberg states of laser-cooled rubidium atoms in a cryostat." In: *Physical Review Research* 2.2 (2020), p. 022032.

[33] Howard Carmichael. *An Open Systems Approach to Quantum Optics.* Springer, 1991.

[34] William B. Case. "Wigner functions and Weyl transforms for pedestrians." In: *American Journal of Physics* 76.10 (2008), pp. 937–946.

[35] Carlton M. Caves. "Quantum-mechanical noise in an interferometer." In: *Physical Review D* 23.8 (1981), p. 1693.

[36] Ulysse Chabaud, Damian Markham, and Frédéric Grosshans. "Stellar representation of non-Gaussian quantum states." In: *Physical Review Letters* 124 (2020), p. 063605.

[37] Christopher Chamberland, Guanyu Zhu, Theodore J. Yoder, Jared B. Hertzberg, and Andrew W. Cross. "Topological and subsystem codes on low-degree graphs with flag qubits." In: *Physical Review X* 10 (2020), p. 011022.

[38] Boris Chesca, Reinhold Kleiner, and Dieter Koelle. "SQUID theory." In: John Clarke and Alex I. Braginski (eds.) *The SQUID Handbook.* John Wiley and Sons, Ltd, 2004, pp. 29–92.

[39] John Clarke. "SQUIDs." In: *Scientific American* 271.2 (1994), pp. 46–53.

[40] J. F. Shimony Clauser and Abner Shimony. "Bell's theorem: Experimental tests and implications." In: *Reports on Progress in Physics* 41.12 (1978), p. 1881.

[41] J. F. Clauser, Michael Horne, Abner Shimony, and R. Holt. "Proposed experiment to test local hidden variable theories." In: *Physical Review Letters* 23 (1969), p. 880.

[42] Aashish A. Clerk. "Optomechanics and quantum measurement." In: *Lecture notes, Les Houches Summer School on Optomechanics* (2015).

[43] Pierre-Francois Cohadon, Jack Harris, Florian Marquardt, and Leticia Cugliandolo (eds.) *Quantum Optomechanics and Nanomechanics: Lecture Notes of the Les Houches Summer School: Volume 105, August 2015.* Vol. 105. Oxford University Press, 2020.

[44] Claude Cohen-Tannoudji, Bernard Diu, and Franck Laloe. *Quantum Mechanics*, Vol. 3. Wiley, 2019.

[45] Claude Cohen-Tannoudji, Bernard Diu, and Frank Laloe. *Quantum Mechanics*, Vols. 1 and 2. Wiley-Interscience, 2006.

[46] Claude Cohen-Tannoudji and David Guery-Odelin. *Advances in Atomic Physics: An Overview*. World Scientific, 2011.

[47] Thomas L. Curtright, David B. Fairlie, and Cosmas K. Zachos. *A Concise Treatise on Quantum Mechanics in Phase Space.* World Scientific Publishing Company, 2013.

[48] Jean Dalibard, Yvan Castin, and Klaus Mølmer. "Wave-function approach to dissipative processes in quantum optics." In: *Physical Review Letters* 68 (1992), pp. 580–583.

[49] Luis Davidovich. "Towards the ultimate precision limits: An introduction to quantum metrology." In: *Invited lecture at College de France* (2016).

[50] H. G. Dehmelt. "Radiofrequency Spectroscopy of Stored Ions". In: *Advances in Atomic and Molecular Physics* 3, (1968), pp. 53–72.

[51] Vincent Delaubert, Nicolas Treps, Mikael Lassen, et al. "TEM 10 homodyne detection as an optimal small-displacement and tilt-measurement scheme." In: *Physical Review A* 74.5 (2006), p. 053823.

[52] Samuel Deleglise, I. Dotsenko, C. Sayrin, et al. "Reconstruction of non-classical cavity field states with snapshots of their decoherence." In: *Nature* 455.7212 (2008), pp. 510–514.

[53] Rafal Demkowicz-Dobrzański, Marcin Jarzyna, and Jan Kolodyński. "Quantum limits in optical interferometry." In: E. Wolf. (ed.), Progress in Optics, Vol. 60. Elsevier, 2015, pp. 345–435.

[54] M. Devoret, Daniel Esteve, C. Urbina, et al. "Macroscopic quantum effects in the current-biased Josephson junction". In: Yu Kagan and A. J. Leggett (eds.) *Modern Problems in Condensed Matter Science*, Elsevier, 1992, pp. 313–345.

[55] M. H. Devoret and R. J. Schoelkopf. "Superconducting circuits for quantum information: An outlook." In: *Science* 339.6124 (2013), pp. 1169–1174.

[56] Michel H. Devoret. "Does Brian Josephson's gauge-invariant phase difference live on a line or a circle?" In: *Journal of Superconductivity and Novel Magnetism* 34.6 (2021), pp. 1633–1642.

[57] Paul Dirac. *The Principles of Quantum Mechanics.* Oxford University Press, 1981.

[58] Paul A. M. Dirac. "On the analogy between classical and quantum mechanics." In: *Reviews of Modern Physics* 17.2–3 (1945), p. 195.

[59] Adrien Dousse, Jan Suffczyński, Alexios Beveratos, et al. "Ultra-bright source of entangled photon pairs." In: *Nature* 466.7303 (2010), pp. 217–220.

[60] Lu-Ming Duan, G. Giedke, J. I. Cirac, and P. Zoller. "Inseparability criterion for continuous variable systems." In: *Physical Review Letters* 84.12 (2000), p. 2722.

[61] Biswadeb Dutta, Narasimhaiengar Mukunda, Rajiah Simon, et al. "The real symplectic groups in quantum mechanics and optics." In: *Pramana* 45.6 (1995), pp. 471–497.

[62] A. Einstein, B. Podolsky, and N. Rosen. "Can quantum-mechanical description of physical reality be considered complete?" In: *Physical Review* 47 (1935), p. 777.

[63] Albert Einstein. "On a heuristic point of view concerning the production and transformation of light." In: *Annalen der Physik* (1905), pp. 1–18.

[64] Albert Einstein. "Über einen die Erzeugung und Verwandlung des Lichtes betreffenden heuristischen Gesichtspunkt." In: *Annalen der Physik* 17 (1905), pp. 132–148.

[65] B. M. Escher, R. L. de Matos Filho, and Luiz Davidovich. "General framework for estimating the ultimate precision limit in noisy quantum-enhanced metrology." In: *Nature Physics* 7.5 (2011), pp. 406–411.

[66] Claude Fabre and Nicolas Treps. "Modes and states in quantum optics." In: *Reviews of Modern Physics* 92.3 (2020), p. 035005.

[67] Claude Fabre, Vahid Sandoghdar, Nicolas Treps, and Leticia F. Cugliandolo (eds). *Quantum optics and nanophotonics.* Vol. 101. Oxford University Press, 2017.

[68] Marc Fischer N. Kolachevsky, M. Zimmermann, et al. "New limits on the drift of fundamental constants from laboratory measurements." In: *Physical Review Letters* 92.23 (2004), p. 230802.

[69] Nicholas Frattini. *Three-wave Mixing in Superconducting Circuits: Stabilizing Cats with SNAILs.* Yale University, PhD thesis, 2021.

[70] Stuart J. Freedman and John F. Clauser. "Experimental test of local hidden-variable theories." In: *Physical Review Letters* 28 (1972), pp. 938–941.

[71] Akira Furusawa, J. L. Sørensen, S. L. Braunstein, et al. "Unconditional quantum teleportation." In: *Science* 282.5389 (1998), pp. 706–709.

[72] Yan Gao, Li Deng, and Aixi Chen. "Optical bistability in an optomechanical system with N-type atoms under nonresonant conditions." In: *Photonics* 7.4. (2020), p. 122.

[73] Raúl García-Patrón, Jelmer J. Renema, and Valery Shchesnovich. "Simulating boson sampling in lossy architectures." In: *Quantum* 3 (2019), p. 169.

[74] Bryan T. Gard, Keith R. Motes, Jonathan P. Olson, Peter P. Rohde, and Jonathan P. Dowling. "An introduction to boson-sampling." In: *From Atomic to Mesoscale: The Role of Quantum Coherence in Systems of Various Complexities.* World Scientific, 2015, pp. 167–192.

[75] Crispin Gardiner and Peter Zoller. *Quantum Noise: A Handbook of Markovian and Non-Markovian Quantum Stochastic Methods with Applications to Quantum Optics.* Springer Science & Business Media, 2004.

[76] Steven M. Girvin. "Superconducting qubits and circuits: Artificial atoms coupled to microwave photons." In: *Lectures delivered at Ecole d'Eté Les Houches*. Oxford University Press, 2011.

[77] M. Giustina, Marijn A. M. Versteegh, Sören Wengerowsky, et al. "Significant loophole-free tests of Bell's theorem with entangled photons." In: *Physical Review Letters* 115 (2015), p. 250401.

[78] Philippe Grangier, Alain Aspect, and Jacques Vigue. "Quantum interference effect for two atoms radiating a single photon." In: *Physical Review Letters* 54 (1985), pp. 418–421.

[79] Philippe Grangier, Juan Ariel Levenson, and Jean-Philippe Poizat. "Quantum non-demolition measurements in optics." In: *Nature* 396.6711 (1998), pp. 537–542.

[80] A. Grimm, N. E. Frattini, S. Puri, et al. "Stabilization and operation of a Kerr-cat qubit." In: *Nature* 584.7820 (2020), pp. 205–209.

[81] Frédéric Grosshans and Philippe Grangier. "Continuous variable quantum cryptography using coherent states." In: *Physical Review Letters* 88.5 (2002), p. 057902.

[82] Gilbert Grynberg, Alain Aspect, and Claude Fabre. *Introduction to Quantum Optics, From the Semi-classical Approach to Quantized Light.* Cambridge University Press, 2010.

[83] C. Guerlin. *Mesure quantique non destructive répétée de la lumière: états de Fock et trajectoires quantiques.* Université Pierre et Marie Curie Paris VI, PhD thesis, 2007.

[84] Biswarup Guha, Silvia Mariani, Aristide Lemaître, et al. "High frequency optomechanical disk resonators in III–V ternary semiconductors." In: *Optics Express* 25.20 (2017), pp. 24639–24649.

[85] Klemens Hammerer, Anders S. Sørensen, and Eugene S. Polzik. "Quantum interface between light and atomic ensembles." In: *Reviews of Modern Physics* 82.2 (2010), p. 1041.

[86] S. Haroche, M. Brune, and J. M. Raimond. "From cavity to circuit quantum electrodynamics." In: *Nature Physics* 16.3 (2020), pp. 243–246.

[87] Serge Haroche. "Nobel Lecture: Controlling photons in a box and exploring the quantum to classical boundary." In: *Reviews of Modern Physics* 85.3 (2013), p. 1083.

[88] Serge Haroche and J.-M. Raimond. *Exploring the Quantum: Atoms, Cavities, and Photons.* Oxford University Press, 2006.

[89] S. E. Harris, M. K. Oshman, and R. L. Byer. "Observation of tunable optical parametric fluorescence." In: *Physical Review Letters* 18.18 (1967), p. 732.

[90] A. Heidmann, R. J. Horowicz, S. Reynaud, et al. "Observation of quantum noise reduction on twin laser beams." In: *Physical Review Letters* 59.22 (1987), p. 2555.

[91] Stefan W. Hell. "Nanoscopy with focused light (Nobel Lecture)." In: *Angewandte Chemie International Edition* 54.28 (2015), pp. 8054–8066.

[92] C. Helstrom. "The minimum variance of estimates in quantum signal detection." In: *IEEE Transactions on Information Theory* 14.2 (1968), pp. 234–242.

[93] Carl W. Helstrom. "Quantum detection and estimation theory." In: *Journal of Statistical Physics* 1.2 (1969), pp. 231–252.

[94] B. Hensen, H. Bernien, A. Dréau, et al. "Loophole-free Bell inequality violation using electron spins separated by 1.3 kilometres." In: *Nature* 526 (2015), p. 682.

[95] M. Hillary, R. F. O'Connell, M. O. Scully, and E. P. Wigner. "Distribution functions in physics: Fundamentals." In: *Physics Reports* 106.3 (1984), pp. 121–167.

[96] Chong-Ki Hong, Zhe-Yu Ou, and Leonard Mandel. "Measurement of subpicosecond time intervals between two photons by interference." In: *Physical Review Letters* 59.18 (1987), p. 2044.

[97] Ryszard Horodecki, Paweł Horodecki, Michał Horodecki, and Karol Horodecki. "Quantum entanglement." In: *Reviews of Modern Physics* 81.2 (2009), p. 865.

[98] Randall G. Hulet, Eric S. Hilfer, and Daniel Kleppner. "Inhibited spontaneous emission by a Rydberg atom." In: *Physical Review Letters* 55 (1985), pp. 2137–2140.

[99] Marc Thierry Jaekel and Serge Reynaud. "Quantum limits in interferometric measurements." In: *Europhysics Letters* 13.4 (1990), p. 301.

[100] E. T. Jaynes. "Clearing up mysteries – The original goal." In: J. Skilling (ed.) *Maximum Entropy and Bayesian Methods*. Springer, 1989, pp. 1–27.

[101] E. T. Jaynes. *Probability Theory: The Logic of Science*. Edited by G. Bretthorst. Cambridge University Press, 2003.

[102] Earle H. Kennard. "Zur quantenmechanik einfacher bewegungstypen." In: *Zeitschrift für Physik* 44.4 (1927), pp. 326–352.

[103] Gerhard Kirchmair, B. Vlastakis, Z. Leghtas, et al. "Observation of quantum state collapse and revival due to the single-photon Kerr effect." In: *Nature* 495.7440 (2013), pp. 205–209.

[104] Daniel Kleppner. "Inhibited Spontaneous Emission." In: *Physical Review Letters* 47 (1981), pp. 233–236.

[105] Jens Koch, T. M. Yu, J. Gambetta, et al. "Charge-insensitive qubit design derived from the Cooper pair box." In: *Physical Review A* 76 (2007), p. 042319.

[106] K. Kraus. *States, Effects, and Operations*. Lecture Notes in Physics, vol. 190. Springer, 1983.

[107] Franck Laloë. *Do We Really Understand Quantum Mechanics?* Cambridge University Press, 2019.

[108] Lev Landau. "The damping problem in wave mechanics." In: *Zeitschrift für Physik* 45 (1927), p. 430.

[109] G. G. Lapaire, Pieter Kok, Jonathan P. Dowling, and J. E. Sipe. "Conditional linear-optical measurement schemes generate effective photon nonlinearities." In: *Physical Review A* 68 (2003), p. 042314.

[110] Julien Laurat, T. Coudreau, N. Treps, A. Maître, and C. Fabre. "Conditional preparation of a quantum state in the continuous variable regime: Generation of a sub-Poissonian state from twin beams." In: *Physical Review Letters* 91.21 (2003), p. 213601.

[111] Julien Laurat, K. S. Choi, H. Deng, C. W. Chou, and H. J. Kimble. "Heralded entanglement between atomic ensembles: Preparation, decoherence, and scaling." In: *Physical Review Letters* 99.18 (2007), p. 180504.

[112] Michel Le Bellac. *Quantum Physics*. Cambridge University Press, 2011.

[113] Ulf Leonhardt. *Measuring the Quantum State of Light.* Vol. 22. Cambridge University Press, 1997.

[114] Konstantin K. Likharev. *Quantum Mechanics: Lecture Notes.* Vol. 5. IOP Publishing Ltd. 2019.

[115] Rodney Loudon. *The Quantum Theory of Light.* Oxford University Press, 2000.

[116] L. G. Lutterbach and L. Davidovich. "Method for direct measurement of the Wigner function in cavity QED and ion traps." In: *Physical Review Letters* 78.13 (1997), p. 2547.

[117] Alexander I. Lvovsky and Michael G. Raymer. "Continuous-variable optical quantum-state tomography." In: *Reviews of Modern Physics* 81.1 (2009), p. 299.

[118] P. Magnard, S. Storz, P. Kurpiers, et al. "Microwave quantum link between superconducting circuits housed in spatially separated cryogenic systems." In: *Physical Review Letters* 125.26 (2020), p. 260502.

[119] J. Majer, J. Chow, J. Gambetta, et al. "Coupling superconducting qubits via a cavity bus." In: *Nature* 449.7161 (2007), pp. 443–447.

[120] Stefano Mancini, V. Giovannetti, D. Vitali, and P. Tombesi. "Entangling macroscopic oscillators exploiting radiation pressure." In: *Physical Review Letters* 88.12 (2002), p. 120401.

[121] John M. Martinis and Kevin Osborne. "Superconducting qubits and the physics of Josephson junctions." In: ArXiv cond-mat/0402415 (2004).

[122] Denis V. Martynov, E. D. Hall, B. P. Abbott, et al. "Sensitivity of the advanced LIGO detectors at the beginning of gravitational wave astronomy." In: *Physical Review D* 93.11 (2016), p. 112004.

[123] J. H. McGuire and E. S. Fry. "Restrictions on nonlocal hidden-variable theory." In: *Physical Review D* 7 (1973), pp. 555–557.

[124] D. M. Meekhof, C. Monroe, B. E. King, W. M. Itano, and D. J. Wineland. "Generation of nonclassical motional states of a trapped atom [Phys. Rev. Lett. 76, 1796 (1996)]." In: *Physical Review Letters* 77.11 (1996), p. 2346.

[125] N. David Mermin. "Commentary: Quantum mechanics: Fixing the shifty split." In: *Physics Today* 65.7 (2012), p. 8.

[126] Aaron J Miller, Sae Woo Nam, John M. Martinis, and Alexander V. Sergienko. "Demonstration of a low-noise near-infrared photon counter with multiphoton discrimination." In: *Applied Physics Letters* 83.4 (2003), pp. 791–793.

[127] Z. K. Minev S. O. Mundhada, S. Shankar, et al. "To catch and reverse a quantum jump mid-flight." In: *Nature* 570.7760 (June 2019), pp. 200–204.

[128] Morgan W Mitchell and Silvana Palacios Alvarez. "Colloquium: Quantum limits to the energy resolution of magnetic field sensors." In: *Review of Modern Physics* 92.2 (2020), p. 021001.

[129] W. E. Moerner and David P. Fromm. "Methods of single-molecule fluorescence spectroscopy and microscopy." In: *Review of Scientific Instruments* 74.8 (2003), pp. 3597–3619.

[130] Christopher Monroe, D. M. Meekhof, B. E. King, and D. J. Wineland. "A "Schrödinger cat" superposition state of an atom." In: *Science* 272.5265 (1996), pp. 1131–1136.

[131] Olivier Morin, V. D'Auria, C. Fabre, and J. Laurat. "High-fidelity single-photon source based on a Type II optical parametric oscillator." In: *Optics Letters* 37.17 (2012), pp. 3738–3740.

[132] Warren Nagourney, Jon Sandberg, and Hans Dehmelt. "Shelved optical electron amplifier: Observation of quantum jumps." In: *Physical Review Letters* 56.26 (1986), p. 2797.

[133] Y. Nakamura, Yu A. Pashkin, and J. S. Tsai. "Coherent control of macroscopic quantum states in a single-Cooper-pair box." In: *Nature* 398.6730 (1999), pp. 786–788.

[134] A. Narla, S. Shankar, M. Hatridge, et al. "Robust concurrent remote entanglement between two superconducting qubits." In: *Physical Review X* 6 (2016), p. 031036.

[135] W. Neuhauser, M. Hohenstatt, P. E. Toschek, and H. G. Dehmelt. "Visual observation and optical cooling of electrodynamically contained ions." In: *Applied Physics* 17.2 (1978), pp. 123–129.

[136] Michael Nielsen and Isaac Chuang. *Quantum Computation and Quantum Information.* Cambridge University Press, 2010.

[137] ZY Ou, S. F. Pereira, H. J. Kimble, and K. C. Peng. "Realization of the Einstein–Podolsky–Rosen paradox for continuous variables." In: *Physical Review Letters* 68.25 (1992), p. 3663.

[138] Masanao Ozawa. "Universally valid reformulation of the Heisenberg uncertainty principle on noise and disturbance in measurement." In: *Physical Review A* 67.4 (2003), p. 042105.

[139] Matteo G. A. Paris. "The modern tools of quantum mechanics." In: *The European Physical Journal Special Topics* 203.1 (2012), pp. 61–86.

[140] Wolfgang Paul. "Electromagnetic traps for charged and neutral particles." In: *Reviews of Modern Physics* 62 (1990), pp. 531–540.

[141] David T. Pegg, Lee S. Phillips, and Stephen M. Barnett. "Optical state truncation by projection synthesis." In: *Physical Review Letters* 81.8 (1998), p. 1604.

[142] A. Peres. *Quantum Theory: Concepts and Methods.* Kluwer Academic Publishers, 1993.

[143] Murray Peshkin and Akira Tonomura. *The Aharonov–Bohm Effect.* Vol. 340. Springer, 1989.

[144] Olivier Pinel, Julien Fade, Daniel Braun, et al. "Ultimate sensitivity of precision measurements with intense Gaussian quantum light: A multimodal approach." In: *Physical Review A* 85.1 (2012), p. 010101.

[145] J.-M. Pirkkalainen, E. Damskägg, M. Brandt, F. Massel, and M. A. Sillanpää. "Squeezing of quantum noise of motion in a microme-

chanical resonator." In: *Physical Review Letters* 115.24 (2015), p. 243601.

[146] Max Planck. "On the theory of the energy distribution law of the normal spectrum." In: *Verhandl Deutsche Physikalische Gesellschaft* 2 (1900), p. 237.

[147] John Preskill. "Lecture notes for physics 229: Quantum information and computation." *California Institute of Technology* (1998).

[148] Shruti Puri, Samuel Boutin, and Alexandre Blais. "Engineering the quantum states of light in a Kerr-nonlinear resonator by two-photon driving." In: *npj Quantum Information* 3 (2017), article no. 18.

[149] Robert Raussendorf and Hans J. Briegel. "A one-way quantum computer." In: *Physical Review Letters* 86.22 (2001), p. 5188.

[150] M. G. Raymer. "Uncertainty principle for joint measurement of noncommuting variables." In: *American Journal of Physics* 62.11 (1994), pp. 986–993.

[151] M. Reck, A. Zeilinger, H. J. Bernstein, and P. Bertani. (1994). "Experimental realization of any discrete unitary operator." In: *Physical Review Letters* 73.1 (1994), pp. 58–61.

[152] Philippe Réfrégier. *Noise Theory and Application to Physics: From Fluctuations to Information.* Springer Science & Business Media, 2004.

[153] Margaret D. Reid. "Demonstration of the Einstein–Podolsky–Rosen paradox using nondegenerate parametric amplification." In: *Physical Review A* 40.2 (1989), p. 913.

[154] Howard Percy Robertson. "The uncertainty principle." In: *Physical Review* 34.1 (1929), p. 163.

[155] Ananda Roy and Michel Devoret. "Introduction to parametric amplification of quantum signals with Josephson circuits." In: *Comptes Rendus Physique* 17.7 (2016), pp. 740–755.

[156] Lee A. Rozema, Ardavan Darabi, Dylan H. Mahler, et al. "Violation of Heisenberg's measurement-disturbance relationship by weak measurements." In: *Physical Review Letters* 109.10 (2012), p. 100404.

[157] Wolfgang P. Schleich. *Quantum Optics in Phase Space.* John Wiley and Sons, 2011.

[158] E. Schrödinger. "Discussion of probability relations between separated systems." In: *Proceedings of the Cambridge Philosophical Society* 31.4 (1935), p. 555.

[159] Erwin Schrödinger. "An undulatory theory of the mechanics of atoms and molecules." In: *Physical Review* 28.6 (1926), p. 1049.

[160] Erwin Schrödinger. "Are there quantum jumps? Part I." In: *The British Journal for the Philosophy of Science* 3.10 (1952), p. 109–123.

[161] Marlan O. Scully and M. Suhail Zubairy. *Quantum Optics.* Cambridge University Press, 1999.

[162] Lynden K. Shalm, Evan Meyer-Scott, Bradley G. Christensen, et al. "Strong loophole-free test of local realism." In: *Physical Review Letters* 115 (2015), p. 250402.

[163] Jacob F Sherson, Christof Weitenberg, Manuel Endres, et al. "Single-atom-resolved fluorescence imaging of an atomic Mott insulator." In: *Nature* 467.7311 (2010), pp. 68–72.

[164] Rajiah Simon. "Peres–Horodecki separability criterion for continuous variable systems." In: *Physical Review Letters* 84.12 (2000), p. 2726.

[165] P. H. Souto Ribeiro, C. Schwob, A. Maître, and C. Fabre. "Sub-shot-noise high-sensitivity spectroscopy with optical parametric oscillator twin beams." In: *Optics letters* 22.24 (1997), pp. 1893–1895.

[166] Masao Takamoto, Feng-Lei Hong, Ryoichi Higashi, and Hidetoshi Katori. "An optical lattice clock." In: *Nature* 435.7040 (2005), pp. 321–324.

[167] P. R. Tapster, J. G. Rarity, and J. S. Satchell. "Generation of sub-Poissonian light by high-efficiency light-emitting diodes." In: *Europhysics Letters* 4.3 (1987), p. 293.

[168] Claudia Tesche and John Clarke. "dc SQUID: Noise and optimization." In: *Journal of Low Temperature Physics* 29.3 (1977), pp. 301–331.

[169] Henning Vahlbruch, Moritz Mehmet, Karsten Danzmann, and Roman Schnabel. "Detection of 15 dB squeezed states of light and their application for the absolute calibration of photoelectric quantum efficiency." In: *Physical Review Letters* 117.11 (2016), p. 110801.

[170] Jayameenakshi Venkatraman Xu Xiao, Rodrigo G. Cortiñas, Alec Eickbusch, and Michel H. Devoret. "Static effective Hamiltonian of a rapidly driven nonlinear system." In: *Physical Review Letters* 129 (2022), p. 100601.

[171] R. Vijay, D. H. Slichter, and I. Siddiqi. "Observation of quantum jumps in a superconducting artificial atom." In: *Physical Review Letters* 106 (2011), p. 110502.

[172] John von Neumann. *Mathematical Foundations of Quantum Mechanics.* Princeton University Press, 1955.

[173] John von Neumann. "Probability structure of quantum mechanics." In: *Nachrichten von der Gesellschaft der Wissenschaften zu Göttingen, Mathematisch-Physikalische Klasse* 1927 (1927), pp. 245–272.

[174] Mattia Walschaers, Claude Fabre, Valentina Parigi, and Nicolas Treps. "Entanglement and Wigner function negativity of multimode non-Gaussian states." In: *Physical Review Letters* 119.18 (2017), p. 183601.

[175] Christian Weedbrook, Stefano Pirandola, Raúl García-Patrón, et al. "Gaussian quantum information." In: *Reviews of Modern Physics* 84.2 (2012), p. 621.

[176] B. Wielinga and G. J. Milburn. "Quantum tunneling in a Kerr medium with parametric pumping." In: *Physical Review A* 48 (1993), pp. 2494–2496.

[177] Eugene P. Wigner. "On the quantum correction for thermodynamic equilibrium." In: Arthur S. Wightman (ed.) *Part I: Physical Chemistry. Part II: Solid State Physics.* Springer, 1997, pp. 110–120.

[178] David J. Wineland. "Nobel lecture: Superposition, entanglement, and raising Schrödinger's cat." In: *Reviews of Modern Physics* 85.3 (2013), p. 1103.

[179] Howard M. Wiseman and Gerard J. Milburn. *Quantum Measurement and Control.* Cambridge University Press, 2009.

[180] Ling-An Wu, H. J. Kimble, J. L. Hall, and Huifa Wu. "Generation of squeezed states by parametric down conversion." In: *Physical Review Letters* 57.20 (1986), p. 2520.

[181] Kai Yang, William Paul, Soo-Hyon Phark, et al. "Coherent spin manipulation of individual atoms on a surface." In: *Science* 366.6464 (2019), pp. 509–512.

[182] B. Yurke and D. Stoler. "The dynamic generation of Schrödinger cats and their detection." In: *Physica B+C* 151.1 (1988), pp. 298–301.

[183] Bernard Yurke, Samuel L. McCall, and John R. Klauder. "SU(2) and SU(1,1) interferometers." In: *Physical Review A* 33 (1986), pp. 4033–4054.

[184] T-C Zhang, Jean-Philippe Poizat, Philippe Grelu, et al. "Quantum noise of free-running and externally-stabilized laser diodes." In: *Quantum and Semiclassical Optics: Journal of the European Optical Society Part B* 7.4 (1995), p. 601.

[185] Zlatko Minev. *Catching and Reversing a Quantum Jump Mid-Flight.* Yale University, PhD thesis, 2019.

[186] W. H. Zurek. "Pointer basis of quantum apparatus: Into what mixture does the wave packet collapse?" In: *Physical Review D* 24 (1981), pp. 1516–1525.

[187] Wojciech Hubert Zurek. "Decoherence, einselection, and the quantum origins of the classical." In: *Reviews of Modern Physics* 75.3 (2003), p. 715.

Index